VAULT GUIDE TO THE
TOP TECH EMPLOYERS

EDITED BY MICHAELA R. DRAPES AND NICHOLAS R. LICHTENBERG

Library of Congress CIP Data is available.

ISBN 13: 978-1-58131-545-5

ISBN 10: 1-58131-545-7

Printed in the United States of America

ACKNOWLEDGMENTS

We are extremely grateful to Vault's entire staff—especially Laurie Pasiuk, Mary Sotomayor, Elena Boldeskou and Marcy Lerner—for all their help in the editorial, production and marketing processes for this guide. Vault also would like to acknowledge the support of our investors, clients, employees, family and friends. Thank you!

To ensure that our research was thorough and accurate, we relied on a number of people within the companies that we profiled.

To the many employees who took the time to be interviewed or to complete our survey, we could never thank you enough. Your insights about life inside the top technology employers were invaluable, and your willingness to speak candidly will help job seekers for years to come. We also thank the firms who were so helpful in completing this project.

Table of Contents

Visit Vault at **www.vault.com** for insider company profiles, expert advice,
career message boards, expert resume reviews, the Vault Job Board and more.

VAULT CAREER LIBRARY   vii

Visit Vault at **www.vault.com** for insider company profiles, expert advice,
career message boards, expert resume reviews, the Vault Job Board and more.

VAULT CAREER LIBRARY   ix

Introduction

TECHNOLOGY

It's a tech, tech world

As industries go, the tech sector is a fairly new one, but it is revolutionizing the way people conduct themselves socially and in business. Technology has been so rapidly integrated into the fabric of modern life that the semiconductor chip, a device that hardly existed 50 years ago, has become a commonplace part of modern life. In fact, according to the Semiconductor Industry Association, in 2004 humans produced more transistors than grains of rice. The tech sector is fairly broad, and encompasses companies that manufacture hardware—the guts of the computer, like hard drives, chips, circuit boards and wiring—and software, the programs that run on said chips. Other companies offer their customers some combination of hardware and software—like Hewlett-Packard and Dell, who ship computers with operating systems and programs already installed.

Chips off the old block

Microchips are up there with the wheel and domesticated animals in terms of human inventions that have had a dramatic effect on history. The microchip as we know it—which consists of multiple circuits embedded in a semiconductive substrate—was invented by Jack Kilby, an employee of Texas Instruments, in 1958. The first semiconductors, which only had a handful of circuits, were initially used in highly funded government projects, like the Apollo Moon mission and in the guidance systems for Minuteman missiles. As the production of chips became less expensive, they began their inexorable move into the realm of consumer goods during the late 1970s and early 1980s. Unlike those first microchips, which combined tens of circuits on a single chip, microchips produced today can have billions of circuits in an area that can fit on a fingertip.

While the first microchips were soldered together by hand in a lab, today's versions are assembled by building up multiple layers of silicon on a chip in "clean rooms." Chip manufacturing involves hundreds of steps, which gradually build up the microscopic circuitry on the face of the chip. Chips are built on a substrate of pure silicon, called a wafer. A machine cuts wafers—which are about a millimeter thick— off of an enormous cylindrical crystal of pure silicon. After cutting, the surfaces of wafers are polished; any imperfections in the surface of the wafer will result in chips that do not function properly. As chip manufacturing technology improved, wafers increased in size from about four inches in the 1960s to 12 inches (30 cm) today; some memory chips are still made on smaller wafers. Anywhere from a few hundred to 1,000 can fit on a 30-centimeter wafer. After polishing, the wafer is transferred to a machine that coats the surface with an insulating layer of silicon dioxide a few atoms thick, which, in turn, is coated with a light-sensitive material called a photoresist. The wafer is then exposed to light filtered through a lithographic plate, which allows an acid etch bath to remove the photoresist that was exposed to light, as well as some of the silicon underneath it. An atom-thick layer of metal, such as copper or aluminum, can be deposited on these exposed parts of the chip; when the photoresist is removed, only the metal remains to connect the various parts of the chip. Other electrical features can be added to the surface of the chip by implanting the silicon with a smattering of individual atoms, often of boron or arsenic, known as dopants. The boron or arsenic atoms are shot into the crystal structure of the silicon with a particle accelerator. Successive layers of insulating and conducting pathways are gradually built up by repeating these processes, often hundreds of consecutive times.

Since the electrical pathways of modern processor chips are measured in mere nanometers—billionths of an inch—nearly any errant piece of schmutz can short out a circuit. Therefore, chips must be manufactured in what are known as "clean rooms," where the ambient level of dust is significantly reduced by filtration. Workers in these environments also wear bunny suits, which protect the wafers from hair or skin cells that they might shed.

Bumpy road

The semiconductor industry is highly cyclical, prone to vertiginous highs and astonishing lows. This is partly due to the industry's high cost of entry; a fab or foundry, a plant where microchips are made, can take several years—not to mention $2 billion and change—to build. As the price of chips rises, more fabs are built; as they add their capacity to the marketplace, the price of chips falls. This has already happened several times over the course of the industry's history, notably in the 1980s, when Japanese manufacturers flooded the marketplace with inexpensive chips. The market reached a high in 1999—in time to fuel the tech boom—and again in 2004, in order to power the recovery from the tech bust. As of summer 2007, the industry is in a growth phase. Many semiconductor manufacturers are building new factories, and prices for semiconductor products are on the rise. One notable exception is the price for flash memory chips, which suffered a strong downward correction in the spring of 2007; a substantial supply of the chips is still depressing prices. The price of silicon is also rising, as demand for solar cells, which are also made of the same pure silicon wafers that microchips are made of, increases demand for the material. While many major chip manufacturers have locked in contracts for inexpensive silicon, high prices for the material will exert an upward pressure on chip prices.

In Soviet Russia, you power your devices

Tech isn't just chips, however—batteries present their own host of problems for designers of computer gadgets. In addition to adding to the size and weight of the product, they are often taxed by the demands of fueling features like bigger, brighter displays—for days at a time. Engineers are nearly at a loss to provide products that can stand up to the rigors of powering zippy processors, big displays and Internet connections between charges. As batteries are pushed to their limits, they can suddenly fail—or worse, explode.

One explosion-free plan for powering the next generation of cell phones would harvest power from the gadgets' human hosts. While this sounds like a premise for a horror movie, it's really quite clever. Carrying a cell phone while walking—having the device jiggling up and down in a purse or pocket—can generate a small charge, which can be stored in a capacitor for later use. Piezoelectric materials, which generate electricity when compressed, can be put beneath buttons to ease the battery's output during those epic games of *Brick Breaker*. Consumers need not only poke and boogie their way to longer battery life, however. In 2007, two devices emerged that promised to sip power from ambient sources. That July, a company with the literal name of Inflight Power introduced a device that transforms the music distributed through the headphone jacks on airplane seats into power for an iPod or phone. That same month, researchers at the University of Southampton in England introduced a small generator that transforms ambient vibrations into electrical power. While the generator does not produce enough power to fuel electrical gizmos, it can be used to power structural sensors on bridges, where running power cords and replacing batteries is not always possible.

Giving the customers what they want

Speaking of those power-hungry gizmos, the overwhelming trend in the tech industry is to add more features to products. The scheme is a surefire crowd-pleaser: feature-laden gadgets attract customers and sell at a premium, and adding a chip that turns a

Visit Vault at **www.vault.com** for insider company profiles, expert advice, career message boards, expert resume reviews, the Vault Job Board and more.

VAULT CAREER LIBRARY

3

cell phone into a music player doesn't contribute much to its manufacturing cost. This process, known as feature creep, has a downside, however. Consumers only spend about 20 minutes on average fussing around with a new product—and if they can't figure out all the bells and whistles in that period of time, they're likely to return it. In 2006, more than half of all returned electronics had nothing wrong with them. While it may be easy and cheap to add more features, it's extremely difficult to design an elegant, easy-to-use interface for the device, instead of just adding more menus and buttons.

Data, data everywhere

People are also producing more data than ever. Luckily, the cost of data storage is also falling, currently at a rate of around 15 percent every three months. While the cost of data storage has dropped precipitously—from $10,000 per megabyte in 1956 to less than an eighth of a cent in 2004, the amount of data that people and businesses want to store has increased dramatically. One company estimated that humans created 161 billion gigabytes of data in 2006—a number that is forecast to increase five times by 2010. A great deal of this data, especially the portion created by companies—from credit card transactions to funny cat pictures sent to the office via e-mail—must be saved for several years according to regulations like Sarbanes-Oxley. This translates into strong growth for the data-storage products industry--it's expected to expand by 15 percent in 2007—but stand clear of falling margins, since price competition in this sector can be intense.

One laptop per lap

Developing economies are a hot market for tech companies—and not just those who want to manufacture goods there. The wealth of the citizens in these countries is increasing dramatically, pushing people into a growing middle class. Combine that factor with competitive Western markets saturated with products and jaded consumers—and the solutions are obvious. The companies bringing the gospel of the byte to the citizens of growing economies have diverse motives for doing so, from increasing computer literacy among the world's poorest people to seeking profits and gaining market share. Nicholas Negroponte, a former MIT professor, has a plan to bring a $100 laptop—actually, it costs $130—to poor people in developing countries. The laptops, whose design is not yet finalized, can be used as an e-book (with a black-and-white screen that can be read in sunlight), laptop or game device—and can also be networked over a wireless connection, just like conventional machines. The laptops run on a version of Fedora Linux, to encourage children to develop their own software, and also to keep prices down. The governments of Egypt and Argentina are particularly interested in the devices, which are expected to ship in the latter half of 2007.

Microsoft is pursuing customers in the developing world, partly in search of profits, and to curb the growth of free operating systems, which can erode the market share of Windows and its Office suite. In 2007, Microsoft announced that it would begin selling PCs aimed at children in India, which has not adopted the laptops for children initiative. The plan is intended to supplement education in areas where public schools may lack such necessities as running water, electricity, teachers and books. Microsoft's computer, which will cost about $500, or $100 more than some of the less expensive models, will come equipped with a Windows operating system, Word, Encarta and programs tailored to help children cram for school exams. The device isn't designed to get poor children up to speed with the Internet, but will help the company gain market share in the growing Indian economy, where it has the potential to lose ground to less expensive machines running Linux—it hopes to grab one billion customers by 2015. Intel is also looking to sell low-cost computers in Asia in 2007. Aimed at middle-class customers, who make around $5,000 per year, the devices will cost about $300. Company officials say that despite the low prices, they are still making a profit.

An Apple a day

Microsoft is seeking new markets because its dominance at home is increasingly coming under threat. Apple—whose computers were once the purview of a small, rabid fringe of computer users—is now the mainstream. In the younger generation, it's hard to find someone who doesn't own some kind of iPod (and have it permanently welded to his ears). The iPhone has been a roaring success, as well— the company reported that it sold one million of them four months after the launch, and expects to make it an even 10 million by early 2009. However, the company is not just about spiffy consumer electronics—it's been gaining in the computer race as well. Although it still hangs behind industry behemoths Hewlett-Packard and Dell, its share of the pie has been growing. Industry research firms saw Apple's market share creep up by over 1.5 percentage points between January and October of 2007— boosting its share of the market to about 8.3 percent. Industry pundits note that consumers enjoy Apple computers' modern-looking cases, pretty graphics and relative imperviousness to viruses.

WORKING IN THE INDUSTRY

As befits such a large and diverse industry, there are a variety of jobs to please all types—not just shaggy guys with 1970s eyewear who think it's perfectly reasonable to wear socks with sandals. Career options include semiconductor design and manufacture, programmers, systems designers, data storage experts and, of course, those much-maligned IT guys, who are a feature of nearly every sizeable business. The Bureau of Labor Statistics says that the tech sector is experiencing a period of rapid expansion, and will grow 40 percent between 2004 and 2014. While many jobs—including some software development and chip manufacturing jobs—are moving overseas, there will continue to be a demand for IT professionals in the U.S.

Prepare for odd questions

Some companies in the tech industry—notably Google—are notorious for their oddball interviewing tactics. Google actively seeks out techy types that have also written books or started their own companies, and has been known to recruit people by putting brainteasers on billboards. Not every company in the industry uses such off-the-wall strategies, however. Insiders report that interviews usually have several rounds—sometimes as many as three phone screens before a face-to-face interview. Interviews for jobs that require specific skills—like programming expertise—generally have a technical component where the job seeker must demonstrate his skills. Common interview questions of the non-brainteaser sort include the standard behavioral questions—"stories about leadership, teamwork, dealing with ambiguity, multitasking, project management, conflict management," according to one source at a major hardware company.

Getting your foot in the door

Tech companies frequently hire interns for the summer and during the school year. As the number of CS majors shrinks—down 40 percent from 2002 to 2006—the competition for the top candidates is heating up. In addition to the cachet of a big-ticket name on the student's resume, interns are showered with perks as if the tech bubble never burst. Google interns get free nosh—morning, noon and night—at the company's renowned café, while Microsoft interns can get free housing and transportation to and from work. There are also internship opportunities in the field for electrical and mechanical engineers, and IT divisions are frequently in need of younger helpers to patiently teach executives how to send meeting requests in Outlook.

Not moving out, but moving up

Many workers in the tech industry are understandably worried about their jobs moving overseas. An Indian programmer is happy to work for a fraction of the cost of a programmer in Silicon Valley, for instance, and does the same work in about the same period of time. The U.S. is steadily losing tech jobs—60,000 were eliminated in the first three months of 2007. Although these jobs are moving overseas, Americans are moving into positions that can only be done face-to-face—like sales or consulting. IBM uses mixed teams of U.S.-based consultants and programmers in India to design custom software for its consulting projects. The Tata consulting group has even hired 1,000 workers in the U.S., citing the importance of face time with clients. To stay ahead of the outsourcing wave, employees should be flexible and keep their skills up-to-date.

RECOMMENDED READING

There are a very large number of high-quality web sites and online journals for the tech industry.

CNet—www.cnet.com

Electronics Weekly—www.electronicsweekly.com

The Register—www.theregister.co.uk

Slashdot—www.slashdot.org

The Semiconductor Industry Association—www.sia-online.org

Books

Neil Gregory, Stanley D. Nollen and Stoyan Tenev. *New Industries from New Places: The Emergence of the Hardware And Software Industries in China And India.* World Bank Publications, 2007.

Douglas Brown. *The Black Book of Outsourcing: How to Manage the Changes, Challenges, and Opportunities.* John Wiley & Sons, 2005.

EMPLOYER PROFILES

3Com Corporation

350 Campus Drive
Marlborough, MA 01752-3064
Phone: (508) 323-1000
Fax: (508) 323-1111
www.3com.com

LOCATIONS

Marlborough, MA (HQ)
Atlanta, GA • Austin, TX •
Columbia, MD • Durham, NC •
Grand Rapids, MI • Houston, TX •
Miami, FL • Minneapolis, MN • Mt.
Laurel, NJ • New York City •
Reston, VA • Rolling Meadows, IL •
Santa Clara, CA • St. Louis, MO •
West Valley City, UT

DEPARTMENTS

Administrative • Business
Development • Chief Technology
Office • Engineering • Executive •
Finance • HR • IT • Legal •
Marketing • Professional Services •
Sales • Supply Chain

THE STATS

Employer Type: Public Company
Stock Symbol: COMS
Stock Exchange: Nasdaq
Chairman: Eric A. Benhamou
President & CEO: Edgar Masri
2006 Employees: 6,309
2006 Revenue ($mil.): $1,267

KEY COMPETITORS

Avaya
Cisco Systems
Hewlett-Packard

EMPLOYMENT CONTACT

www.3com.com/careers

Visit Vault at **www.vault.com** for insider company profiles, expert advice,
career message boards, expert resume reviews, the Vault Job Board and more.

VAULT CAREER LIBRARY   11

THE SCOOP

Enterprise outfitter

3Com's stable of products includes just about everything a corporation needs to build a network. In addition to computer hardware such as routers, switches and hubs, it also sells the wires, cables and jacks required to deliver a network inside an office building. And for companies that are looking to free themselves from all that wiring, 3Com offers an array of wireless networking products based on the popular Wi-Fi and Bluetooth protocols.

Additional gear offered by the company includes hardware firewalls, redundant power supply systems and equipment to allow voice-over-IP (VoIP) telephony. The company also offers network management software and IT consulting services. The firm manages all these services under three operating units: VoIP and secure networking, which fall under Secure, Converged Networking (SCN), and Huawei-3Com (H-3Com). The company recently gained new ownership, as Bain Capital purchased a majority stake in September 2007.

Networking (pre)history

3Com boasts an illustrious past, having been founded by Robert Metcalfe himself, co-inventor of the Ethernet. The invention was the ancestor of modern local-area-networks, or LANs, as it pioneered the concept of computers communicating with each other. Metcalfe made his discovery while working at Xerox's Palo Alto Research Center (PARC) in 1973.

Six years later, he left PARC to found 3Com, and convinced Xerox to license its Ethernet patent while recruiting other high-tech heavyweights like Digital Equipment Corp. and Intel to support the protocol. Soon, the Ethernet was a PC networking standard. Metcalfe even conceived the 3Com name with the goals of the Ethernet in mind—it is derived from the words "computer," "communication" and "compatibility."

3Com introduced its first product, a PC network interface card called EtherLink, in 1982. EtherLink immediately proved popular, and the company secured a number of influential customers early on, including the White House. In 1983, the Institute of Electrical and Electronics Engineers (IIEE) chose Ethernet as its international networking standard, which boosted its popularity even more, and the technology would ultimately go on to beat out rival networking protocols such as IBM's Token

Ring. 3Com then capitalized on rising PC sales and established itself as one of the early leaders in network interface cards (NICs). By 1984, the company had gone public.

Metcalfe moves on, 3Com grows up

In the late 1980s, 3Com branched out from NICs (which very quickly became low-margin commodities) into more sophisticated networking infrastructure equipment such as routers, hubs and switches. This expansion was aided by a series of acquisitions beginning with Bridge Communications, a router manufacturer, in 1987. In 1990, Eric Benahmou, one of the co-founders of Bridge, ascended to the top spot at his new company as Metcalfe departed. Benahmou oversaw the purchase of a number of smaller companies throughout the 1990s, including BICC Data Networks in 1992, Synernetics in 1993, Centrum Communications and NiceCom in 1994, and Chipcom in 1995, positioning the company to take on infrastructure market leader Cisco Systems.

More than it can chew

In 1997, 3Com took over U.S. Robotics in a huge $7.3 billion deal, but some indigestion came with such a fancy feast. At the time, U.S. Robotics was a leading modem manufacturer and the owner of the Palm line of personal digital assistants. But difficulties in getting 3Com's and U.S. Robotics' combined sales force up to speed resulted first in an inventory backlog and then in layoffs. Dissatisfied investors then sued 3Com for securities violations related to the merger, and the company settled out of court for $259 million in November 2000. The same year as the settlement, 3Com decided to cut its new acquisitions loose, and spun off both U.S. Robotics and the Palm subsidiary in 2000; the latter raised $875 million in an IPO.

To survive a bust, 3Com goes corporate

3Com, like many networking companies, struggled to cope with the sluggish aftermath of the millennial tech bubble bursting—in fact, it has been losing money ever since 2000. Deep cutbacks in corporate IT spending sharply curtailed the company's revenue, and 3Com started cutting jobs at an astonishing rate in 2001, a trend that has continued to the current day. The company changed its direction and strategy several times, re-entering the high-end corporate networking market in 2003 (reversing an earlier decision in 2000 to abandon it) through a partnership with Huawei Technologies of China.

At the same time, 3Com discontinued its consumer-oriented product lines, opting instead to focus solely on the enterprise market. It also moved its headquarters from its longtime home in Silicon Valley to the Boston-metro area, where the bulk of its enterprise networking business is based.

Tipping the balance

Toward the end of 2004, 3Com announced that it was acquiring TippingPoint in an effort to beef up its VoIP security holdings and compete with Cisco. TippingPoint's specialty: Intrusion Prevention Systems (IPS). In fact, TippingPoint invented IPS technology in 2002. The firm's products include automatic VoIP and peer-to-peer protection, made for corporations (3Com's core customer base), service providers, government agencies and educational clients.

The $430 million cash buyout made TippingPoint a division of 3Com once the acquisition was completed early the next year. But in June 2007, 3Com reversed course again, announcing its intentions to spin the company off in an IPO. At the time, analysts Andrew Braunberg and Charlotte Dunlap of Current Analysis criticized 3Com's plan to "effectively divest itself of one of its information security crown jewels" as "both financially risky and competitively questionable."

All the IT in China

3Com has not reversed itself in one crucial area: Asia. Many analysts have predicted that the company could reap huge rewards from H-3C, its partnership with the Chinese Huawei Technologies. And 3Com agrees: it purchased Huawei's 49 percent of the venture in November 2006. In doing so, 3Com beat out several would-be buyers, including Texas Pacific Group and Bain Capital. After the $882 million buyout was completed, the computing firm got complete ownership of H-3C, its R&D facilities, and its profits.

What's so attractive about it? Peter Chai, VP and general manager of 3Com Asia-Pacific, elaborated in an August 2007 interview: "3Com has the 'China advantage,' which gives it opportunities that many vendors do not have. Our investments in China allow us to employ low-cost manufacturing, research and development, as well as support services." It goes without saying that the venture also brings 3Com into play in one of the fastest growing (tech) economies in history.

Changes at the helm

After six straight years of losses, CEO Bruce L. Claflin retired in January 2006, and R. Scott Murray, the former CEO of supply chain management firm Modus Media International, took over soon after. However, Murray left eight months later, citing a concern that travel to China for work with Huawei would take away from time with his family. Veteran venture capitalist and 3Com manager Edgar Masri took his place, also becoming H-3C's CEO. The parent company chose a China head, Robert Y.L. Mao, to take some of the burden off its American chief. But would these boardroom changes affect the bottom line?

Profits ahead?

3Com's profitless streak came to an end in 2006, with the incorporation of H-3C networking and security product profits into the end of 3Com's fiscal year. The firm's revenue rose by 22 percent over 2005, to $795 million, and the company earned $8 million in profits.

At the beginning of 2007, 3Com announced further restructuring: it expects to cut employee numbers and 21 plants around the world, and double R&D spending. At the end of June, 3Com released its fiscal 2007 results, and they were good: $1.2 billion in sales, and a further rise in profits, to $38 million. The firm raised another $2.2 billion in September (but lost its independence), when it agreed to a buyout offer from Bain Capital, with Huawei Technologies taking a minor stake in the company. Analysts credited the company's recent Chinese success as a major motivation for the deal; it is unclear whether management changes are to follow.

GETTING HIRED

Network at 3Com

Though 3Com continues to make workforce cuts, jobs are always available—for the right candidate. Listings may be found on the company's careers web site at www.3Com.com/careers. Applicants can job hunt by creating a login, and searching by keyword, location, schedule and job type, and department. Administrative, business development, chief technology office, engineering, executive, HR, IT, legal, marketing, professional services, sales and supply chain positions are offered.

In addition to full-time posts, 3Com offers internship positions to both undergraduate and MBA students. Interns work on specific projects that match their academic specialty with one of 3Com's business needs. Contact information for students interested in either internship or full-time, entry-level positions is posted at the company's web site. If 3Com is interested in an interview, a company representative will contact each applicant within 10 days of his/her resume submission.

Benefits still tops

3Com has instituted a pay-for-performance compensation model that aims to compensate all of its employees fairly according to their contributions to the company's success. It offers performance-based incentives consisting of either cash or equity to reward outstanding results. In addition, 3Com employees also have twice yearly salary reviews to ensure that pay levels are both internally equitable and competitive with what other companies are offering.

The company's stock ownership plan is another way that 3Com attempts to give its employees a stake in the success of the business. All new employees receive a grant of company stock upon their hiring, and workers may purchase additional shares for 85 percent of their market value. The company's standard benefits package includes a choice of health plans, life, disability and dental insurance, matching contributions to retirement savings accounts, and access to professional financial planning advice. Other perks enjoyed by 3Com employees include on-site fitness centers in certain locations and tuition reimbursement.

OUR SURVEY SAYS

Becoming a part of the team

The interview process with 3Com usually involves several rounds, involving meetings with both managers and potential colleagues in groups of four to six, insiders tell us. Interviews usually last two to three hours, depending on the position. Although one insider says that while "some managers like to use brainteasers or riddles," this practice seems uncommon.

Be ready for technical questions, though, even if you're not an engineer. "As a technology company, we are very concerned with having people who are technology-literate even in non-technology positions, says one manager. "That does not mean that an individual needs to have a strong technical background, simply technology and possibly industry awareness." Another insider adds, "Some of the [interviewers] are pretty technical, whereas others try to get a feel of your attitude more than current technical ability. I believe that the ability to learn new things is more important than what you already know."

"3Com really likes engineers," says one former marketing MBA intern. "Know your technical stuff before interviewing." And don't try to fake it. Says one contact, "I've seen some people lie about their technical skills, they put something on the resume and when asked, they couldn't back it up. Managers really drill on that." And, adds one product manager, "learn about the product before you interview. A little knowledge goes a long way."

Insiders also stress that when it comes to personal traits, the company values teamwork highly. "There are two aspects they look at: how you deal with yourself (motivation and enthusiasm), and your interpersonal skills working in a group," says one. "Be sure to have examples of each, especially the working in a group part. The big companies usually look for team players." Says one techie, "They mainly asked me about past jobs and experiences, different systems I have knowledge of, and about working in a group atmosphere." Sources tell us that 3Com is good about making employment decisions soon—interviewees usually hear in a week or so.

Uneven culture

Many surveyed note that 3Com's company culture is difficult to define in large part because it has grown through acquisitions, not organically. "The acquired company is turned into a division and tends to keep their pre-acquisition culture," explains one employee.

However, 3Com's work environment is marked by communication and teamwork. "3Com has tried to minimize the class divisions of more traditional companies. Everyone in the company, including the VPs and the CEO, has cubicles," says one contact. "They want to reinforce an open-door policy for all employees. Understandably, the executives' cubicles are larger, but still without a door." "The environment is very friendly with a hearty teamwork-driven flavor," says one contact. "Corporate culture is very open all the way to the top. I have had personal conversations with our COO and CEO about things affecting my ability to manage well, and when I was an individual contributor." Says another, "Open communication is a very important part of our culture." "The culture is very open and honest, and exhibits great team effort," reports another insider. "The people here tend to be smart, open and happy," says one contact at company headquarters. "Friendly, fun, hardworking real people," sums up a Boston-based colleague.

The absence of hierarchy, which one employee describes as "fraternalistic rather than paternalistic," does have its drawbacks: "You need to fend for yourself, but you will be supported in what you are trying to do, rather than having your manager responsible for steering your career," says one respondent. "It is a hard-driving place, so anyone coming here should be ready to hit the ground running, and running hard and fast," says another. "It's a big company," says a source. "That means you don't work in a small, family-like environment. But you may feel a little more secure compared to small startups. Companies this size don't go away easy."

Says one insider of his tenure with 3Com that spans more than a decade, "This [length of employment] is somewhat unheard of for high tech, but 3Com is a company that works hard to retain its employees." However, a newer contact reports something of a disheartening hierarchy between old-timers and recent hires: "It is hard for the newcomers to connect at times with the inner circle that has been around for a while. There seems to be some favoritism along that line, but a meritocracy really reigns as the overall guiding principle." Adds another, "This industry has a higher-than-normal turnover rate due to the constant opportunities at neighboring companies, so every employee is seen as a valued resource."

While sources say there is "no special treatment for women or minorities, that does not mean that 3Com looks like a male WASP club. There is a great deal of diversity within the workforce." "The women here seem to be very respected, employees and managers alike," says one respondent. "I've seen representation of women in good numbers at all levels." "I felt very welcome here since day one," says one minority contact. "I do know many females who held high positions within our company. My old manager was a female and the current director of my new job is also female." Adds another, "There is a wide variety of cultures here, and being a white male, I'm almost a minority around here." "This is definitely a multicultural company," says another. "You are quite likely to end up working with people with different nationalities and cultures."

Part of the reason employees may stay at 3Com is that they get to spend a lot of time away from the company. Every four years, staffers get an extra one-month paid "sabbatical," which insiders say is "mandated." This is in addition to an already whopping vacation schedule that includes a "companywide shutdown between Christmas and New Year's." To encourage time off, 3Com "doesn't let employees accrue more than 120 hours. If you earn more you lose it because you don't take what you've got."

While they're there, 3Com employees often work quite hard. One contact says, "You will need to consistently put in 12 to 14 hours a day except when working on something 'mission critical.'" Although many report working nine- to 10-hour days, and some complain about long hours, others remark that "unlike the unwritten rule at many high-tech companies who really want 50 to 60+ each week, you work hard while you're here and then go home or go play. You're expected to work 40 hours a week." And even during the workday, many employees seem to take time off: "Many of us even exercise at lunch and then eat—it is assumed we are responsible enough to get our jobs done." As one source explains: "They want well-rounded employees who have lives outside of work." This does not seem to be a problem, according to another contact, who notes that, "Most people here are hardworking but are able to draw the line between work and outside activities like family and friends, recreation, etc."

While 3Com may have the amenities of other Silicon Valley companies, it hasn't gone quite as far toward casual dress days. "The dress code is generally dress casual—by this, I mean nice clothes—Dockers and maybe a button-down shirt for men," reports one employee. However, "on Fridays, you can wear just about anything you want." Engineers report casual dress; sales and marketing employees report wearing suits like the rest of the world. Says one product manager, "It's dressier than jeans, but less than 'business dress.' When we meet to do a customer briefing, I usually do the suit thing complete with power tie." As for pay, most respondents report receiving above what they believe to be the industry average. "Few companies will offer more," says one engineer.

Adaptec, Inc.

691 South Milpitas Boulevard
Milpitas, CA 95035
Phone: (408) 945-8600
Fax: (408) 262-2533
www.adaptec.com

LOCATIONS

Milpitas, CA (HQ)
Austin, TX • Durham, NC • Foothill
Ranch, CA • Grapevine, TX •
Houston, TX • Laguna Hills, CA •
Longmont, CO • Minneapolis, MN •
North Andover, MA • Orlando, FL •
Plymouth, MN • Raleigh, NC •
Redmond, WA • Seattle, WA •
Tampa, FL

30 locations in 9 countries
worldwide.

DEPARTMENTS

Administration • Applications
Engineering • Communications •
Design Engineering • Executive •
Facilities • Finance & Accounting •
Hardware Engineering • HR •
Information Systems • Legal •
Manufacturing • Marketing •
Materials • Operations •
Performance Engineering • Quality •
Sales & Support • Software
Engineering • Technical Writing •
Technicians • Test Engineering

THE STATS

Employer Type: Public Company
Stock Symbol: ADPT
Stock Exchange: Nasdaq
Chairman: D. Scott Mercer
President & CEO: Subramanian
Sundaresh
2007 Employees: 598
2007 Revenue ($mil.): $255

KEY COMPETITORS

Agere Systems
Broadcom
LSI Logic

EMPLOYMENT CONTACT

www.cytiva.com/cejobs/templateAdp
tc.asp

THE SCOOP

Adaptable Adaptec

Based in Milpitas, California, Adaptec makes hardware and software that expedites data storage and transfer between computers, networks and peripherals. It leads the market for SCSI technology; pronounced "scuzzy," this technology connects peripheral devices to computers and is used by servers and workstations to move large files.

The company's bandwidth management solutions increase the speed of data transfer, while its network attached storage (NAS) and fast Ethernet networking devices help direct information through computer systems. The company also has several storage chip and RAID (redundant array of independent disks) lines. Adaptec's input/output software is compatible with all of the leading processor platforms and operating systems.

SCSI-Wuzzy was a success

Larry Boucher created the SCSI interface at Shugart Associates (later known as Seagate Technology) in the late 1970s. He formed Adaptec in 1981, the company sold its first SCSI products in 1983, and it went public three years later. SCSI was a niche hit; simply put, it controlled the flow of data through a computer's operating system, and generally made computer use quicker and easier.

But Boucher was more of an engineer than a businessman, and left a year after the company went public to start up another tech firm; he has stayed friendly with his first company, though, serving on the board of directors and even stepping in as interim CEO for a brief period in 1998. Big tech firms increasingly adapted to Adaptec's SCSI offerings into the 1990s: In 1989, Adaptec developed the advanced SCSO programming interface (ASPI), which Novell adopted for its NetWare Networking software in 1991, and IBM followed a year, later using it for its OS/2 operating system.

Adaptec goes gobble-gobble

Like many successful high-tech enterprises, Adaptec grew largely by acquiring other software companies. In 1995, it purchased Future Domain, which makes CD-ROM connection devices, and Trillium Research, which makes Apple-compatible connection devices. In 1996, the company bought Western Digital's Connectivity

Systems Group and Corel's CD creator software. It also purchased software maker Data Kinesis and Cogent Data Technologies, which makes networking devices. In early 1998, Adaptec tried to acquire Symbios, but abandoned those efforts after federal regulators opposed the deal (it would have given Adaptec a monopoly in the SCSI market).

In the Corel acquisition, Adaptec picked up the award-winning, CD-recording software Easy CD Creator Deluxe, which home PC users could use to record music, images and videos onto CDs. The company also offered a similar product for Mac users called Jam for Macintosh and a DVD software application called DVD Toast. However, the company let go of those products after it spun off its software products group, Adaptec SPG, in 2001.

Ooooh, Snap!

After the software products unit sale, Adaptec focused most of its attention on networking and storage solutions. The firm snatched up Snap Appliance for $100 million in 2004, and the computer storage unit has been a big part of Adaptec's business ever since. In May 2005, Adaptec proudly announced that its Snap Server 4500 line surpassed products from rivals like HP and Iomega, as it won *PC Magazine*'s Editor's Choice award for its high performance at a low price.

Adaptec adapts to the marketplace

But although Adaptec was receiving plaudits in 2005 for some products, it was not receiving any of the pecuniary kind: its profits plummeted from $62 million in 2004 to a loss of $145 million the next year. The company decided it needed to streamline its businesses in order to focus on building its SCSI market share, and sold its systems business unit in October, along with many of its NAS holdings.

In December, Adaptec unveiled some products with mass-market appeal. The company introduced the GameBridge system, which allowed avid gamers to connect video game consoles to PCs or TVs and also record and replay their games—for posterity, of course. GameBridgeTV users could even record epic battles on DVD!

Snappy solutions

In an effort to expand one of its core businesses into new markets, Adaptec updated the Snap line throughout 2006. By the end of that year, three new versions, the Snap

Server 110, 210 and 410, had hit the market. Small and low-priced, the 110 to 410 models made sophisticated storage more accessible to small and midsized businesses.

Although its revenue in 2007 was the lowest in at least five years, the company returned to profitability that year, earning over $30 million. Of course, if sales are lower and profits higher, the money isn't coming from what the company is producing so much as what it's taking away. Indeed, it reduced its headcount from 2006 to 2007 by nearly 50 percent—a staggering reduction from 1,128 to 598 employees in the span of a single year. The company has announced further restructuring; it plans to cut another 20 percent of its employees by the close of its 2008 fiscal year.

GETTING HIRED

Connect at Adaptec

Adaptec's careers site, www.cytiva.com/cejobs/templateAdptc.asp, gives prospective employees two search options: they can search for jobs by department or location or view a complete list of openings. Each listing may be clicked on to access an online application. The main page also has a "Submit Resume" link for general interest applications.

In terms of benefits, the company offers medical and dental, with co-pay, on the first day of employment. It also offers various insurance services, a 401(k) and flexible spending accounts. Time off includes 12 paid holidays and 12 vacation days, with an additional day for each year added per year after the second year of employment. It offers an educational assistance program, and if you refer someone to the company, Adaptec will pay you a bonus of $1,000 to $2,000! The company web site also mentions on-site services, such as dry cleaning and oil changes, stating that they vary by location.

Adobe Systems Incorporated

345 Park Avenue
San Jose, CA 95110-2704
Phone: (408) 536-6000
Fax: (408) 537-6000
www.adobe.com

LOCATIONS

San Jose, CA (HQ)
Arden Hills, MN • McLean, VA •
Newton, MA • New York, NY • San
Diego, CA • San Francisco, CA •
Seattle, WA • Amsterdam •
Bangalore • Barcelona • Bucharest •
Chatswood, Australia • Copenhagen
• Diegem, Belgium • Dublin •
Edinburgh • Hamburg • Hong Kong
• Kista, Sweden • London •
Longueuil, Québec, Canada • Milan
• Mumbai • Munich • Noida, India •
Oslo • Ottawa • Paris • Ruben
Dario, Mexico • Saggart, Ireland •
São Paulo • Seoul • Singapore •
Taipei • Tokyo • Toronto •
Uxbridge, UK

DEPARTMENTS

Administration • Business/Corporate
Development • College Internships •
College MBA Internships • College
MBA Positions • College New
Graduate • Engineering •
Engineering Program Manager •
Finance • HR • IT • Legal •
Marketing • MBA Internships •
Operations • Product Management •
Professional Services • Sales •
Technical Support/Customer Service
• User Interface

THE STATS

Employer Type: Public Company
Stock Symbol: ADBE
Stock Exchange: Nasdaq
Co-Chairmen: Charles M. Geschke,
   John E. Warnock
CEO & Director: Bruce R. Chizen
2006 Employees: 6,082
2006 Revenue ($mil.): $2,575

KEY COMPETITORS

Microsoft
Quark
Xerox

EMPLOYMENT CONTACT

www.adobe.com/aboutadobe/career
opp

THE SCOOP

The graphics software of choice

Big in publishing, famed for its free web downloads and a favorite platform among photo and graphics geeks everywhere, Adobe software is practically ubiquitous. Many of the graphics you see in newspapers and magazines are created by Adobe programs like Illustrator and Photoshop, special effects in movies often get Adobe's After Effects treatment, and many of the web sites you see are created using Adobe software. Adobe offers portable document files (PDFs), which are universally readable, under its Acrobat brand; and since the acquisition of Flash in 2005, it provides quick video applications everywhere from the Web to the cell phone.

In 2006, the company was ranked 31st on *Fortune*'s 2006 100 Best Companies to Work For list (12th among midsized firms) and placed on Canada's Top 100 Employers list; and in 2007 *BusinessWeek* listed the company at No. 98 on its InfoTech 100.

Acrobat software is also used to display the text and art for e-books, and the company has donated equipment to Yahoo!'s book-scanning project, the Open Content Alliance. PDF Merchant is used for encrypting and selling files online, and Web Buy inhibits piracy by allowing authors to control the distribution of their works that are downloaded. This technology can be used in electronic distribution of novels, how-to guides and business reports.

Finding their calling

Adobe Co-Founders (and current Chairmen) John Warnock and Chuck Geschke first met at Xerox's Palo Alto Research Center (PARC), where they developed PostScript, a computer language that translates code into printable pages of type. After failing to convince Xerox to market the application, the two left to start Adobe in 1982. Canon came out with a cheap laser printer before they could, forcing them to alter their path. Luckily, Steve Jobs approached them to develop PostScript technology for Apple's Macintosh. The result: the Apple LaserWriter and the desktop publishing revolution.

A solid foundation made of Adobe

In the 1980s, Adobe's ascension went hand in hand with the rise of the PC, specifically the Apple Macintosh. It was a neat partnership, as Adobe's laser-printing

Visit Vault at **www.vault.com** for insider company profiles, expert advice, career message boards, expert resume reviews, the Vault Job Board and more.

V∧ULT CAREER LIBRARY   25

technology complemented the Mac's many innovations in the field of computer design. By 1986, Apple accounted for 80 percent of Adobe's sales, and Adobe released its own design and illustration software, the aptly-named Illustrator, the next year.

Just as Adobe's PostScript had become the industry standard for desktop publishing, the company's Acrobat suite from 1993 became a universal document viewing file, with its trademark PDF files. But Adobe was a pioneer in the industry, and had to guard its innovations carefully. It won a landmark legal decision against Southern Software in 1997, which ruled that fonts must be classified as intellectual property and subject to the same laws. Adobe won a similar lawsuit against Southern Software in 1999, and the company maintains an antipiracy initiative.

Territorial dispute

Adobe continues to deal with the ramifications of its products being so widespread as to lose their market cache. Most recently, it entered into a dispute with Microsoft in June 2006. Adobe's beef? Microsoft claimed that its Vista system would automatically save PDFs, and Adobe threatened to sue, arguing that the system would block PDF program competition. Acrobat 8 debuted that September and Microsoft held off on PDF saving programs, although it did release PDF-free Expression Web in December, to compete with Adobe's design software.

Battling through the late 1990s

In fact, when Adobe's financial health was far from secure in the late 1990s, other companies aping its design was a major problem. The company suffered an $11.8 million loss in 1995, caused largely by the problematic $500 million acquisition of software maker Frame Technology Group. Adobe's problems compounded over the next few years; its old friend Apple seemed to have lost its way, its Asian customers stopped spending money during the Asian financial crisis (sales in Japan fell about 40 percent from 1997 to 1998) and it lost business to Hewlett-Packard, which had designed a clone of the PostScript software.

The Quark killer

In November 1998, principal rival Quark made an offer to buy the weakened Adobe, but dropped it after Adobe made signs it would not fight any takeover attempt. Months later, Adobe introduced its page-layout program InDesign, which quickly earned the title of "Quark-killer" because its functions and features far exceed the capabilities of Quark Xpress. Adobe's formidable rival in desktop publishing software; it also sells for less than a Quark upgrade.

The Chizen one

Much of Adobe's late-1990s recovery and recent success has come under the leadership of its Brooklynite CEO Bruce Chizen, who has held the position since 2000. He has steered the company to record revenue and profits, gobbling up numerous other companies along the way, including a huge 2005 merger with erstwhile competitor Macromedia.

In a January 2007 article with the consulting firm Accenture, he characterized his management style, and Adobe's operating process as "managed anarchy—giving our technical, engineering and research community a lot of flexibility to experiment and to play, but with a layer of discipline on top." In 1998, he said, the company didn't have a firm grasp on its own performance, as its international systems weren't integrated—its European database was an Oracle model, and its American and Japanese were two different models of SAP, so the company had to wait for weeks to calculate its own financial performance.

But he also had a commitment to the founding ideals of the company; he even admitted to the Wharton Business School in a 2004 interview that he was "afraid" of taking over management from the company's founders Warnock and Geschke "without destroying the culture that John and Chuck built," because "at the end of the day, innovation is about the employees. We're an IP (intellectual property) company – it's what's in people's heads that makes a difference." Chizen's reforms took note of the employees; he instituted an annual review at the company, which had never been a widespread practice, and he consulted with Adobe customers about the Quark takeover attempt at the Seybold publishing show, noting that "I don't think anybody (wanted) Quark to take over Adobe. That is because our customers like what Adobe does and stands for."

A Macro merger

The acquisition of Macromedia in April 2005 gave Adobe footholds in the networking and mobile markets as well as a defense against the encroachment of Microsoft programs into the electronic document management sector. The $3.4 billion buyout increased Adobe's workforce by about one third and Adobe also acquired all of Macromedia's software programs, including web design program

Dreamweaver and web animator Flash, used by the likes of Google and YouTube for page displays and videos.

After the buyout, Adobe began packaging its own programs with Macromedia applications in suites targeted at different niches. Macromedia's Breeze Meeting software—meeting and web conferencing software that provides virtual meeting sites and multimedia platforms for business customers—became Adobe Acrobat Connect Professional. Adobe also integrated Flash technology into its ColdFusion business presentation software, in the form of portable documents in PDF and FlashPaper versions.

Prime pickups

Adding to its holdings, Adobe picked up Trade and Technologies France in April 2006. The buyout gave Adobe more 3D visualization software, which the company announced it would use to reach more technical markets, such as manufacturers. That June, Adobe got another imaging firm under its belt, Pixmantec ApS. Pixmantec's holdings gave Adobe more imaging software, with an emphasis on raw image processing technology. Finally, in October 2006 the software firm acquired Serious Magic Inc., a video communications company whose products Adobe plans to peddle along with its own video programs.

Picture perfect

While Adobe had to absorb some of the costs of its December 2005 Macromedia acquisition, fiscal 2006 marked its fourth straight year of record profits. Overall, 2006 revenue reached $2.6 billion, a 31 percent increase over 2005. Profits clocked in at $505 million, less than the $602 million of 2005, but a vast improvement over 2002's $191 million. Although shares of Adobe went down after the close of the fiscal year, analysts predicted that sales of Acrobat and Flash Lite mobile applications would grow over the course of 2007.

Executive acrobatics

Adobe's recent history hasn't been all upward success, though, as its May 2006 appointment of ex-Sanmina SCI exec Randy Furr to the CFO position ended on a sour note. Furr had left Sanmina just as questions surfaced over improper stock practices during his tenure. He left Adobe for "personal reasons" in November, just six months after joining, but he was vindicated when the sad news broke that his wife had contracted cancer. CEO Chizen asserted that Adobe's finances were in order regardless of Furr's departure, and a new CFO, EMC vet Mark Garrett, arrived in January 2007.

Options inquiries

Adobe began a voluntary review of its stock options practice in October 2006 and found that no executive stock grants were noted as backdated. The company did find, however, that it had underpaid employees in dividends according to their options grant dates. The company wrapped up its inquiry in January 2007 with the decision to provide employees with up to $170,000 in back payments.

Adobe au naturel

In December 2006, Adobe was awarded three platinum LEED (Leadership in Energy and Environmental Design) certifications from the U.S. Green Building Council for its redesigned headquarters—a first for both Adobe and the agency. What makes the award even more noteworthy? Rather than receiving props for building green towers, Adobe has been recognized for a more difficult feat: renovating existing structures with an eye on the environment.

GETTING HIRED

Flash a smile at Adobe

Adobe's careers site, www.adobe.com/aboutadobe/careeropp, is busy but thorough. The page has links for registering a user ID and profile, repeat logins, searching for jobs, uploading resumes and viewing company benefits policies. By following the "search for jobs" link, prospective employees can view a list of openings or search by keyword, department and locale. Examples of positions offered are business architect, account manager, computer scientist, technical evangelist, quality assurance engineer and web engineer. Each job listing contains a link to Adobe's online application process.

In addition to positions for experienced professionals, Adobe also lists internship and entry-level positions for college students and recent grads. Position descriptions and openings for college-level MBA, technical and intern positions may be found at www.adobe.com/aboutadobe/careeropp/college/index.html. Finally, all applicants may submit a resume to the company via a link on the main careers page just below the "search for jobs" link, in (what else?) PDF form. The link also leads to a general employment application form that may be printed and completed prior to interviews.

OUR SURVEY SAYS

Employee-centric culture

According to Adobe sources, in general the firm is an "innovative," "employee-friendly" place with "excellent managers." Repeatedly ranked as one of *Fortune*'s Best 100 Companies to Work For, Adobe does its best to accommodate workers, offering flexible schedules, telecommuting and "a great many training classes to get up to speed with the products." Employees also enjoy "a great cafeteria," access to a company gym ("they even have aerobics classes"), and free fresh-squeezed orange juice, soda, Starbucks coffee and "various other goodies" every day.

Toil well compensated

If you work best under pressure, Adobe is the company for you— "most of the time is pressure time," says one stressed Adobe inmate. "Our hours fluctuate between 40 and six million a week," quips an engineer, "but the marketing, sales and administrative people have different work patterns." Though they work a lot, engineers and other technical specialists "pretty much come in when they want if they're working on a long-term project," and get to "dress how they want." Plus, Adobe engineers' salaries are "typical of the geek scale" (that means high). Tech professionals can start "anywhere in the high 30s," and "if you know C++, you can make up to $50 an hour!" The corporate employees who "get away with working 40 hours a week" also enjoy good salaries, but "they have to come in by 9, and dress professionally."

Benefits are "excellent," including "very generous" stock purchase and profit-sharing options, quarterly bonuses, full insurance, and a 401(k) matched 50 percent. Adobe employees also get up to five weeks of vacation, eight paid holidays and paid sick/personal days. Insiders also feel fortunate to have "access to cutting-edge computers and software."

Brushes with greatness

Finally, insiders say Adobe is a "progressive," "laid-back company" that "knows its people are its strongest asset." "The level of intelligence of the people here" is also a big draw. "We have some incredibly artistic people working here," says one programmer, "and just being able to talk to some of the engineers who wrote Photoshop is a perk to me." Working at Adobe is also extremely challenging: "It's a good place to work, and work you will," says one insider. "We run a lean company— every single one of us has the job of at least two—but it keeps us on our toes."

Visit Vault at **www.vault.com** for insider company profiles, expert advice, career message boards, expert resume reviews, the Vault Job Board and more.

V/\ULT CAREER LIBRARY　　**31**

Advanced Micro Devices

1 AMD Place
P.O. Box 3453
Sunnyvale, CA 94088-3453
Phone: (408) 749-4000
Fax: (408) 749-4291
www.amd.com

LOCATIONS

Sunnyvale, CA (HQ)
Austin, TX • Bellevue, WA •
Boxborough, MA • East Fishkill, NY
• Fort Collins, CO • Houston, TX •
Miami Lakes, FL • Portland, OR •
Raleigh, NC • Santa Clara, CA

11 locations in 9 countries
worldwide.

DEPARTMENTS

Administrative/Clerical • Applications
Engineering • CAD Engineering •
Customer Sales Support • Design
Engineering • Device Engineering •
Device Tech Engineering • Finance •
Hardware Design Engineering • HR •
IT • Legal • Manufacturing
Engineering • Material Control •
Operations Management • Process
Engineering • Procurement •
Product Engineering • Product
Marketing • Production Control •
Reliability & Test Engineering •
Software Engineering • Technical
Marketing • Technical Sales Support
• Technician

THE STATS

Employer Type: Public Company
Stock Symbol: AMD
Stock Exchange: NYSE
Chairman & CEO: Hector de J. Ruiz
2006 Employees: 16,500
2006 Revenue ($mil.): $5,649

KEY COMPETITORS

Intel
Freescale Semiconductor
Samsung

EMPLOYMENT CONTACT

careers.amd.com

THE SCOOP

The other chip maker

Advanced Micro Devices (AMD) is the second-largest PC chipmaker in the world, although it is much smaller than the largest, industry giant Intel. AMD also makes flash memory devices and support circuitry for communications and networking applications. Based in Sunnyvale, California, the company is truly a global operation—it has manufacturing plants in the U.S., Europe and Asia, and collects 70 percent of its revenue from overseas markets.

The company has released a slew of microprocessors, including its AMD Athlon and AMD Duron. Memory products include flash memory devices and erasable, programmable, read-only memory (EPROM) devices. The company's PCS products include embedded processors, platform products (which primarily consist of chipsets) and networking products. The company ended 2006 with a 25 percent share in the chipmaking market, its largest coverage of the sector in 11 years.

The Jerry Sanders show

Advanced Micro Devices began in 1968 after Jerry Sanders left his lofty perch as Fairchild Semiconductor's director of worldwide marketing to found his own company. It started out small (it had 53 employees in 1970) but it stayed true to its name: it focused on all sorts of tiny (i.e., micro) devices, which it received from other semiconductor companies, and improved (i.e., advanced) their speed and efficiency. Also, befitting a small company with a sales-conscious CEO, the company devoted itself to the twin ideals of quality products and salesmanship. It took root in the semiconductor industry, going public in 1972, and employing nearly 1,500 people by 1974.

As CEO, Sanders grew his company through overseas expansion and increased investment in research and development, and took care to have fun along the way. He celebrated his company's five-year anniversary in 1974 with a blowout street party, where he gave away TVs, barbecue grills and other goodies to his valued employees; and he held a company raffle for 20 years, awarding $12,000 annually. After 33 years at the helm, the flamboyant Sanders stepped down as CEO in April 2002; Hector Ruiz, who was appointed president and chief operating officer in 2000, took over as CEO the same month. Ruiz became chairman two years later in 2004, when Sanders relinquished the title.

Visit Vault at **www.vault.com** for insider company profiles, expert advice, career message boards, expert resume reviews, the Vault Job Board and more.

V∧ULT CAREER LIBRARY

33

Price wars

With the economic doldrums of 2001 and the processor market down a record 32 percent, both AMD and Intel faced sudden drops in chip demand, leading to unexpectedly high inventories. The companies could have settled into a duopoly, with Intel controlling most of the market and AMD taking up a good amount of the rest, but Intel opted to cut its prices in a bid to claim the microprocessor sector for itself. In response, AMD cut 2,300 jobs (15 percent of its workforce at the time), closed two factories and engaged Intel in a "price war."

Microsoft was heavily invested in keeping the price war going, as low microchip prices could only help PC sales. Therefore, Microsoft bought its more advanced chips from Intel (Itanium) and bought more basic chips from AMD (x86), which it used in consumer PCs. In the last quarter of 2003, AMD finally broke a nine-quarter string of losses, marking the first time it had posted a profit since 2001. The company also announced that its 2003 sales had increased by 30 percent from the year before. Intel started losing market share to AMD at this point, which has only increased in recent years.

Dealing with Dell

After spending almost five years beefing up its R&D in the hopes of grabbing a foothold in the Intel-dominated chip market, AMD's tactic paid off in 2006. That May, longtime Intel-devotee Dell announced that it was switching to AMD for its chip provisions, but in a way nearly opposite from Microsoft.

AMD had come out with the Opteron chip, a high-end item designed exclusively to compete with Intel's Itanium, and Dell signed on to use the chip exclusively for its high-end servers, and keep using Intel's chips for the majority of its PC line. But the deal marked the end of a 22-year exclusive relationship between Dell and Intel, and a significant victory for AMD in the continuing price wars.

Taking the fight to court

In 2005, AMD broadened the scope of its price war, suing Intel in Delaware federal court, alleging that it paid computer manufacturers not to buy AMD's chips. (The trial is currently scheduled for 2009.) And in 2007, the European Commission got into the Intel/AMD fight as well, charging Intel with trade violations of the same sort.

AMD has created a web site devoted to the legal battle; you can check it out at breakfree.amd.com.

In the meantime, diversification

In the last decade, AMD made some acquisitions to become more web-friendly. In February 2002, it acquired Alchemy Semiconductor, a company that designs, develops and markets low-power, high-performance microprocessors for the non-PC Internet appliance market. Alchemy's embedded processors became a part of AMD's PC and PDA products, and the following November AMD announced its first product for the wireless local area network (LAN) market, the Alchemy Solutions Wireless LAN Am1772 chipset. In August 2003, the company acquired National Semiconductor's information appliance business unit, which focuses on providing easier access to the Internet with its Geode product family of silicon and system solutions.

Ex"spansion"

In July 2003 AMD and Fujitsu Limited announced the integration of their flash memory business to form Spansion, which is now the largest flash memory company in the world, with a value of about $3 billion and approximately 8,900 employees. Its flash memory solutions are available worldwide, and the company is headquartered in Sunnyvale, Calif., while also operating a Japanese headquarters in Tokyo. AMD and Fujitsu spun the company off in late 2005, but due to falling memory chip prices it raised less money than expected—AMD and Fujitsu still own large portions of the company's stock.

Sweet motherboard! AMD expands in the Fatherland ...

In October 2005, AMD opened a new plant in Dresden, Germany. Dresden, where the company had already established semiconductor production, was cited as a center of technological expertise. The 300 millimeter Fab 36 chip factory, also known as a Fab 36, was expected to double AMD's production. The move was seen as part of an effort to increase investment in Germany; the company plans to build two more plants there and raise investment to $2.5 billion by 2008. The company marked its 10th anniversary in Dresden, where it first established microprocessor Fab 30, in October 2006; it has over 3,000 employees in that city. It should be the beginning of a long and happy collaboration, as this factory has already proved invaluable in AMD's competition with Intel.

... and sets up shop upstate

In June 2006, the company announced that it would build a new semiconductor plant—the most technologically advanced in the world—in Saratoga County, New York. The 1.2 million-square-foot, $600 million project is expected to create 1,200 high-tech positions, more than 3,000 other positions, and thousands more construction jobs. Construction of the plant, which will produce 300 mm wafers, will be completed between 2012 and 2014.

AMD gets animated

AMD's chips aren't just about business. In fact, they've been used in the processors that produce some of the most sophisticated film effects around. In April 2005, AMD announced that Opteron processors were used in the filming of comic noir-inspired *Sin City*, one of the first films to use a completely digital green screen. That same month, AMD made a three-year deal with DreamWorks Animation SKG. The two companies agreed to collaborate on computer-generated (CG) filmmaking, using AMD processors. That agreement produced the children's films *Madagascar* in 2005 and *Flushed Away* in 2006. The chip company has another studio alliance, too. In October 2005, AMD and Lucasfilm Singapore announced that AMD processors would be a part of the studio's digital animation center. Not too shabby.

Getting graphic

AMD beefed up its interests in graphics with the July 2006 purchase of ATI Technologies for $5.4 billion. The deal transformed the semiconductor game overnight, extending AMD's already formidable reach into the graphic chipsets that power high-end personal computer gaming, computer animation and chipsets in cell phones and PDAs. (Chipsets connect a computer's processor to its memory and other parts.)

Analysts explained that the deal put AMD on more of an equal footing with rival Intel, and also takes AMD beyond computers into the handheld device market. ATI proved attractive to AMD for several reasons, mainly for its experience with making chips for new forms of TV and image processors to display video on cell phones and mobile video game players. At the time of the sale, ATI ranked No. 25 in a 2005 list of chip suppliers compiled by iSuppli, with about $2 billion in revenue.

Looking ahead

The price war with Intel is not over, although AMD has scored many unexpected successes to date. In August 2007 AMD lost its worldwide sales chief, Henri Richard, who departed "on his own accord and on completely amicable terms," which nevertheless set off speculation that AMD is headed for trouble. The bottom line of late hasn't been encouraging—the company has lost $1.2 billion in the first two quarters of 2007. The firm is pinning hopes on the Barcelona, an upgrade of its x86 line of microprocessors.

GETTING HIRED

Advance at AMD

Although the two merged in 2006, job listings for AMD and ATI are listed separately on the AMD careers site (careers.amd.com). The labyrinthine AMD site allows prospective employees to search by region; while some regions have their own job listings page others, like that of the U.S., have their own search engines and different ways of creating online profiles and applications. The U.S. site also lists regular, intern and co-op positions. The ATI site, in contrast, has one search engine that filters openings by location, department and keyword. The site also allows applicants to register at an online career center in order to submit applications and sign up for a job alert service.

Benefits at AMD include health, dental and disability plans, life and death insurance, travel and accident plans, profit-sharing and stock purchase plans, and tuition assistance. In addition, the chip company offers its employees tickets to movies and sports events and access to fitness centers.

A wide net

According to insiders, the company finds new employees through a wide variety of channels. "Recruiting at AMD is done through campus interviews, through job fairs conducted at AMD and elsewhere, and through resumes sent directly through the placement office," instructs one. AMD accepts resumes through regular mail, fax and e-mail—and also through an online form available at its web site. The site also contains job listings.

Applicants who land an interview may have difficulty figuring out what to expect ahead of time. "The nature of the interview depends on various factors," says a source, "but in general it is 'medium.' [That is], I have seen both more relaxed and more stressful situations." Those who interview on campus "may get invited for an interview at our site, and that's when you'll be interviewed by about seven or eight people." It is true that "there are many rounds and lots of technical questions if you are applying for a technical position." On the other hand, interviewers "don't try to deliberately stress interviewees out simply in order to gauge their reactions." Prospects for landing a job appear bright, explains one respondent: "If you are smart and talented, we certainly want and need you."

OUR SURVEY SAYS

Swift interviews

Ladies and gentlemen, start your spell checkers: "AMD tends to put a tremendous amount of emphasis on the resume," notes one source.

An insider describes his experience: "The interview process was rather painless. It was far easier than another experience I had with a related technology company (six different interviews, lasting eight hours). I went through a 30-minute phone interview, then a two-hour, in-person interview with three people and that was followed up with an hour phone conversation with a director. The interview questions were normal/typical, e.g., what are your strengths, what's do you enjoy doing, etc. I also received lots of questions about working on a team and being able to navigate through different teams/organizations." His co-worker cautions that "Many AMD positions are contract-to-hire positions. Sometimes a contract is extended for up to three years before a full employee position is offered."

Hired already? One source reports that "HR does have tools for career advancement, and even has a very generous tuition reimbursement program."

Mixed satisfaction reviews

Although AMD is frequently cited in *Fortune*'s 100 Best Companies to Work For, insider reviews are somewhat mixed overall. One contact agrees with the external assessment, saying "I love working at AMD and I especially enjoy my team."

On the other hand, another source says, "there have been quite a few people [who were] dissatisfied with their jobs in the last year and have left our group … One left because she could not get along with a certain individual—who also left."

Part of the reason for the varied responses is that AMD employees are subject to the vicissitudes of their work. "It's a bit of a roller coaster financially, but that's the nature of the industry," according to one. "When times are good, the perks are good." This up-and-down nature is especially true when it comes to work hours and intensity. "I've put in 17-hour days and I've put in five-hour days," says one respondent. Another says he spent most of a year working 9 to 5, but as his group approaches a target date, "sometimes I am up 'til 1 [a.m.] or 2 [a.m.]." "There isn't an expectation to be at work at 9 a.m. sharp but you are expected to work as late/long as it takes to get work done," adds another insider. However, most say their bosses are flexible with their hours, as long as they get their work done. A source agrees: "AMD values family time … For most departments, the work can get done any time of day as long as it does get done, so AMD offers tremendous flexibility." His gizmo-deprived co-worker adds, "getting a BlackBerry for work purposes is … difficult (they don't want to disturb your home life)."

A people's company

Although a few sources describe AMD as "a little frantic," respondents generally gave good reviews of their work atmosphere. "It's less stressful than many other high-tech companies I am familiar with," says one. "Although we are in a competitive industry, I don't feel unduly pressured beyond what is reasonable." Another observes that the "corporate culture is largely one of an underdog. There is immense pride but there still is a subtle tone of second best." Yet another says his co-workers are "not workaholics." The "people are very committed to the job, (but) the work environment is very relaxed most of the time." One insider describes a "tight community that's very friendly."

As far as bosses go, one AMDer says he's "proud" to report that "upper management really cares what its employees think." Another AMD resident comments that, "We are viewed as individuals, rather than cogs in a machine." According to another source, "It seems that AMD places a lot more value on people than other companies do." The people-oriented atmosphere extends to the dress code, which is business casual to casual. Many report wearing jeans and T-shirts. One insider points out that the "dress code is nonexistent. It's not uncommon to see executives walking around in jeans." "We dress very casual," says one. "Finance is the only department I know of (that) wear ties every day." "Today I am wearing cowboy boots," says another. "Yee-haw."

Harmonious environment

The vast majority of those surveyed say AMD is a diverse company when it comes to women and ethnic minorities. Any variations on this theme are chalked up to the ways of the world itself, not AMD. "I have noticed that most of the senior executives—as with many Silicon Valley companies—are white males," says one insider. "As far as treatment of women and minorities, they have taken huge strides to address this issue. As with all companies, however, all is not perfect," notes one female insider. Her co-worker in marketing adds, "I don't think AMD can claim to be diverse. African-Americans are woefully underrepresented, [though] some women hold high-ranking positions". "I've known some folks who felt discriminated against but rarely if ever have I seen this feeling justified," observes a manager.

Make some dough and rock out with the Stones

Our contact reports that AMD pays industry standard salaries. One engineer remark, "I guess they could pay me more, but then I guess I could stand to win the lottery too." Another source cautiously notes, "I'm not sure it this is true, but there's a saying around AMD [that] 'it's near impossible to obtain a raise unless you leave the company and then come back a year later, at which point your salary will double.'"

Aside from the quirky wage structure, AMD has "a good profit-sharing program" and "1.5 percent match on our 401(k)." An employee stock purchase plan may turn out to be a special plus from a company "on the horizon of greatness." Other perks include "discounts for amusement parks," "free workout programs," an on-site day care center, computer purchase reimbursement, and "a fully paid MBA program." The corporate headquarters in Sunnyvale "are beautiful," and include a fitness center with two basketball courts. But the most popular perk is the company's sabbatical system, which offers AMD vets two extra months of paid vacation after they complete seven years.

Affiliated Computer Services Incorporated

2828 North Haskell
Dallas, TX 75204
Phone: (214) 841-6111
Fax: (214) 821-8315
www.acs-inc.com

LOCATIONS

Dallas, TX (HQ)
Atlanta, GA • Austin, TX • Baltimore, MD • Baton Rouge, LA • Boston, MA • Chicago, IL • Cincinnati, OH • Cleveland, OH • Denver, CO • Des Moines, IA • Detroit, MI • El Paso, TX • Fort Lauderdale, FL • Fort Worth, TX • Honolulu, HI • Houston, TX • Indianapolis, IN • Kansas City, MO • Lexington, KY • Los Angeles, CA • Milwaukee, WI • Minneapolis, MN • Nashville, TN • New Orleans, LA • New York, NY • Oklahoma City, OK • Orlando, FL • Philadelphia, PA • Phoenix, AZ • Pittsburgh, PA • Portland, OR • Reston, VA • Sacramento, CA • Salt Lake City, UT • San Antonio, TX • San Francisco, CA • Seattle, WA • St. Louis, MO • Tallahassee, FL • Tempe, AZ • Trenton, NJ • Tucson, AZ • Washington, DC

700 other locations in the US and worldwide.

THE STATS

Employer Type: Public Company
Stock Symbol: ACS
Stock Exchange: NYSE
Chairman: Darwin Deason
President & CEO: Lynn Blodgett
2006 Employees: 60,000
2006 Revenue ($mil.): $5,353

DEPARTMENTS

Administration • Commercial Solutions • Corporate Servicesv • Education Solutions • Employee Service & Global Community • Executive • Finance & Accounting • Government & Community Solutions • Government Healthcare Solutions • HR • IT • Information Services • Legal • State & Local Solutions • State Healthcare Services • Transportation Solutions • Vertical Markets

KEY COMPETITORS

Accenture
Computer Sciences Corp.
EDS
IBM

EMPLOYMENT CONTACT

www.acs-inc.com/career

Visit Vault at **www.vault.com** for insider company profiles, expert advice, career message boards, expert resume reviews, the Vault Job Board and more.

VAULT CAREER LIBRARY

41

THE SCOOP

Darwinian Beginnings

Affiliated Computer Services is an outsourcing company with over $5.35 billion in annual revenue. It offers a suite of IT-oriented services such as data center management, disaster recovery, network management, security services, storage solutions and transition services for human resources. The brainchild of Darwin Deason—who rallied IT pros to work with clients on IT outsourcing projects and integration services—ACS has built quite a name for itself since its founding in 1988. In fact, ACS claims to have pioneered the concept of providing business process outsourcing (BPO) services on a grand scale.

The firm first gained clients in the financial industry, followed by ones in the communications, education, energy, government, health care, insurance, manufacturing, retail and transportation sectors. The company has amassed a client roster of top commercial and government organizations and specializes in BPO, IT outsourcing and systems integration. ACS joined the exclusive ranks of the Fortune 500 in 2003, and maintains a slot at No. 424. From its headquarters in Dallas, Texas, ACS employs over 60,000 people worldwide, serving customers in 100 countries.

Wheeling and dealing

ACS executed several strategic buys in the 1990s to become king of the jungle that is business outsourcing. With the 1996 purchase of Genix Group for $135 million, ACS became the largest outside provider of data processing services in the nation. Other major acquisitions included BRC Holdings in 1998, a developer of computer and management systems for health care organizations and local governments, and Medicaid software and services specialist Consultec in 1999. The firm had consumed around 50 companies by early 2001.

Beginning in 2000, though, ACS changed direction somewhat, selling off units to balance its business portfolio with commercial and government clients. The strategy paid off, and ACS landed long-term, multimillion-dollar contracts with high-profile customers like Blue Cross, American Express, General Motors and the Department of Education.

But it didn't stop spending entirely. As part of a plan to expand into the student loan processing business, ACS acquired FleetBoston's education services unit for $410 million in June 2002. The early half of the decade also saw several transactions with defense technology giant Lockheed Martin, from whom ACS purchased IMS Corporation for $825 million. Two years later, ACS and Lockheed Martin swapped IT service units; Lockheed purchased ACS's federal government tech services business for $658 million, and ACS bought Lockheed's commercial information technology business for $107 million.

Medicaid aid across the USA

In 2003, ACS divested a good portion of its federal business division, including its Air Force contracts, but the firm retains a solid relationship with many state agencies. It helps them handle the tech side of programs such as Medicaid (for which it processes more than 450 million claims annually), and it has been able to rev up its business in this area. ACS inked a $65 million deal to develop New Hampshire's Medicaid system in January 2005, followed by an $82.5 million contract with New Mexico's Human Services Department that October. The company swept up more state health department contracts in 2006, including $11.6 million from Louisiana in January, $100 million from Mississippi in May and $230 million from Texas in December.

Superior acquisitions

Expanding its offerings within the health care industry, ACS bought Superior Consultant Holdings in late 2004. The resulting division, ACS Healthcare Solutions, serves hospitals and health systems, other care providers, payers and health plans, as well as state and federal government agencies. Since its inception, ACS Healthcare has been raking in lucrative contracts in both the private and public sectors. In particular, it won contracts in May 2006 from OhioHealth and New Mexico's Presbyterian Healthcare Systems, both nonprofit state health care providers.

In mid-2005, ACS also completed its purchase of Mellon Financial's HR consulting and outsourcing business for $405 million. The firm made a few more acquisitions in 2005, including LiveBridge, a global customer services company, for $32 million, and the $104 million transport revenue division of Ascom AG, a Swiss communications company.

Tickets to ride

Always eager to explore new avenues, ACS derives a healthy portion of its revenue from its transport revenue group. The firm is the largest worldwide provider of transportation services to governments. It provides toll collection, transit fare

Visit Vault at **www.vault.com** for insider company profiles, expert advice, career message boards, expert resume reviews, the Vault Job Board and more.

VAULT CAREER LIBRARY

43

collection, parking fare collection, transportation safety enforcement, transportation violation collection and port management, among other services.

France hired ACS in February 2007 to deliver a contactless ticketing system for public transport in Marseilles, a contract worth $14 million. Later that month, the company was tapped to provide similar services for Jerusalem's first tramway. The firm has done similar jobs using "smartcard" technology across the globe, in such international locales as Mexico City, Melbourne and Zurich. Following a partnership with MasterCard, ACS introduced contactless parking-fee payment at six U.S. airports for those bearing MasterCard's PayPass credit card, in June 2007.

Say it ain't so! It ain't.

Rumors began to swirl in late 2005 and early 2006 that ACS might be put up for sale. The firm broke its silence in January 2006, announcing that it had called off "unsolicited discussions" with a group of private equity investors angling to buy it. However, ACS said, it was still considering ways to boost value for shareholders. A few weeks later, the firm said it would borrow money to buy nearly half of its outstanding shares. The move delighted investors, who sent ACS stock skyward.

Pumping up the bottom line

In other cost-cutting moves, ACS filed papers in February 2006 with the Securities and Exchange Commission, indicating that it would cut 1,700 jobs and reduce operations in one of its Mexican nearshore services centers. According to the documents, the positions to be cut were mainly in "offshore processors and related management." The firm said it would further reduce its bottom line by selling a corporate jet.

At the same time, the company hustled to bring in new revenue, closing a $56 million BPO agreement in January 2006 with Aetna, a client since 1999, covering document preparation, imaging, storage, management and retrieval services for medical, dental, disability and other types of claims. Another BPO win came that same month when the firm announced a contract to process student loans with KeyBank National Association. A whopping IT outsourcing deal with MeadWestvaco brought in $200 million in March, and the contracts continue to roll in—a February 2007 agreement to provide GlaxoSmithKline with HR BPO services brought $171 million into company coffers.

World-class company

Even as ACS cuts its forces in some global locations like Mexico, it continues to open new international offices. In January 2005, the firm opened a new global center for BPO and IT services in Tianjin, China, with an initial employee count of around 200. In May 2006, ACS opened a technical development center in Kuala Lumpur. The facility employs 700 workers and provides network and desktop engineering solutions, system engineering services, mainframe support, application management systems, customer care and HR services. On the other side of the world, the company announced in January 2007 that it will expand its presence in Jamaica with a new 65,000-square-foot office and call center in Montego Bay, adding another 600 jobs to the 1,300-strong workforce in Jamaica.

Murky maneuvers ...

Like many other companies, ACS has undergone scrutiny in the recent wave of stock options backdating scandals that have left so many captains of corporate America stranded on the rocks. In response to an inquiry from the SEC and a grand jury subpoena from New York's Southern District U.S. Attorney, the company announced in August 2006 that it would conduct an internal investigation into its stock option practices during the past 12 years.

The result? CEO Mark King and CFO Warren Edwards stepped down from their posts in November 2006 after the internal audit found that they had violated the company's code of ethics in connection with the options. Correcting the incorrect accounting will cost the company more than $51 million. Moving forward, Lynn Blodgett assumed the CEO job, while John Rexford was appointed to the CFO post.

Down the slippery slope

Blodgett and Rexford have ample experience with company—more than 10 years each—but they will have to turn the company around a little bit. While 2006 revenue was up $1 billion from 2005 (the year saw an impressive leap from $4.4 billion to $5.4 billion), profits were down $51 million, from $409 million in 2005 to $358 million the next year.

This is part of a continued downward slide in profits, which totaled $521 million in 2004, although profits were mired in the $200 million range at the beginning of the decade, so all recent results have shown marked improvement. More ominously, the company's debt load has increased 333 percent since 2004, shooting up from $372 million to $1.6 billion in 2006. Company officers are nevertheless optimistic,

Visit Vault at **www.vault.com** for insider company profiles, expert advice, career message boards, expert resume reviews, the Vault Job Board and more.

VAULT CAREER LIBRARY   45

blaming deflating profits on ACS's shedding of non-core business segments and pointing to new contracts that should help the company show profits in a few years.

... but we love 'em anyway

Despite the bad publicity that comes with executive shake-ups and less-than-optimal financials, ACS continues to gather kudos as an industry leader. The firm was ranked No. 248 on *Forbes*' Platinum 400 for the seventh consecutive year in 2006 and also graced *Fortune*'s Most Admired Companies list for the fifth consecutive year. The company received numerous other awards for leadership in its industry in 2006, boosting optimism that the days of scandal and upset are largely behind it.

Helping humans serve humans

ACS acquired a cute little software company, Albion Inc., in April 2007, for $25 million. Aside from giving ACS more of a stake in the health and human services sector of local and state government, Albion brought its software suite @Vantage into the ACS catalog. While similar to the business process software already developed by ACS, @Vantage's advantage is its focus on programming that smooths out the process of deploying state services like food stamps and Medicaid.

GETTING HIRED

Seeking experienced hires

The ACS careers site (acs-inc.com/career) allows you to search openings within the company by title, category, division, keyword and location, including jobs at international offices. Applicants can also build a profile and post a resume online. Though the firm seeks mostly experienced hires for consulting roles, it offers internships for undergraduate and graduate students (with a minimum 3.0 GPA) who are working toward a degree in business, computer science, management information systems or computer engineering. These internships typically run from May through August. Benefits include comprehensive insurance, 401(k) and a stock purchase plan. The firm also seeks to benefit its employees' mental growth by providing training in several areas, including project management, leadership management, interpersonal skills and client care.

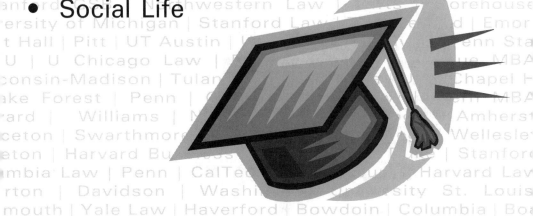

Agilent Techologies, Inc.

5301 Stevens Creek Boulevard
Santa Clara, CA 95051
Phone: (408) 345-8886
Fax: (408) 345-8474
www.agilent.com

LOCATIONS

Santa Clara, CA (HQ)
Alpharetta, GA • Anaheim, CA •
Andover, MA • Arlington Heights, IL •
Austin, TX • Bellevue, WA •
Bethlehem, PA • Boise, ID • Budd
Lake, NJ • Carmel, IN • Cary, NC •
Chesterfield, MO • Colorado Springs,
CO • Columbia, MD • Corvallis, OR •
Durham, NC • Eagan, MN • El
Segundo, CA • Englewood, CO •
Everett, WA • Fishkill, NY • Folsom,
CA • Ft. Collins, CO • Germantown,
MD • Greensboro, NC • Huntsville, AL
• Kokomo, IN • Livonia, MI •
Loveland, CO • Melbourne, FL •
Melville, NY • Miami, FL • Milpitas,
CA • New Castle, DE • Newark, CA •
Orlando, FL • Overland Park, KS •
Paramus, NJ • Pleasanton, CA •
Portland, OR • Richardson, TX •
Roseville, CA • San Diego, CA • San
Francisco, CA • San Jose, CA •
Santa Clara, CA • Santa Rosa, CA •
Schaumburg, IL • South Plainfield, NJ
• Spokane, WA • Tempe, AZ •
Washington, DC • Westlake Village,
CA • Wilmington, DE

THE STATS

Employer Type: Public Company
Stock Symbol: A
Stock Exchange: NYSE
Chairman: James G. Cullen
President & CEO: William P. Sullivan
2007 Employees: 19,400
2007 Revenue ($mil.): $5,420

DEPARTMENTS

Administration • Communications •
Customer Service • Finance • HR • IT
• Legal • Manufacturing • Marketing •
Quality • Research & Development •
Sales • Support/Service • Workplace
Services

KEY COMPETITORS

Tektronix
Teradyne
Thermo Fisher Scientific
Varian Inc.

EMPLOYMENT CONTACT

www.agilent.com/about/careers

THE SCOOP

One testy company

Agilent Techologies, headquartered in Santa Clara, California, is one Silicon Valley preteen with a distinguished family lineage. Tracing its roots back to the founding of Hewlett-Packard in 1938, the company was spun off in 1999, when HP decided to concentrate solely on computer hardware and software for the consumer market. That left Agilent with three main areas of business: test and measurement equipment, semiconductor products, and life sciences and chemical analysis equipment.

The company sold off most of its semiconductor sectors in December 2005, with the remaining semiconductor test solution division to follow 10 months later. Now leaner and focused on bioanalytic and electronic measurement, the company sells testing gadgetry to such diverse markets as the communications industry, drug manufacturers and the U.S. Department of Homeland Security. Agilent is the No. 1 manufacturer of test and measurement equipment, and holds many of the No. 1 slots in the life sciences and chemical analysis industry.

Agilent has an enormous catalog of products; in all, the company offers more than 20,000 test, measurement and monitoring devices and chemical analysis tools. Through its electronic measurement division, Agilent sells equipment such as oscilloscopes, multimeters and spectrum analyzers to the communications, electronics and semiconductor industries. Some of Agilent's test and measurement customers include Verizon, Nokia, IBM and, unsurprisingly, Hewlett-Packard. Its life sciences and chemical analysis division manufactures products such as DNA microarrays and "lab-on-a-chip" devices that are used in biotech and pharmaceutical research, as well as tools for chemical and materials science research like gas chromatography, liquid chromatography and mass spectrometry equipment.

Once upon a time ...

In 1938, Bill Hewlett and Dave Packard teamed up to create an oscillator used to test sound equipment. Their company, aptly named Hewlett-Packard, grew rapidly during WWII due to the U.S. government's rising demand for electronic instruments. In the late 1950s, the company went public and expanded into Europe. During the following decade, HP branched out into related fields, including medical electronics and analytical instrumentation, and created HP Laboratories, which later became Agilent Labs. The company went through a dramatic change during the 1980s, as the computer revolution transformed everything from HP's manufacturing processes to its product line.

All on its own

Hewlett-Packard realigned its corporate structure in 1999, and Agilent was one of the divisions affected. HP spun it off in a $2.1 billion IPO—the largest in Silicon Valley history at that time. Hewlett-Packard (which would subsequently merge with the large PC manufacturer Compaq) had decided to position itself as a consumer-oriented company, while Agilent took on all of its former industrial and scientific equipment businesses. The new company became fully independent in June 2000, when Hewlett-Packard distributed its Agilent shares to HP stockholders.

Growing pains

The future looked promising for the high-born company. With CEO Ned Barnholt at the helm, Agilent posted healthy revenue of $10.8 billion in 2000, supporting 47,000 employees. Unfortunately, the company was cut loose from HP just as the 1990s tech boom was coming to a close. Large corporations accordingly trimmed their technology investment budgets, and Agilent saw its sales fall in tandem. The company reacted with a massive restructuring project while shedding excess divisions to focus on its three main sectors.

In 2001 the company cut 10,000 jobs, on top of 10 percent pay cuts throughout the workforce, and sold its health care solutions group to Royal Philips Electronics for $1.7 billion. But these efforts couldn't keep 2002 from wreaking havoc on Agilent: that year, the company posted a loss of more than $1 billion. Agilent then cut another 7,000 jobs, bringing the 2003 headcount down to 29,000 from its peak of 47,000 only three years earlier. This less-than-triumphant year did bring one gift to the corporation—induction into the Fortune 500, with a comfortable top-half position at No. 212.

A little bit "agile," a little bit "diligent"

Agilent continued its downward spiral in 2003, posting a loss of $2 billion at the year's end. But the fourth quarter of that year signaled that better times were coming, as the company cautiously rejoiced over a profitable quarter and the turnaround it might signal.

Sure enough, like a jumbo jet pulling out of a steep nosedive, Agilent steadily but surely leveled out and climbed back to profitability in 2004. The previous years' restructuring efforts finally paid off, leaving the now-lean and mean corporate machine with a $349 million net income. In 2004, Agilent acquired Silicon Genetics, a life sciences software producer, to round out the company's offerings in its increasingly profitable life sciences and chemical analysis equipment division.

Genomic essentials

Indeed, while Agilent's test and measurement and semiconductor divisions muddled their way through a slack economy, the company's life sciences business quietly flourished like so much bacteria in a Petri dish or, more aptly, genes in a microarray. Microarrays are tools that help researchers study gene expression by looking at human (or mouse or elephant or what-have-you) genes. In June 2003, the company introduced a microarray of human genes that, combined with another human array introduced the previous year, allowed researchers to display more than 36,000 human genes. Both microarrays include a customizable region so that scientists can include newly discovered genes that are of particular interest to their own research.

The company topped itself in October 2003 with a new microarray that included 44,000 human features and then, in August 2004, rolled out a microarray that captured the entire mouse genome, made up of more than 41,000 features. The whole rat genome followed in March 2005. What's the use of this technology? In October 2006, three cancer research centers won grants totaling $35 million to use Agilent's microarrays, hardware and data analysis tools to study gene aberrations that are believed to play a role in the development of cancer.

Have sensors, will travel

In November 2004, Agilent introduced its "lab-on-a-chip" to adoring industry fans. The innovation was the first fully automated system for DNA and protein sample analysis, thus saving lab time and allowing for research on a grander scale. The little chip had a big impact, and won an international product design award, the iF Gold Award, in March 2005; it also ranked on *R&D Magazine*'s top-100 significant new technology products in August of that year (joining such past winners as the fax machine and the high-def TV).

Another market that spurred Agilent's turnaround is homeland security. In October 2003, the company unveiled its "lab-on-wheels," a mobile bioterrorism detector capable of sensing biological and chemical agents such as anthrax, arsenic, Ebola

Visit Vault at **www.vault.com** for insider company profiles, expert advice, career message boards, expert resume reviews, the Vault Job Board and more.

VAULT CAREER LIBRARY   **51**

virus and sarin nerve gas. Two years later, the company won a research grant from the U.S. Department of Homeland Security to work on a machine that the company hopes will be able to detect toxic compounds and chemical warfare agents within closed spaces.

To the East

As the fast-growing market for pretty much anything marketable, China became Agilent's second-largest customer base in 2004, trailing only the company's home U.S. turf. The company and China go way back to the HP days, and Agilent continued to do business in China after its spin-off from HP in 1999.

In response to the demands of this burgeoning market, Agilent consolidated its interests in the region into the Agilent Technologies China Holding Company in January 2005. Agilent opened a research and development center in Beijing in February 2007, bringing the company's tally of facilities in the nation to 17. The company also operates facilities in Singapore, Taiwan, India and Malaysia.

Change of guard

Having steered the company into calmer (not to mention profitable) waters, CEO Ned Barnholt stepped down from his post in 2005, and Executive Vice President Bill Sullivan, who joined Hewlett-Packard in 1976 and has been with Agilent from the start, was promoted to the CEO position. That year saw another big change in the company, as Agilent shed its semiconductor businesses to focus on its renamed bioanalytical measurement and electronic measurement divisions in December. Revenue and income for 2005 declined slightly from 2004, but the company stayed in the black.

When Agilent was seven, it was a very good year

In mid-2006, Agilent spun off the remaining vestiges of its semiconductor days, grouping its semiconductor test solutions business under the new name Verigy. Acquisitions abounded in the now-prospering company, and in April 2006 Agilent snagged SynPro Corporation, a manufacturer of synthetic nucleic acids, followed by Acqiris, a provider of high-speed digitizers and analyzers, that November. Additionally, Agilent's joint venture with Japan's Yokogawa Electric Corp., an alliance dating back to 1992, became a wholly owned subsidiary of Agilent in February. Still more acquisitions followed in 2007, with the purchase of life sciences research technology firm Stratagene in April, and the buyout of test-product manufacturer Adaptif two months later.

A few big sales made year-end profit figures soar in fiscal 2006. The November 2005 sale of Agilent's interest in Lumileds Holding to Philips for $949 million and the December 2005 sale of Agilent's semiconductor business for $2.6 billion fattened fiscal 2006 income to $3.3 billion. Lump sums aside, net revenue was up 6 percent from 2005. Electronic measurement, the larger of the company's two divisions, saw a 5 percent increase in revenue despite a slowdown in Agilent's wireless monitoring business. The bioanalytic division enjoyed record revenue in 2006, up 9 percent from 2005.

Sales were especially brisk in Europe and Asia, where new food and environmental testing regulations spawned greater demand for Agilent's chemical analysis products. In June 2006, Agilent completed a stock repurchase program of $4.4 billion of its common stock—six months ahead of schedule—and got a head start on another stock repurchase blitz the following September (this one is for $2 billion and is slated to end in 2008).

Being all that it can be

The efforts of 2006 paid off with industry kudos in 2007. In February Agilent was dubbed Company of the Year by the tech newsletter *Instrument Business Outlook*, which gave a thumbs-up to the company's bio-analytics testing arm. Agilent won Test Product of the Year the following month from readers of *Test & Measurement World* magazine, for its latest series of oscilloscopes.

The company's good behavior also brought awards of the monetary kind in March, when Agilent scored a $94 million contract to provide radio testing equipment to the U.S. Army's aviation and missile command arsenal. Under the six-year engagement, Agilent's AN/PRM-35 Radio Test Set will put the Army's radios through harsh environments and rigorous tests in an ongoing effort to modernize the armed forces' crucial field equipment.

GETTING HIRED

It's an Agilent world after all

Although headquartered in Santa Clara, California, Agilent Technologies has U.S. branch offices in more than 50 cities. The company also has a strong overseas presence with facilities in 30 foreign countries. Of the company's 19,000 employees,

Visit Vault at **www.vault.com** for insider company profiles, expert advice, career message boards, expert resume reviews, the Vault Job Board and more.

V∧ULT CAREER LIBRARY

53

approximately 11,300 are based in the United States. Major U.S. employment centers outside of California include Delaware, Illinois, New York, New Jersey, Massachusetts, Idaho and Colorado. The Agilent web site posts all open positions—both in the U.S. and abroad—searchable by location, department and level of experience.

OUR SURVEY SAYS

U of Agilent

Agilent recruits for new talent at colleges throughout the nation, including Stanford, Colorado State University, Penn State and the University of Illinois, as well as at job fairs sponsored by professional organizations such as the National Black MBA Association, the Society of Women Engineers and the National Society of Hispanic MBAs. Internships offer another way for students to get their foot in the door. According to Agilent's web site, "The goal of [the internship] program is to hire students into regular jobs after graduation."

The 10-week program is open to students pursuing undergraduate or advanced degrees in the following disciplines: electrical, mechanical, industrial, computer or chemical engineering; chemistry; biology; computer science; physics; materials science; IT/IS; and business. All internships are paid, and many offer relocation, housing and transportation allowances in addition to medical insurance. Internships are offered in research and development, manufacturing, marketing, quality control, materials, facilities, IT, finance and HR.

Living up to its name

For more traditional hires, an engineer describes the interview thusly: "The telephone interview is with the hiring manager and it is to assess if there is a fit between the company and the interviewee. If there is a fit, the hiring manager will invite you on site for a full day of interviews. This gives you a chance to evaluate the company and it gives the company a chance to evaluate you. The questions are usually behavioral-based and there are technical questions if the interview is for a technical position." One insider notes that "There are very few opportunities for advancement. The company has gone through some major restructuring over the last five years and most new opportunities are being offered in Asia." Another employee disagrees, noting that "If you have a lot of self-generated initiative, Agilent might the place for you."

But it's not all negative reviews. Sources find that Agilent is agile, and praise the company's "flexibility." "The dress code is very casual. In fact, some people wear shorts and sandals in the summer," points out an engineer. His colleague goes on to add, "They treat [you] like adults with very [few] rules!"

"There is a generous 401(k) program with company match. There is a full set of health benefits and a bonus program that is paid out in relation to how well the company is doing." His co-worker in Malaysia adds that he thinks it's "a great culture with great people."

Akamai Technologies, Inc.

8 Cambridge Center
Cambridge, MA 02142
Phone: (617) 444-3000
Fax: (617) 444-3001
www.akamai.com

LOCATIONS

Cambridge, MA (HQ)
Atlanta, GA • Bellevue, WA •
Chicago, IL • Dallas, TX • Denver,
CO • Long Beach, CA • Mountain
View, CA • New York, NY •
Reston, VA • San Mateo, CA • San
Diego, CA

Bangalore • Beijing • London •
Madrid • Munich • Paris • Singapore
• Sydney • Tokyo

DEPARTMENTS

College
Corporate Services
Engineering
Finance
Human Resources
Information Technology
Marketing
Network Operations
Product Management
Professional Services
Quality Assurance
Sales
Sales Engineering
Services & Support

THE STATS

Employer Type: Public Company
Stock Symbol: AKAM
Stock Exchange: Nasdaq
Chairman: George H. Conrades
President & CEO: Paul L. Sagan
2006 Employees: 1,058
2006 Revenue ($mil.): $428

KEY COMPETITORS

Blue Coat Systems
Mirror Image Internet
SAVVIS

EMPLOYMENT CONTACT

www.akamai.com/html/careers

All about Akamai

If you've downloaded a song, watched the Live Earth concerts or booked a flight over the Internet, then chances are you've benefited from Akamai Technologies' offerings. This Cambridge-headquartered company "provides services for accelerating and improving the delivery of content and applications over the Internet from live and on-demand streaming videos to conventional content on web pages to tools that help people transact business." In other words, Akamai helps its customers direct web traffic so it doesn't get jammed up; the company estimates that it delivers between 10 and 20 percent of all web traffic.

Akamai's diverse group of clients has included Audi, NBC, Victoria's Secret, the U.S. Department of Defense and NASDAQ. It topped *Business 2.0*'s 2007 list of the 100 Fastest-Growing Tech Companies. The firm has more than 1,000 employees, more than half of them in Massachusetts. In 2006, Akamai's revenue was nearly $429 million, a 51 percent increase from 2005's total of $283 million. It joined the S&P 500 in July 2007.

Intelligent design

"Akamai" is a Hawaiian word that means "smart" or "intelligent." And some very smart people were involved in Akamai's creation. In fact, the company's origin is directly tied to Dr. Tim Berners-Lee, one of the founding fathers of the World Wide Web. In 1995, Dr. Berners-Lee's second year at MIT, he challenged his computer science colleagues to come up with improvements on his design, specifically for the speed of transferring online content. Remember how slow the Internet was in the 1990s? The challenge was to make it faster and more efficient.

Perhaps other MIT professors took the bait, but Dr. Tom Leighton, a professor of applied mathematics, came up with a solution. He thought algorithms could improve the flow of online information, as they could analyze all the random information floating around the Web in a more specific and complex way. If this still doesn't make sense, think of Google, which has similarly used algorithms to improve the routing of online information. In the hands of companies such as Akamai and Google, the algorithm has become sort of a 21st century "secret recipe," like Coca-Cola for the Web.

Visit Vault at **www.vault.com** for insider company profiles, expert advice,
career message boards, expert resume reviews, the Vault Job Board and more.

VAULT CAREER LIBRARY   57

Team Akamai—assemble!

Dr. Leighton soon recruited the Akamai team, whose principal members were still graduate students. One was algorithm writer extraordinaire Danny Lewin, who would become the company's chief technology officer, and another was MBA student Jonathan Seelig, now a managing director at the venture firm Globespan Capital Partners. In 1998 the team won the annual MIT $50K Entrepreneurship Competition, through which it obtained an exclusive license to some of MIT's intellectual property; and it began the development of the company in earnest.

Within a year, Akamai was incorporated and had attracted a number of high-flying tech professionals. Among them were Paul Sagan and George Conrades, Akamai's current CEO and chairman, respectively. Sagan, who boasted a founding position with Road Runner cable/modem, a former presidency at Time Inc. New Media and a founding role with 24-hour news channel NY1 News, came aboard as chief operating officer. Conrades, previously the chairman and CEO of technology company BBN and a senior vice president at IBM, stepped into the CEO position.

How may we serve-r you?

Basically, the Akamai system works like so: it has assembled a huge network of servers, literally reaching around the world (hardware manufacturers even donated them in early days, once they got wind of the MIT project), and when someone requests information from anywhere on Earth, the famous Akamai algorithms kick in and route that content to the server nearest the requesting party, making for as quick and painless an info transfer as possible.

For instance, web sites and blogs occasionally attract more users than their servers can support—like when they are conducting a live chat with a celebrity—and invariably stop functioning, citing "server problems" as the reason. Akamai works to prevent such things occurring, sort of like an Internet traffic signal coordinator. Of course, Akamai's services are not available to all the obscure blogs and sites out there; companies have to contract for its offerings.

In CDN heaven

But due to its auspicious beginnings at MIT and the clutch of MIT patents that came along with its research project status, the company has grown to dominate the content delivery network (CDN) sector, although it is facing increasing competition as the Internet grows quicker and juicier content for delivery (like online videos) become easier propositions. Indeed, the company had a head start in distributing online video and audio, as it started offering streaming media and content delivery services all the way back in 1999.

The year before, it had enjoyed one of the best IPOs ever. In their first day of trading, Akamai's shares increased fivefold. The success was astounding, as many of the company's employees were still MIT students, and co-founder Daniel Lewin, only 28 years old at the time, was suddenly worth $240 million after the offering.

CEO Conrades told the reporter Gavin McCormack about running into a 21-year-old Akamaier at company offices on the Sunday after the IPO. "When Conrades asked him how much he was worth, the young [man] turned flame red before stammering a figure in the double-digit millions. 'And how does that make you feel?' Conrades asked. The kid responded, 'Why do you think I'm working a Sunday?'" The CEO was pleased with the attitude, saying "We've got a lot of people working their buns off to make sure that our stock's undervalued."

Not so fast there!

But in the years since, Akamai's stock price has fluctuated wildly. At one point during the tech boom, its price hit $327, but shares plummeted to less than a dollar a share after the tech bubble burst a few years later. Akamai was well positioned in the CDN market and recovered decently from the bursting, trading at around $60 per share by February 2007, but it fell to below $30 per share six months later.

Wherefore the travails of Akamai's stock price? Well, in the decade or so since the company's inception, the CDN market has become increasingly commoditized—the Internet has grown as a presence in people's lives and Akamai has simply not gotten CDN contracts for every web site out there. Competitors such as Limelight and Speedera (sometimes called the "anti-Akamai") entered the market, and Akamai was no longer able to set the terms of its services. Analysts began to talk of a price war for CDN services, as rival firms started to affect the number of Akamai's clients. Investors stopped buying as much of Akamai's stock because there were simply more options.

Smart buys ...

To its credit, Akamai has adapted to the changing marketplace by shoring up its CDN offerings, increasingly through acquisitions. The most dramatic example occurred when, in the middle of a nasty lawsuit with Speedera, its archrival and main competitor for market leadership, Akami whipped out its checkbook and purchased

Speedera for 142.2 million. CEO Paul Sagan even admitted in the post-acquisition conference call that he expected some Speedera customers to defect following the merger.

Akamai added more companies in recent years. In December 2006 it purchased Nine Systems, and it acquired a private company called Netli in early 2007. Both firms should add to its repertoire of CDN solutions.

... and smart guys

It's not all acquisitions for Akamai, though. In August 2007 the company announced the launch of a new large file download optimization technology, which should enhance users' ability to download large files, such as high-definition videos. In the company's 2006 annual report, CEO Sagan outlined its commitment to innovation: "Akamai will continue to work on developing future cutting-edge solutions that serve the needs of our enterprise customers in a rapidly changing landscape." Like other tech companies, Akamai is also trying to expand its presence in countries like India and China.

Remembering a founder

The company's continuing innovation is, sadly, a tribute to its co-founder Daniel Lewin, who died in the September 11th attacks. He was a passenger on one of the hijacked airplanes that crashed into the World Trade Center, and the company observed the fifth anniversary of his death in 2006. CEO Sagan wrote, in the firm's 2006 annual report, "Danny's entrepreneurial spirit of innovation and excellence continues to permeate all of our efforts at Akamai. We remain faithful to Danny's ideal and are committed to building on his vision for applying complex solutions and novel approaches to solving seemingly intractable technical problems."

Nurturing nerdiness

Akamai founded and provides financial support to the Akamai Foundation, which aims to promote excellence in math and mathematics education among a new generation of technology innovators. The foundation sponsors the US Math Olympiad and the International Math Olympiad, and it also awards college scholarships to top high school math students. In addition, through an initiative called "Magic of Math," the Akamai Foundation tries to teach kids in grades K-12 that mathematics can be "magical and fun."

GETTING HIRED

Committed to having fun

Akamai says the company will always work closely with customers to understand their needs. It also promises to conduct business ethically, deliver on its commitments to stakeholders and strive for excellence. However, another one of the firm's governing principles is to "have fun."

The extensive list of employee benefits at Akamai includes medical insurance, prescription drug coverage, vision benefits, dental insurance, a 401(k) retirement and savings plan, life insurance, an employee stock purchase plan, a pretax commuter benefit and an educational assistance plan. Akamai's people also enjoy fitness programs and get discounts on entertainment.

Akamize yourself

Akamai participates in college recruiting events at school including Carnegie Mellon, MIT, McGill, Brown, Harvard, Northeastern, Stanford and UCLA. Additionally, the firm looks for new hires at several professional recruiting events. A list of upcoming events is online at www.akamai.com/html/careers/events.html.

Potential applicants can view current job openings and apply for them online at www.akamai.com/html/careers. Recently, the company was hiring network operations engineers in Cambridge, Mass.; software engineers in San Diego, Calif.; network operations technicians in Reston, Va.; product marketing specialists in Bangalore, India; and more. The firm's web site also features video testimonials from several current employees, including CEO Paul Sagan and an intern named Joseph Judge.

Visit Vault at **www.vault.com** for insider company profiles, expert advice,
career message boards, expert resume reviews, the Vault Job Board and more.

V/\ULT CAREER LIBRARY

61

Alliance Data Systems, Inc.

17655 Waterview Parkway
Dallas, TX 75252
Phone: (972) 348-5100
Fax: (972) 348-5555
www.alliancedata.com

LOCATIONS

Dallas, TX (HQ)
Atlanta, GA • Bend, OR •
Columbus, OH • Dallas, TX • Ennis,
TX • Jacksonville, TX • Johnson
City, TN • Kennesaw, GA • Lenexa,
KS • Milford, OH • New York, NY •
San Antonio, TX • Scottsbluff, NE •
St. Louis, MO • Tampa, FL •
Thornton, CO • Tulsa, OK •
Wakefield, MA • West Monroe, LA
• Westminster, CO

DEPARTMENTS

Account Management •
Accounting/Finance • Administration
• Commercial/Industry Relations •
Communications • Credit Operations
• Customer Relations • Engineering
Services • HR • Information
Systems • Internal Audit • Legal •
Lending • Management •
Manufacturing • Marketing •
Operations • Planning • Product
Engineering • Purchasing • Research
& Development • Sales

THE STATS

Employer Type: Public Company
Stock Symbol: ADS
Stock Exchange: NYSE
Chairman & CEO: Mike Parks
2006 Employees: 9,300
2006 Revenue ($mil.): $1,998

KEY COMPETITORS

First Data
MasterCard
Total System Services

EMPLOYMENT CONTACT

www.alliancedata.com/careers.html

THE SCOOP

Credit card kings

Wherever gift cards are given, Alliance Data Systems is there. Wherever a gas bill is paid online, Alliance Data Systems is there. Wherever consumer loyalty is rewarded with free movie tickets, Alliance Data Systems ... well, you get the idea. A strong believer in the purchasing power of that thin slice of plastic, Alliance Data Systems (better known by the abbreviated Alliance Data) handles all things transactional, including payment authorization, processing, underwriting, risk management and database marketing for more than 600 clients in the retail, utility and financial industries.

The company processes more than 3.8 billion transactions annually—that's a whole lot of trips to the convenience store. Paired with its marketing services segment— which promotes consumer loyalty through incentives like the company's Air Miles rewards program, offered only in Canada—Alliance Data delivers a double whammy that makes it stand out in its industry. The company not only handles all the details of customer purchases for its clients, but it also creates and manages the little perks that hopefully keep those same customers coming back.

Child of the 1990s

Alliance Data was born in 1996, when investment firm Welsh, Carson, Anderson and Stowe merged two entities it had purchased, BSI Business Services (JC Penney's transaction services arm) and World Financial Network National Bank (the credit card banking segment of The Limited). Under the leadership of CEO J. Michael Parks, Alliance Data embarked upon several strategic acquisitions to fill out the company's abilities in the world of transaction, credit and marketing services.

In 1998, Alliance Data acquired the Loyalty Management Group Canada, bringing the popular Air Miles rewards program (for Loyalty's Canadian clients) into the Alliance Data fold. The program allows Canadian consumers to earn "reward miles" when they shop at participating companies; the reward miles can, in turn, be redeemed for goodies like travel discounts, movie passes, CDs and DVDs at more than 100 Canadian companies. The acquisition of SPS Payment Systems, a firm providing electronic payment services to mass transit systems and tollways, followed in 1999, with utilities billing firm Utilipro entering ADS's lineup in 2001.

More recently, Alliance Data has acquired a number of loyalty marketing-related businesses, including Epsilon Data Management (Epsilon), Bigfoot Interactive, Abacus and DoubleClick Email Solutions. These companies provide unique data-driven solutions that help clients internationally better manage their customer relationships and improve profitability through database management, direct mail, sophisticated analytics and permission-based e-mail marketing services.

A Limited clientele

Initially, a healthy chunk of Alliance Data's business came from providing private-label credit card services to the company from whence it sprung, The Limited. After Alliance Data's IPO in 2001, The Limited continued to hold approximately 20 percent of the firm and constituted almost as much of its business; the firm eventually sold its stake in 2003.

The two companies signed a contract in December 2005 that extended Alliance Data's credit card services to Limited Brands (including The Limited, Bath & Body Works, Express, Victoria's Secret and more) through 2012.

Out of the blues and into the black

Unwilling to "limit" all its eggs to one basket, ADS began inking agreements with other retail chains as well: Williams-Sonoma, AnnTaylor Stores and Crate and Barrel all signed onto Alliance Data's private label credit services in 2002. ADS also started new consumer loyalty programs with the giant convenience chain 7-Eleven and launched an online mall through its Air Miles rewards service. Alliance Data's utilities outsourcing business was bolstered with the purchase of Orcom (then the leader in utilities billing outsourcing) in 2003.

A pattern was forming: pair a smattering of annual acquisitions with aggressive pursuit of new clients, and wait patiently for the fermentation to begin. Through this modus operandi Alliance Data's revenue increased by leaps and bounds each year. Revenue for 2003 reached the $1 billion mark, almost twice as much as the company was pulling in just four years earlier. More importantly, the company was starting to show a profit after several years of million-dollar losses. Following the lead of The Limited, Welsh, Carson, Anderson and Stowe cut the umbilical chord at this high point, beginning in 2003 with a reduction in its holdings and ending with a full redistribution of shares in October 2006.

Alliance Data's list of clients (and, in turn, profits) continued to swell throughout 2004 and 2005. Design Within Reach, The Buckle and New York & Company all signed on for private-label credit card services in 2004, with Maurices and The Dress Barn to follow in 2005. In the loyalty marketing arena, Alliance Data continued to expand its offerings in the Air Miles program by penning agreements with WestJet in 2004, while, through its Epsilon business, also extending its agreements with Hilton Hotels.

This period saw growth in Alliance Data's business with utilities companies as well with the acquisition of Capstone Consulting Partners, a provider of technical services to the energy industry, in November 2004, and subsequent contracts with Pepco Energy Services in July 2005 and First Choice Power, of the Lone Star State, in November 2005.

... and on into the green

In 2006, a $600 million stock repurchase program, announced that October, brought smiles to the faces Alliance Data stockholders, while the acquisitions of e-mail marketing firms' iCom Information & Communications and DoubleClick Email Solutions brought deal-closing handshakes all around. These additions to ADS's Epsilon segment beefed up the company's e-mail marketing services and helped make Epsilon the leading generator of growth at ADS. Citibank signed up for ADS's services in creating a points-based customer reward program in March 2006 and private-label credit card deals abounded, with new and renewing clients including Abercrombie & Fitch, Friedman's Jewelers and Pamida Stores.

What resulted was a year of revenue just a smidge under $2 billion (up 29 percent from the previous year) and a healthy forecast for continued growth based on the steady expansion of the company's credit and transaction services and the budding potential of its marketing services division. Alliance Data's bright forecast for the future was best summed up by CFO Ed Heffernan in late 2006: "We're going to have a fun run."

A little privacy, please

As it turns out, the folks at Alliance Data will be having that fun behind closed doors; in May 2007, the company agreed to be taken private by The Blackstone Group. The $7.8 billion deal—which includes Alliance Data's $1.1 billion debt—is expected to close in Q4 2007, pending customary closing requirements and regulatory approvals.

CEO Mike Parks and his cohorts expect to keep their positions in the company hierarchy after the deal goes through. Citing Blackstone's history of investing in the companies it acquires, Parks expects that the sale will ultimately benefit Alliance Data's customer base, and in the meanwhile promises that it will be "business as usual" at Alliance Data headquarters.

GETTING HIRED

Reward your career at Alliance Data

You can get the goods on opportunities at Alliance Data through its career site, at www.alliancedata.com/careers. There you will find listings for all open positions at the company, searchable by location, department, full- or part-time status and, for the truly discriminating applicant, by desired salary. Upon applying, you will required to set up an account on Alliance Data's career web page, allowing you to upload and edit your resume (you also have the option of applying without using a resume), in addition to filling out online forms regarding work experience and education history.

Since Alliance Data is in the loyalty marketing business, it can be assumed that it knows a thing or two about fostering employee loyalty, as well. The company offers the usual health benefits, including reimbursement for gym membership. Working parents will be attracted by the backup care program, which grants 80 hours of emergency child or dependant care to all associates. Wannabe working parents will appreciate the adoption assistance program, which reimburses employees up to $2,500 per adoption. Other niceties include a 401(k) plan with company matching and a stock purchase program.

Altera Corporation

101 Innovation Drive
San Jose, CA 95134-2020
Phone: (408) 544-7000
Fax: (408) 544-6403
www.altera.com

LOCATIONS

San Jose, CA (HQ)
Broomfield, CO • Chelmsford, MA •
Duluth, GA • Ellicott City, MD •
Kenilworth, NJ • Minneapolis, MN •
Portland, OR • Raleigh, NC •
Richardson, TX • San Diego, CA •
Schaumburg, IL • Sussex, WI •
Willoughby, OH

DEPARTMENTS

Administrative
Applications Engineering
CAD/Layout Design Automation
Facilities
Field Applications
Finance
HR
IC Design Engineering
Information Systems
Legal
Marketing
Operations
Product Engineering
Purchasing
Sales
Software Engineering
Test Engineering

THE STATS

Employer Type: Public Company
Stock Symbol: ALTR
Stock Exchange: Nasdaq
Chairman & CEO: John P. Daane
2006 Employees: 2,654
2006 Revenue ($mil.): $1,285

KEY COMPETITORS

Actel
Lattice Semiconductor
Xilinx

EMPLOYMENT CONTACT

www.altera.com/corporate/jobs

Visit Vault at **www.vault.com** for insider company profiles, expert advice,
career message boards, expert resume reviews, the Vault Job Board and more.

VAULT CAREER LIBRARY   67

THE SCOOP

A leader of PLDs

If the Internet relies on device manufacturers such as Lucent and Cisco Systems to supply technological components, those companies rely upon Altera to keep the components humming along. The enigmatically-named Altera is one of the largest producers of specialized computer chips used by companies in a wide range of industries, including the networking, telecommunications and industrial machinery fields.

The acronym-friendly firm specializes in the manufacture of PLDs, or programmable logic devices, and SOPCs, or systems-on-a-programmable-chip. Because these logic devices are programmable, Altera's clients can craft the chips to meet their own needs—using Altera software, of course. The majority of Altera's chips are being sold overseas, increasingly in Asian markets; in 2006, Japan constituted 23 percent of sales, with the rest of Asia Pacific garnering another 27 percent. North America and Europe made up about a quarter of sales each.

Chips get smaller, company gets bigger

Four semiconductor entrepreneurs who foresaw great things in chips founded Altera in 1983 (their names are Robert Hartmann, Michael Magranet, Paul Newhagen and Jim Sansbury). The company went public in 1988, introducing a new generation of chips that year, as they would in 1991 and 1992. Altera acquired Intel's PLD line in 1994, thereby increasing its market share to 20 percent.

From there, things only got better—Altera's sales had more than doubled in 1995, from $199 million to $402 million, and its stock price had jumped from about $10 per share to $25. In fact, 1997 proved so profitable that the company handed out $10 million in profit shares and bonuses. The following year, during which the overall PLD industry suffered, Altera still managed its sixth-straight profitable fiscal year, bringing in $654 million in revenue.

Downs followed by ups

Altera finished out the century strong, with sales growing to $1.3 billion by the end of 2000. But the new century brought with it new woes. Due to a general slowdown in the semiconductor industry, revenue for 2001 fell by 39 percent. Sales fell further in 2002, to $711 million, but the brisk sales of that year's new products—the Stratix,

Stratix GX and low-cost Cyclone families of PLDs—promised to bring the company out of the slump.

Sure enough, revenue was again on the rise by 2003, giving Altera something to cheer about on its 20th birthday. Its focus on innovation and new product offerings—especially in the rapidly expanding realm of field programmable gate arrays (FPGAs)—brought 2004 revenue back to the levels Altera had once enjoyed—that year, the company pulled down $1 billion in sales.

Altera's good fortune continued through 2005. In July of that year, the company celebrated the milestone of its thousandth patent. Indeed, having learned that new products could provide a cushion against industry turmoil, Altera filed more than 300 patent applications in 2005 alone. This business model, based on innovation, made Altera the fastest-growing PLD company that year.

Back that date up

In May 2006, Altera began an internal review of historical stock options practices in response to federal probes into backdating schemes at other Silicon Valley firms. This investigation caused a delay in Altera's filing of its quarterly report—a violation of Nasdaq's marketplace rules—and the association threatened to delist the company. Meanwhile, a special committee of board members found seven instances of backdating between 1996 and 2001, forcing Altera to restate its earnings for those years by $47 million.

The attentions of the SEC soon followed, as did the resignation of Altera CFO Nathan Sarkisian in October. The company received conditional Nasdaq listing in August 2006, subject to the firm's compliance with federal investigations into is stock option pricing practices, followed by an extension of the deadline to late October, with which Altera complied.

Forward momentum

The SEC completed its investigation in February 2007 without recommending enforcement action against the chip producer. Revenue and income for 2006 gave the company every reason to believe it would recover from the bad publicity, with sales increasing 14 percent and profit rising 16 percent, even in that tumultuous year. FPGAs constituted 71 percent of revenue that year, reaching a record $909 million in sales, and the company enjoyed double-digit growth in all its market segments—industrial, computer and storage, and communications.

Looking for a clean break with its somewhat tarnished past, Altera hired Timothy Morse, coming off a 15-year stint with GE, as its new CFO in January 2007. When COO Denis Berlan announced his retirement the following February, Altera announced the elimination of that position entirely.

Tapping China

It is increasingly common for tech firms in the U.S. to shop outside their homeland's borders for new talent. Altera joined the fun in April 2007, when it opened its 30th joint laboratory and training center at the South China Normal University. Designed to recruit and train the next generation of engineers to fit Altera's particular needs, such centers have been popping up all over China since 2004. Because the sites function as labs as well as schools, Altera not only benefits from the new workers, but also from their work. The company also sponsors a Nios II Embedded Processor design contest in China every year to attract students to the company. As one of the company's most popular products, Nios II processors are used to reduce bottlenecks during the PLD development process. It recently won the 2006 Innovation Award from industry rag EDN.

GETTING HIRED

The kids are alright (with Altera)

Altera is eager to woo the fresh-from-college crowd, and on its careers web site (www.altera.com/corporate/jobs) you will find a plethora of information on how new college grads can get in with the firm. At press time, however, the "campus calendar" neglected to list any event, but other sections make up for this area's out-of-date-ness. One more helpful portion of the site gives a rundown of the college majors of particular use to the company, with a summary of typical job functions and responsibilities for each. From here you can also explore internship opportunities—though these are typically limited to engineering positions. Once you find something that piques your interest, Altera even helps you through the hiring process with its "Interview Tips" section, complete with a list of typical interview questions and a guide to proper apparel.

Experienced hires will find goodies on the careers site as well, with a jobs database searchable by department and location (Altera has facilities throughout North America, Europe and Asia). Upon uploading your resume and filling out a short application online, you will be considered for current and future opportunities with the company. An exhaustive listing of benefits—for U.S. employees only—reveals the usual spate of medical, dental and vision insurance, in addition to paid time off (10 holidays plus 12 vacation days per year), child care assistance and a flexible health care spending account. If you want to keep your poodle out of the pet cemetery, you can opt for the veterinary pet insurance program, one of the voluntary benefits offered along with a college savings plan and auto and home insurance. The company also provides 100 percent education reimbursement for full-time employees and an employee stock purchasing program.

Visit Vault at **www.vault.com** for insider company profiles, expert advice, career message boards, expert resume reviews, the Vault Job Board and more.

V∧ULT CAREER LIBRARY

71

Analog Devices Incorporated

1 Technology Way
Norwood, MA 02062
Phone: (781) 329-4700
Fax: (781) 461-3638
www.analog.com

LOCATIONS

Norwood, MA (HQ)
Austin, TX
Beaverton, OR
Boston, MA
Cambridge, MA
Fort Collins, CO
Greensboro, NC
Longmont, CO
Nashua, NH
Raleigh, NC
San Diego, CA
San Jose, CA
Somerset, NJ
Tucson, AZ
Richardson, TX
Vancouver, WA
Wilmington, MA

DEPARTMENTS

Administration
Engineering
Finance
HR
Information Systems
Legal
Manufacturing
Marketing
Sales
Technician

THE STATS

Employer Type: Public Company
Stock Symbol: ADI
Stock Exchange: NYSE
Chairman: Ray Stata
President & CEO: Jerald G. Fisherman
2006 Employees: 9,800
2006 Revenue ($mil.): $2,573

KEY COMPETITORS

Cirrus Logic
Linear Technology
Maxim Integrated Products
National Semiconductor
STMicroelectronics
Texas Instruments

EMPLOYMENT CONTACT

www.analog.com/jobs

THE SCOOP

Circuiting the globe

Analog Devices Incorporated makes analog, mixed-signal and digital signal processing integrated circuits (IC). One of the world's leading semiconductor companies, ADI technology routinely converts "real-world phenomena" such as temperature, light and sound into electrical signals used in a range of devices, from CAT scanners and digital cameras to cars and radar systems.

The firm has a portfolio of over 10,000 products, and employs nearly as many people, including over 3,000 engineers, at its design centers, sales offices and manufacturing facilities across the globe. ADI was listed as one of America's most admired companies by *Fortune* in 2000, 2001, 2005 and 2006; it has also made five appearances on the Forbes Platinum 400 list, most recently in 2005 at No. 347.

The year was A.D. 101

Ray Stata and Matthew Lorber met in Cambridge, Mass. as engineering students (and roommates) at MIT in the 1950s. They first went into business with each other in 1961, founding a gyroscope-testing venture called Solid State Instruments. Stata would later call the venture "a failure in all respects but one,"—the sale of the company to New England-based Kollmorgen Corp., which provided Stata and Lorber with the funds to launch Analog Devices in 1965.

For ADI's first product, the two moved into a market where they wouldn't have to compete with the almighty IBM—amplifiers that receive and strengthen electrical signal. The company's first product was a small, modular amplifier of this kind, called the AD101. The next year, ADI opened its first international sales subsidiary in the U.K. and began publishing *Analog Dialogue* in 1967, now the longest continuously-published corporate technical journal (recently supplemented by the online offering *Analog Diablog*).

An amplified breakup

Two years later, Stata took the company public in order to raise funds for an expansion into analog-to-digital (A/D) converters, as well as high-quality integrated circuits (ICs), which could perform several different functions on a single chip at a much lower cost than the modules ADI was producing.

With the direction of the company changing, Lorber opted out of the firm in 1969. Lorber has since founded numerous other firms, including Printer Technology in 1971, Torque Systems in 1973 and Copley Controls Corp. in 1984. He continues making power amplifiers, as well as motion-control drives at Copley, located in Canton, Mass., steering the company as president and CEO.

A *Fortune* profile of ADI, circa 1985, detailed the Stata/Lorber split—it cited a strained, "two-headed monster" relationship where Stata felt like a "superclerk" and Lorber didn't like the idea of working for a big company (even if it was his own creation). Stata, a self-described "workaholic," didn't share the sentiment. He would devote so much energy to ADI that, in the same *Fortune* profile, he admitted consciously trying to see more of his family during weekends.

The first monolith

In 1970, ADI introduced the AD550mDAC, its first monolithic integrated circuit, a design that would carry much of the company's innovation throughout that decade. That same year, ADI broke ground on what is today the company's headquarters in Norwood, Massachusetts. The first Japan-based sales office also opened that same year, establishing an Asian Pacific presence for ADI.

Current CEO Jerry Fishman came aboard in 1971, as ADI advanced its design and manufacture of linear semiconductors through the acquisition of Nova Devices. By the end of the 1970s, the company had joined the New York Stock Exchange and rolled out a bunch of firsts, including a high-speed operational amplifier IC, a CMOS digital-to-analog converter and its analog-to-digital counterpart.

How ADI made its first billion

Analog opened IC test and assembly plants in Japan and the Philippines in 1981. The company continued product development throughout the decade, releasing several industry debutantes to beef up its reputation. ADI introduced 33 new ICs in 1987 alone. By 1992, ADI was a Fortune 500 company (it has since fallen to No. 721), and by 1995, it had cracked the top half of *Fortune*'s list of the 1,000 most valuable companies. By 1998, sales topped $1.2 billion, and for the second year in a row the company was named one of *Fortune*'s 100 Best Companies to Work For; that same year, ADI opened a wafer fab in Cambridge, Mass.

ADI made its first headway into the automotive industry in the 1990s, when the company's accelerometer, a sensory chip that reads movement, was first employed by Saab to make its 1994-model cars detect crashes and deploy airbags. The company's products also found use in consumer electronics in this time, as Sony Corporation began using its analog-to-digital converter chips for its digital camcorders.

A change in leadership also took place in the 1990s, as Ray Stata decided to step down as president of the company in order to assume the role of chairman of the board and CEO. Thus, Jerry Fishman was appointed president and COO in 1991. In 1996, Stata passed the reigns of CEO to Fishman.

... and then its second

ADI started the 21st century with a string of acquisitions, including the Northern Ireland-based electronic components-maker BCO Technologies and Chiplogic of Santa Clara, California, a developer of high-level communications processing tools. Although its 2000 performance beat analysts' estimates—with sales topping $2 billion for the first time—the company had trouble in 2001. Despite a number of new products, due to the market environment, orders fell, causing ADI to report lower-than-expected sales and earnings. Profit dropped to $356 million, just over half of what the company had made the year before. Despite the slump, *The Wall Street Journal* still ranked ADI's stock among the top-20 growth stocks from among the world's 1,000-largest companies.

Technology for the future

In response, ADI cranked out some new products, investing $512 million in R&D in 2004 alone. The company not only launched a new Blackfin electronic processor for everything from cameras to car computer systems, but also introduced new performance-leading converter technology that reinforced ADI's leadership position in that product category. At the International Fire and Security Exhibit and Conference held in Birmingham, England, in May 2004, ADI released a number of surveillance and security products, including microchips to run security camera innovations.

As hoped, the investment in R&D brought a turnaround in the company's fiscal fortunes, as sales climbed back above $2 billion. The resurgence was spurred on in part by increasing consumer demand for high-performance electronics such as DVD players and digital TVs and cameras, all of which require ADI's analog components—indeed, consumer electronics would soon become ADI's fastest-growing market. In 2005 the company celebrated its 40th birthday, and, while year-end revenue was down slightly to $2.4 billion from the record $2.6 billion of the previous year, profits reached $26 billion.

Back is the new black.

Like so many of its technology cohorts, ADI was pegged in late 2005 as a potential violator in the wave of stock options backdating scandals that poured out of corporate America. The initial inquiry into ADI's stock option granting practices by the Securities and Exchange Commission came in 2004. Seeking to avoid a protracted battle with the SEC, a tentative settlement was reached in November 2005, by which ADI would pay a $3 million penalty, in addition to reducing the options granted to its executives. So far, so good.

Then, in May 2006, the U.S. Attorney for the Southern District of New York subpoenaed company documents relating to granted stock options, as it also did with several other corporations. While the company cooperated with the investigation and will probably not have to restate its financial results from years past, the affair brought unwanted publicity to the company and has raised stockholder awareness of excessive executive compensation practices throughout the industry.

Still admired

ADI managed to put the unpleasantness aside and make a good year out of 2006. The company was again listed as one of *Fortune*'s Most Admired Companies, ranking fifth in the semiconductor industry for 2006. In June, the company acquired Korean mobile communications technology firm Integrant, followed in September with the purchase of AudioAsics A/S, a Danish company specializing in microphone and audio signal technology.

A major win for the company in 2006 came through its contribution to Nintendo's popular Wii video game console, the key component of which is ADI's 3-axis accelerometer chip, the ADXL330. The application of the technology in consumer electronics is opening new doors for the chipmaker; it is already becoming useful to cell phones and remote controls, allowing users to simply gesture or flick their wrist instead of poring over rows of buttons. ADI intends to be at the forefront as the accelerometer realizes its full potential in everyday appliances. Sales for 2006 reflected the company's healthy future and solid standing, climbing 8 percent to $2.6 billion. This rise reflected growth of 16 percent in the consumer market and 15 percent in the industrial market, which makes up the majority of ADI's business.

Jack-of-all-Blackfins

ADI continued on its consumer electronics-bender in 2007, when several of its products found new applications in the latest tech must-haves. Mobile television, improved cell-phone cameras and improved cell-phone sound were all features ADI showed off at the annual 3GSM World Conference in Spain in February. The company's real MVP of 2007, though, was its Blackfin processor, which is designed to power the multifunctional electronics that are increasingly vogue.

In January 2007, the Blackfin found its way into the Litecomputer, a low cost PC (about $150 a pop) that offers Internet, a music and video media player, instant messaging, and Internet-based calling without a high-maintenance hard drive. X-Digital Systems chose the Blackfin in April for its satellite radio receivers, used to convert digital signals to an audio format for radio broadcasters. The Blackfin processor is also used in the medical technology field; in May it started powering the display and analysis functions of the AfibAlert, an atrial fibrillator that allows homebound patients to monitor their heart rates.

GETTING HIRED

It's all about the benefits

ADI's career web site (www.analog.com/jobs) offers the standard job search functions, allowing you to casually browse its extensive listings or narrow your focus to specific locations and departments. Following the "Engineering Jobs: What and Where?" link will get you a handy chart letting you know what facilities employ what kind of engineers. Applications engineers can forget about finding work with ADI in the Windy City, but perhaps they could learn to love Beaverton, Oregon. Layout engineers, on the other hand, are welcome pretty much everywhere. ADI does much of its business outside the U.S.—approximately 75 percent of revenue is earned abroad—and accordingly more than half of the company's workforce is located at facilities in Australia, Ireland, India, Taiwan, Japan, the Philippines and the U.K.

ADI considers its employees to be its greatest asset, and as such goes to great lengths to curry their favor. The company offers extensive medical, dental and vision insurance, in addition to a slew of life insurance options. There's also education assistance covering 100 percent of tuition expenses (for eligible programs), a retirement plan with company match and a generous dollop of vacation time. In 2005, ADI was recognized as one of the 100 Best Places to Work in the Bay Area—

Visit Vault at **www.vault.com** for insider company profiles, expert advice, career message boards, expert resume reviews, the Vault Job Board and more.

V/\ULT CAREER LIBRARY

77

an especially telling honor; considering the legendary perks offered by other Silicon Valley firms, the company has some tough competition in terms of attracting and keeping employees.

Will solve for cash

The web site also features a section for college job opportunities, which are found at ADI locations throughout the U.S. Internships and co-ops are also available at most U.S. locations. ADI also sponsors the Analog Circuit Design Contest, which runs from November to December and is now entering its fifth year. Participating college students (U.S. and Canada only) are given a series of engineering conundrums to solve—submitting both their answers and their resume by e-mail. First prize is $1,000, but more important than money is the opportunity to get your foot in the door with the firm, which uses the contest as a way to connect with potential employees.

In fact, ADI has a history of reaching out to college engineering programs dating back to the 1980s, when it established engineering chairs at MIT (its founders' alma mater), Northeastern and Lowell University. (Stata has also been a frequent donor to MIT, at one point donating a $25 million building to the institute, then the largest such gift in the organization's history.)

Student or not, the application and selection process is the same across the board. Should you find an opening to your liking, you will be asked to fill out some basic information in an online form before e-mailing your resume to human resources. If ADI likes what it sees, you will be contacted for a phone interview before a face-to-face meeting.

Apple Incorporated

1 Infinite Loop
Cupertino, CA 95014
Phone: (408) 996-1010
Fax: (408) 974-2113
www.apple.com

LOCATIONS

Cupertino, CA (HQ)
Austin, TX
Sacramento, CA
Cork, Ireland
Singapore

DEPARTMENTS

Apple Care
Applications
Facilities
Finance
HR
Information Systems & Technology
iPod Engineering
Legal
Mac Hardware Engineering
Marketing
Operations
Retail
Sales
Software Engineering

THE STATS

Employer Type: Public Company
Stock Symbol: AAPL
Stock Exchange: Nasdaq
Director & CEO: Steve Jobs
2006 Employees: 17,787
2006 Revenue ($mil.): $19,315

KEY COMPETITORS

Dell
Hewlett-Packard
International Business Machines
Microsoft

EMPLOYMENT CONTACT

www.apple.com/jobs

Visit Vault at **www.vault.com** for insider company profiles, expert advice,
career message boards, expert resume reviews, the Vault Job Board and more.

VAULT CAREER LIBRARY

79

THE SCOOP

Thirty years of thinking different

Apple has come a long way since the day in 1976 when the Steves (Jobs and Wozniak, that is) first built their own computer. Today, the Cupertino, California-based company makes a line of constantly evolving laptop and desktop computers, which, equipped with the Mac OS X operating system, vlogger-friendly QuickTime and—in a first for the company—Intel processors running the show, have generated a zealous cult following. The Macintosh has also long been popular in the graphic arts and publishing worlds. Apple ranked No. 159 on the 2006 list of the Fortune 500, and sixth on *BusinessWeek*'s 2007 edition of the InfoTech 100 (ahead of competitor Microsoft, which placed ninth).

The bulk of the company's profits, however, come from its personal digital products rather than its computers. At the outset of the 2000s, Apple co-founder and CEO Jobs radically revised the company's core identity, drawing the firm's attention to high-end electronic devices and services that draw on digital technology. Most notably, the iPod—which works with both PCs and Macs—with its attendant iTunes program and online music store, has revolutionized the music industry, and changed the way people listen to (and record labels distribute) music.

A series of ups and downs

Steve Wozniak and Steve Jobs built the first Apple computer in 1976, working out of Jobs' garage; it's been said that they chose the name "Apple" for their new creation and company because it was Jobs' favorite fruit. The following year, the fledgling tech firm released the Apple II, which proved so popular that it helped launch the personal computer revolution. In 1980, with revenue in excess of $100 million, the company went public.

The company pioneered the PC in 1983 with the release of the "Lisa," the first commercially successful computer with a mouse and GUI (graphical user interface) system, the most popular example of which is Microsoft's Windows. The next year, Apple released the more affordable and accessible line of Macintosh computers. Jobs abruptly resigned in 1985, following a power struggle with then-CEO John Sculley, causing some shareholders to lose confidence in the company. However, Apple quickly redeemed itself in investors' eyes by introducing its LaserWriter printer, PageMaker desktop publishing program and the Mac II.

In 1988, Apple shocked the PC world by filing a copyright lawsuit against its main rival, Microsoft, claiming that the recent Windows 2.03 was too "Mac-like." The lawsuit also revealed some juicy details, like a "secret pact" between the two companies in 1985 to share details of the GUI system, which resulted in the Mac and Windows (the subject of the lawsuit). In a decision still debated to this day, judges in both lower and appellate courts found in favor of Microsoft; a 1994 decision handed Microsoft victory, and Apple's subsequent appeal to the Supreme Court was denied.

A big bite out of the apple

While the litigation was pending, Apple followed up on its earlier successes throughout the early 1990s, producing the first-generation PowerBooks and PowerMacs, and licensing the Mac OS to other companies. But the 1994 conclusion of the Microsoft lawsuit shook it to its core—business suffered in 1995 when the company faced a parts shortage, leading to $1 billion in backorders. Apple also made product missteps, however noble, as with the Newton personal digital assistant, a Palm Pilot-like device that was woefully ahead of its time.

Microsoft truly had its comeuppance with the launch of Windows 95, luring customers away from Apple with a graphical user interface that closely (and legally, although perhaps unethically) mimicked that of the Mac. Windows became the nearly universal OS between 1994 and 1996, and Apple's market share shrank from 18 to 6 percent in the period, prompting some industry insiders to predict the company's imminent demise.

Jobs gets his job back

As it turned out, reports of Apple's demise were greatly exaggerated. In an effort to return to the company's roots, CEO Gil Amelio acquired software company NeXT and announced that its CEO, a guy named Steve Jobs, would return to Apple. The next year, Jobs was appointed CEO after Amelio was ousted in a boardroom coup.

Jobs began to turn the company around, first by ruthlessly cutting departments he deemed unprofitable (for a while, getting fired from Apple was called "getting Steved"), but later, in one of his most crucial moves, by calling an effective truce with Microsoft in 1997. The two firms signed a five-year patent cross-licensing pact and finally settled the longstanding GUI dispute. Microsoft agreed to allow its Internet Explorer web browser onto the Apple OS, purchase $150 million of nonvoting Apple stock and develop versions of MS Office software for Apple.

In following years, Jobs also focused on eliminating the production of Mac clones, which were draining sales from Apple's own line. In May 1998, Jobs unveiled the G3 computer, which was touted as running up to two times faster than PCs with Pentium processors. The G3 arrived in three incarnations: the sleek black PowerBook laptop, the PowerMac desktop and the visually arresting, one-piece iMac.

Apple's eye-popping iMac

The iMac, with its bright colors and bulbous, translucent shell, was a runaway hit with consumers. In promoting the new product, Jobs said, "The iMac is next year's computer for $1,299; not last year's computer for $999." The company sold 1.8 million iMacs in 1999, and sales had surpassed the six million mark by 2002. But the iMac didn't just look different; it was different. For one, it had one-step Internet access.

Its designers also made the unusual decision (at the time, anyway) not to include a floppy disk drive. Jobs justified the move by saying that most information was being transferred via the Internet, CD-ROMs and Zip drives; Apple has been proven totally justified in this assumption. The iBook, the iMac's portable equivalent, arrived in mid-1999—and, by the same time following year, Apple's iBook and PowerBook lines claimed about 10 percent of the retail laptop computer market. On the desktop side, Apple controlled only 4 to 5 percent, behind PC giants like Dell, Compaq and Gateway.

The iPod cometh

Apple, along with other computer, tech and Internet companies, was hurt by the industrywide slowdown in 2001; following a sharp decline in sales, the company reported a net loss of $25 million for the year. Luckily, though, Apple had an ace up its sleeve. In November, the company unveiled the iPod, its disarmingly simple and easy-to-use portable MP3 player with the signature "click wheel."

Consumers snapped up 125,000 of the devices in its first two months on the market. By 2006, 70 million had been sold, in ever-shrinking varieties, including the Shuffle (the smallest and most inexpensive, without a screen), the Mini (the best-selling iPod so far, which has since been replaced by the Nano) and the Video (the largest version, capable of playing movies, TV shows and more). The tech firm has also used the iPod as a limited edition vehicle, producing the "U2 edition" in 2004—a black and red iPod that came pre-loaded with U2 songs—as well as a red Nano to support AIDS research in Africa, unveiled in 2006.

iTune in

Since launching iTunes, the iPod's music player and online store complement, Apple has seen its value—as well as its estimation in the eyes of the public—rise significantly. With a premise as simple as it is revolutionary, the iTunes Music Store grants both Mac and PC users access to songs and albums from all five major record labels, as well as a growing selection of independents. After downloading the program, consumers can listen to a 30-second preview of any song and then purchase a full copy for 99 cents, or the album it's on for $9.99; the store sold over 70 million songs in its first year.

Apple also secures exclusive and unreleased recordings for its store, as well as TV shows, news and entertainment podcasts, and movies. In the years following its U.S. launch, iTunes has sprouted specialized versions from Europe to Japan to New Zealand, as well as sections (such as iTunes Latino) devoted to different genres. By December 2006, the iTunes Store's song-selling feature, once viewed as a stagnant enterprise, was growing at a higher rate than in years past, with sales increasing by 84 percent over the course of the year.

Some industry pundits have suggested that the lucrative iTunes library could be even more lucrative if Apple raises prices on the more popular music, TV shows and movies that it sells. So far, though, Apple has fought price-rises like an online Robin Hood (or Wal-Mart), keeping all of its prices level regardless of popularity. In late August 2007 NBC-Universal cancelled its iTunes contract in a huff when Apple refused to raise prices on popular shows like *The Office*. Apple accused NBC of being greedy and the network responded that Apple's retail strategy is self-serving, seeking to promote the iPod and iTunes at the expense of the content they play/sell. Neither can much do without the other: NBC is in last place among the broadcast networks and Apple relies on it for 30 percent of its iTunes TV show sales. As of fall 2007 the two parties were working on a new contract.

CellTunes (and BMWTunes)

Music-loving cell phone addicts got the best of both worlds in 2005, when iTunes and Motorola unveiled a technology that allows songs purchased through the iTunes Store to play on certain Motorola models. The phones, equipped with Apple software called iTunes Mobile Music, allow transfer of songs from a PC or Mac to the phone over a cable or wireless connection. Also, through a deal between Apple and the BMW group, drivers of "Beamers" and Mini Coopers can plug their iPods directly into their cars.

None too Zune

Microsoft's rival MP3 player, the Zune, debuted in fall 2006 to widespread skepticism from industry watchers. Though it came in a 30GB model and featured radio as well as photo and song sharing (for up to "three days or plays"), nothing it offered seemed a good enough reason to make the switch from the iPod. The Zune was rather obviously viewed as a joyless, business-minded product, the beginning of Microsoft's assault on the iPod-dominated market. And, while the Vibez, a German-engineered player with a slick click wheel, was seen as a better alternative, it wasn't widely available in the U.S. Apple kept 70 percent of the MP3 player market, and the iPod captured Christmas yet again.

iPhone mania

Just as the Zune was introduced, Apple had enigmatic commercials on TV that made its new product look like a magical instrument from the future. This product was the iPhone, and the hype surrounding its subsequent June 2007 release was almost deafening. The Apple faithful started lining up outside the company's flagship store on Fifth Avenue in New York City days before the phone's release; crafty consumers in other parts of the country scoped out less-trafficked AT&T stores for a chance to grab an iPhone away from the crowds gathering at suburban mall Apple stores.

Despite worries that the status level and steep price ($500 for the 4 GB model and $600 for the 8 GB model) of the iPhone would lead to inflated prices for the accompanying wireless service plans from AT&T, the carrier's reasonable plans didn't deviate from existing offerings, with plans priced at the $60, $80 and $100 per month. Consumers have also grumbled that the iPhone's sole affiliation with AT&T in America will deter them from buying the product, although an anonymous computer hacker revealed a free open-source procedure in mid-2007 to "unlock" the iPhone, and accommodate other wireless providers.

Regardless, iPhones have been flying off the shelves. Despite investor worries that demand was sinking, the one millionth iPhone was sold in September 2007, only three months after the gadget's incipience. The figure was reached a week after Apple slashed the 8 GB model price by a third, which spiked demand for the phones. And don't expect sales to slow anytime soon—CEO Jobs said that the iPhone would represent one of the company's major cash cows during the 2007 holiday season, next to Macintosh computers and the ubiquitous iPod.

iPhone Euro

And even before the season arrives, the iPhone will head overseas, set for release in the U.K. and Germany in November. The devices will be sold across the pond for a starting price of £269 and €556, respectively. Domestically Apple has linked the iPhone with AT&T coverage, to some consumers' chagrin, and abroad it is using Telefonica's O2 wireless unit in Britain and Deutsche Telekom's T-Mobile in the Rhineland. Days later, France joined in the fun, selecting France Telecom's Orange subsidiary as service carrier.

Some skeptics wonder why the European iPhone is the same model as the American version, which will make it unable to harness high-speed European wireless to its full advantage. Apple has countered the complaints, saying that the battery power necessary to accommodate the third-generation wireless networks in Europe would be exorbitant and impractical.

Cisco courts the iPhone

A scuffle over ownership of the iPhone name with Cisco Systems resulted in a February 2007 settlement that allows both companies to use the iPhone trademark on their products. Cisco, which has been using the iPhone name since 1997, craftily used the lawsuit to horn in on the Apple iPhone's predicted popularity—the companies are currently looking into making the iPhone compatible with Cisco's business and consumer phone systems.

New iTerritory

The Apple TV, originally dubbed iTV, might be the next new Apple gadget to explode into retail stores. Shipped to stores in March 2007, and priced at about $300, the TV-top box (which looks like the Mac Mini) is able to stream any media playing on a Mac onto a TV screen. And what's even more aggressive than the Apple TV's potential television takeover? The fact that it will be compatible with Windows.

Options questions

Just as Steve Jobs was earning praise for these exciting new products, he and his company were under investigation for stock options backdating. An internal

Visit Vault at **www.vault.com** for insider company profiles, expert advice, career message boards, expert resume reviews, the Vault Job Board and more.

VAULT CAREER LIBRARY

85

investigation instigated by Apple's board in June 2006 turned up something fishy, and soon the SEC and the U.S. attorney's San Francisco office began investigations into Apple's practices regarding executive compensation.

While backdating options isn't illegal, failing to report it—and then profiting from their sale, as Jobs did—is quite illegal. It remains unclear if Jobs will personally be implicated in the federal probe, but already Apple has agreed to pay $84 million to make up for the improper backdating of 6,400 options granted between 1997 and 2001.

Staying stylish, targeting big business

Ultimately, though, money talks, and Apple has every right to shout its triumph from the rooftops after a strong year of financial growth. The firm finished 2006 with profits up by a whopping 46 percent, as the company outpaced the expansion of rivals like HP. All is not perfect in the land of Apple, however, as it will be difficult to maintain the iPod's incredible growth, and the stock options brouhaha has twice-delayed the company's SEC filing.

On the other hand, predictions that Apple would begin its assault on the business world—and sell up to nine million Macs in 2007—reassured devoted fans that Apple would continue to fly high, extending its two-year streak of increasing its value and unveiling new, much-loved products for every market sector. To this end, the 2007 releases of iPhone, Apple TV and Leopard, the latest generation of the Mac OS X operation system should go a long way toward reiterating Apple's position as everybody's favorite tech firm. So far, Apple's 2007 releases have significantly boosted the bottom line; quarterly financial results released in July detailed $5.41 billion in revenue and $818 million, an increase from $4.37 billion and $472 million from the same quarter the previous year.

GETTING HIRED

Take a bite out of Apple

Unsatisfied with driving innovation only in the tech and music industries, Apple also wants to lead the pack in the benefits it grants its employees. Through its FlexBenefits program, Apple-ites can choose benefits that best fit their unique little lives. The company also offers 401(k) with company match, an employee stock purchasing program, and on-site fitness centers. And you don't have to be a born-and-raised Apple fan to fit in here; in fact, the company says it appreciates diversity in employee background, which extends to personal computer history. Taking diversity one step further, the company also partners with several organizations to ensure a nonhomogeneous workforce, including National Black MBAs, the National Society of Hispanic MBAs and the Society of Women Engineers.

With job openings for staid tech workers as well as recent grads in areas like hardware engineering, software engineering, applications, finance, operations and marketing, Apple is looking for candidates with skills. Just what skills, exactly, are outlined on the company's job web site (www.apple.com/jobs), which provides a breakdown of necessary abilities in its overview of each job category. When you've determined the area of operations that's right for you, a listing of all available jobs for that category is available for your perusal. Found something you like? Interested parties can set up a profile on Apple's job page and submit their resume online.

Apple, the next generation

For the soon-to-graduate set, Apple makes an appearance at several campus events throughout the year, including pit stops at Stanford, Harvard, Duke University, the University of California at Irvine and New York University (to name a few—a full schedule can be found at www.apple.com/jobs/us/pro/college/campus_events.html). The company also offers full-time and part-time paid internships in its engineering divisions; while internships are occasionally offered in nontechnical areas like finance, marketing, or graphic design, these positions often go to candidates with some technical experience.

Along with learning about the company and industry, interns enjoy more immediately rewarding perks like discounts on Apple products, free health insurance and a casual dress code—not to mention that rarest of internship compensations: money. Internships are offered at Apple's headquarters in Cupertino and generally last for three months during the summer, although there are a limited number of six-month co-op positions available throughout the year.

Visit Vault at **www.vault.com** for insider company profiles, expert advice,
career message boards, expert resume reviews, the Vault Job Board and more.

VAULT CAREER LIBRARY

87

OUR SURVEY SAYS

Great place to work, with some reservations

Talk to some Apple employees, and you might come away with the impression that there's no better place to work than Apple. "Funny, brilliant, relaxed" co-workers and "modern, spacious, beautiful" offices filled with comfortable couches and huge picture windows make work time a pleasure." I can't imagine any reason not to work here," coos one insider." The atmosphere is extremely relaxed and open, with a very friendly culture," gushes a source in the finance department. "It is not uncommon to see people playing Frisbee, volleyball or basketball on the campus."

While most admit "Apple has changed over the past 10 years," employees stay put. "Apple believes strongly in hiring from within the company, so opportunities to advance are plentiful. There are also numerous opportunities to shift horizontally within the company as well ... Employees are generally very dedicated, and it's easy to find people who have worked here for 20+ years who are still very happy with their jobs," notes another source. Comments like "absolutely no downers," "everyone is an essential part of the team," and "Apple remains my dream company," could make anyone want to pack their bags for California.

It's not all sunshine and sweet dreams in Cupertino, however. "Yeah, Apple! Right? Well maybe," says one insider. "If you put your heart into the 'Mac vs. Windows' jihad, it's very frustrating to see an inferior OS getting all the attention," says one programmer. The downside of the unusually creative atmosphere is that sometimes trying to enforce a corporate strategy is like "herding cats." The pay is "not bad, but not the best in the Bay Area," though some contacts add "the people make up for the pay."

Still want to hightail it to Cupertino? Vault sources report going through two (or more) rounds of interviews. One intern pointed out that his interviewer "asked me one question about my resume, then opened the rest of the time for me to ask him questions. Coming prepared with five or six good, open-ended questions is extremely important."

Most of our contacts say they have no plans to leave Apple anytime soon. A dress code that is casual "with a capital C" and a uniquely Californian "work hard/play hard" culture make up for the hours toiling at the office. "Working at Apple is a wonderful experience," one source says, noting that the environment is "intellectually challenging, inspiring and rewarding." "The culture is the best," another gushes.

The main R&D campus in "Cupertino is pretty posh, with a huge lawn in the middle, a very nice cafeteria and a big indoor atrium with a cafe." Perks at the liberal icon include "same-sex domestic partner medical benefits." And while the company is mostly white and male," Apple does look for women and minorities." Insiders say "it's common knowledge that Apple's progressive nature is a large part of its image."

Applied Materials, Inc.

3050 Bowers Avenue
Santa Clara, CA 95054-3299
Phone: (408) 727-5555
Fax: (408) 748-9943
www.appliedmaterials.com

LOCATIONS

Santa Clara, CA (HQ)
Albuquerque, NM • Allentown, PA •
Arlington, TX • Austin, TX • Austin,
TX • Beverly, MA • Boise, ID •
Louisville, KY • Burlington, VA •
Camus, WA • Chandler, AZ •
Colorado Springs, CO • Dallas, TX •
Eugene, OR • Hayward, CA •
Hillsboro, OR • Hopewell Junction, NY
• Hudson, MA • Lake Oswego, OR •
Manassas, VA • Mountain View, CA
• Orlando, FL • Phoenix, AZ •
Portland, OR • Poughkeepsie, NY •
Richmond, VA • Rio Rancho, NM •
San Antonio, TX • Tempe, AZ •
Tucson, AZ • Vancouver, WA

Additional offices in China, India,
Malaysia, France, Israel, Taiwan,
Germany, Japan and South Korea.
Applied Materials has locations
throughout Canada, China, Europe,
Israel, India, Malaysia, Singapore,
Japan, Korea, Taiwan and the United
States.

THE STATS

Employer Type: Public Company
Stock Symbol: AMAT
Stock Exchange: Nasdaq
Chairman: James C. Morgan
President & CEO: Michael R. Splinter
2006 Employees: 14,000
2006 Revenue ($mil.): $9,167

DEPARTMENTS

Administration • Corporate Affairs •
Environmental Health and Services
(EHS) • Finance • HR • Information
Systems • Legal • Manufacturing •
Marketing • Materials Management
Operations• Quality & Safety • Real
Estate & Facilities • Sales • Technical
Publications • Technical Support •
Technical Training • Technology

KEY COMPETITORS

ASML Holding
KLA-Tencor
LAM Research

EMPLOYMENT CONTACT

www.appliedmaterials.com/careers

THE SCOOP

Time to make the chips

Applied Materials leads a large niche of the semiconductor industry. It prides itself on being the world's largest purveyor of nanomanufacturing technology, which means that it sells the machines that other companies use to make semiconductor chips, solar modules and flat panel displays. When Applied first came on the scene in the late 1960s, semiconductor machines were produced by the large companies that made the machines themselves, such as IBM and Texas Instruments. By catering to this specific part of the market, Applied Materials has made itself a Silicon Valley mainstay, and the largest company of its kind.

Through its four business segments—silicon, fab solutions, display and adjacent technologies—the company strives to meet all the demands of its wide-ranging customer base. Applied Materials' silicon arm makes the manufacturing equipment that makes semiconductors. The fab solutions department works in tandem with both the silicon and solar businesses, providing services to make the manufacturing process more efficient. The display segment is a bit more self-explanatory—here Applied Materials manufactures equipment used to build liquid crystal displays (LCDs) for televisions and computer monitors. Through its adjacent technologies, the company applies its display technology to solar photovoltaic cells (which harness the awesome power of the sun to run your blow dryer or coffee grinder) and energy-efficient glass.

The Applied Materials empire increasingly stretches far beyond U.S. borders; in addition to its corporate offices in Santa Clara, it has sales and service offices in Europe, Israel, Japan, Malaysia, South Korea, Taiwan, Singapore and China. Its research, development and manufacturing centers are based in the U.S., Israel and Europe, while its technology centers are located in Japan, South Korea and Taiwan. As many semiconductor clients are based in Asia, that corner of the globe accounts for the company's largest amount of orders (70 percent of Applied Materials' sales), with Taiwan alone leading the pack in 2006 with a 23 percent share. North America is next with 19 percent, followed by Europe with 11 percent. Applied Materials' biggest customers include Intel, Advanced Micro Devices, IBM, Samsung, Toshiba and Freescale Semiconductor.

Visit Vault at www.vault.com for insider company profiles, expert advice, career message boards, expert resume reviews, the Vault Job Board and more.

VAULT CAREER LIBRARY   91

The chips are stacked

Michael McNeilly, a scientist with a background in chemistry, co-founded a company called Apogee Chemicals in 1964, at the age of 25. Very shortly, he realized an opening in the semiconductor industry for the competent handling of all the chemicals involved in producing nanomanufacturing machines. In a 2004 interview with the Semiconductor Equipment and Materials International Foundation (SEMI), he said "it became obvious to me that the industry was populated by guys that … really didn't understand chemistry (and) that the future of semiconductor manufacturing was going to be based on their ability to utilize a variety of hazardous chemicals and other materials for device production and that that was not in place."

Semiconducting an orchestra of machinery

McNeilly left his job at Apogee Chemical, and in November 1967 joined with others to create Applied Materials Technology, Inc., focused on the manufacturing and sales of semiconductor materials and equipment, primarily gases and gas flow panels. Within five years, the company had established operations in Europe and gone public. McNeilly quickly realized that going public would change the way his company operated, and he hired James Morgan in 1976, a former executive at Bank of America, to handle the business aspect of operations.

Within a year, Morgan replaced McNeilly as CEO, and identified the manufacturing of semiconductor equipment as the firm's best growth opportunity. It was a wise choice, allowing the firm to capitalize on the just-beginning revolution in personal computers. In 1987, Applied Materials revolutionized chip manufacturing with its Precision 5000 single-wafer, multichamber processing systems. After further expansion into Europe and Asia in 1992, Applied Materials became the world's largest manufacturer of semiconductor equipment, a title it has retained throughout the many fluctuations of the cyclical microchip industry.

Growing through buying the three O's

Since the late 1990s, much of Applied Materials' growth has been driven by acquisitions. Some of its biggest deals included the 1997 purchases of Opal (for $175 million) and Orbot Instruments ($110 million), which gave Applied Materials entry into the metrology (measurement) and inspection equipment markets, respectively. In 1999, the company added chemical-mechanical polishing machines to its line via the $135 million purchase of Obsidian.

The slowdown

In 2001, an industrywide slowdown in the semiconductor sector began to chip away at many companies' top and bottom lines. When orders began to trickle back in 2002, the company recalled some employees it had laid off and opened a new facility in Sunnyvale, California.

But in 2003, Applied Materials' sales fell again and income fell further, albeit at a slower pace, prompting analysts to suggest that a full recovery in the chip equipment market would take longer than expected. In an effort to bring a fresh perspective to the company, Applied named a former Intel executive vice president, Michael R. Splinter, as its new president and CEO. James Morgan stayed on as chairman of the board of directors.

Finally moving on up

In 2004, Applied's sales soared to $8 billion, a high that hadn't been attained in three years. The three-year drought—one that management considered the worst in its nearly 40-year history—was over, as Applied Materials reveled in a profit of more than $1 billion. With its newfound riches, the company purchased Torrex Equipment, a developer of multi-wafer technology, in June 2004, followed by Metron Technology, a San Jose, California-based provider of marketing, sales, services and support for semiconductor manufacturers and equipment suppliers, for $85 million that December.

On average, Applied invests more than $1 billion in research and development annually. As a result, the firm introduced 10 new products in 2005, including a couple of industry firsts—the OPC Check (a tool that allows automated qualification of chip designs) and the UVision system (a laser 3-D inspection tool that locates chip defects). To foster more of these kinds of breakthroughs, Applied Materials launched a subsidiary called Applied Ventures in November 2005; its purpose is to invest in small tech companies that could bring new innovations into the parent organization.

Another year in the clear

The maneuverings of 2005 proved wise, as sales in 2006 exceeded $9 billion for the first time in six years. A stock repurchase plan was put into effect, bringing 20 percent of company stock back home by September 2006, at which time the board of directors announced yet another repurchase program—this one for $5 billion and slated to occur over the next three years.

Applied also continued to cut costs, as it announced the selling of several facilities it had gained through mergers over the years, including sites in California, Massachusetts, South Korea and Japan. On the other hand, a new 106,000-square-foot engineering and software service center in Xi'an, China extended the company's reach in that region. (Applied Materials had been the first semiconductor company to establish a presence in China, opening a Beijing outlet in 1984.)

Cleanliness is a very nice thing

Applied Materials returned to its tried-and-true method of growth through acquisition, buying a number of cleaning technologies. In 2005, it helped itself get cleaner by purchasing EcoSys, a part of ATMI Inc., which treats gas emissions produced by semiconductor manufacturing to minimize their environmental impact, and some single-wafer cleaning technology from SCP Global Technology. The company was serious about getting clean, as it also bought ChemTrace Precision Cleaning for $22 million in 2005; Applied's Metron Technology subsidiary purchased the Singapore-based parts cleaning and recycling firm UMS Solutions for $10 million.

Who loves the sun?

Applied Materials put its new thin film division to work right away. Long an active participant in corporate and social responsibility, as evidenced by its numerous awards—most recently, in February 2007, it was named one of the 100 Best Corporate Citizens in an annual survey by *Corporate Responsibility Officer* magazine.

Indeed, the company announced that March the installation of a nearly two-megawatt solar power plant on the grounds of its Sunnyvale campus. Upon completion in 2008, the generator will produce enough electricity every year to juice 1,400 homes, and will be one of the largest solar installations on any corporate campus in the nation—now there's a good corporate citizen! The company seems enamored of solar power; in July 2006 it bought Applied Films, a manufacturer of thin film equipment used in making flat panel displays, solar cells and energy-efficient glass, and later it invested in Solaicx, a firm that makes single-crystal silicon wafers used in the photovoltaic industry. Following the announcement of its solar venture, the company won a contract with Spanish-based T-Solar Global to provide Europe with its first ultra-large solar panel production line.

Bringing on Brooks

Applied Materials gave its fab solutions arm a little love as well, when it purchased Brooks Software, a provider of factory management software, for $125 million in 2007. Formerly a division of Brooks Automation, the new software firm brings its knowledge of semiconductor and flat panel display manufacturing processes to the already-thoroughly versed Applied Materials. Although Brooks Software might not be able to teach the folks at Applied Materials anything they didn't already know, its addition to the company ranks makes Applied Materials's portfolio of factory management software the largest in the industry.

GETTING HIRED

Extensive info has been applied online

Applied Materials has extensive hiring information at www.appliedmaterials.com/careers. Job listings are sorted by region and resumes may be submitted online. For those suffering from a resume deficiency, the company even provides an online resume builder. The "job agent" function allows you to enter and save search criteria; the company will contact you when open positions matching your interests come up. Applied's careers site also hosts a special section for exiting military personnel; recognizing the skills gained through duty in the armed forces, the company works with local military transition centers to match servicemen and women with opportunities. It's because of efforts like this that the company earned a spot on *G.I. Jobs'* 2006 Top 50 Most Military-Friendly Employers list.

From the classroom to the boardroom (or at least a cubicle)

Recent grads should look into the company's "College to Corporate" development program, a full-time paid opportunity to get your foot in the Applied Materials door. Designed to train and develop graduates on the ins and outs of nanomanufacturing technology, the program welcomes applicants with technical backgrounds (materials science, physics, computer science and engineering majors) as well as those with business degrees in supply chain management, finance, accounting, business administration and the like. A graduate fellowship program and internship and co-op opportunities accommodate those who are still in school but looking to add a little corporate flavor to their lives. Details for all these opportunities are listed under the

"College Recruiting" portion of the careers web site, while details on Applied Materials' myriad of college campus visits can be found under the "Job Fairs" link.

Nice perks

Once you're in, Applied Materials wants to keep you happy. Some of the benefits on offer—apart from the usual health and dental insurance—include a 401(k) with company matching, training and development through on-site learning centers and academic partnerships, wellness programs and athletic facilities, profit sharing, employee stock purchase plans, credit union membership, service awards, community and volunteer giving opportunities, and tuition reimbursement.

ASML Holding N.V.

De Run 6501
5504 DR Veldhoven
The Netherlands
Phone: +31-40-268-3000
Fax: +31-40-268-2000
www.asml.com

LOCATIONS

Veldhoven, Netherlands (HQ)
Albany, NY • Albuquerque, NM •
Austin, TX • Bloomington, MN •
Boise, ID • Colorado Springs, CO •
Eugene, OR • Hillsboro, OR •
Hopewell Junction, NY • Hudson,
MA • Irvine, CA • Manassas, VA •
Pleasant Grove, UT • Portland, ME •
Portland, OR • Richardson, TX •
San Jose, CA • Sandston, VA •
Santa Clara, CA • South Burlington,
VT • Tempe, AZ • Wilton, CT

50 locations in 14 countries
worldwide.

DEPARTMENTS

Administration • Communications •
Facility Management • Finance &
Accounting • General Management
• HR • IT • Legal • Manufacturing •
Marketing & Sales • Planning &
Logistics • Procurement • Quality •
Research & Development • Service
& Maintenance • Training &
Publications

THE STATS

Employer Type: Public Company
Stock Symbol: ASML (ADR)
Stock Exchange: Nasdaq
Chairman: Arthur van der Poel
President & CEO: Eric Meurice
2006 Employees: 6,000
2006 Revenue ($mil.): $4,111

KEY COMPETITORS

Canon
Nikon
Ultratech

EMPLOYMENT CONTACT

www.asml.com/careers

Visit Vault at **www.vault.com** for insider company profiles, expert advice,
career message boards, expert resume reviews, the Vault Job Board and more.

VAULT CAREER LIBRARY   97

THE SCOOP

Stamp of approval

When it comes to advanced lithography and semiconductor solutions, ASML is king. The firm makes and distributes photolithography and scanning products needed to print the circuitry patterns found on silicon wafers. These machines are the heart of the manufacturing process to produce integrated circuits (aka ICs or chips) and can cost up to €30 million apiece.

It is the largest chip equipment maker in Europe, and a global market leader with 63 percent market share, well ahead of rivals Nikon and Canon. It boasts customers such as Taiwan Semiconductor Manufacturing Company, Samsung and Intel. As a producer of top-shelf systems, the company has relied on superior engineering, a specialized approach and innovative business strategies, such as outsourcing, to stay competitive.

High-profile parents

ASML, initially known as ASM Lithography Holding, began in 1984 as a joint venture between Advanced Semiconductor Materials (ASMI), a global semiconductor equipment manufacturer, and Royal Philips Electronics, one of the world's biggest electronics firms.

ASMI, which is active in semiconductor equipment outside lithography and does not compete with ASML, sold its 50 percent stake in the venture to Philips in 1990. ASML went public in 1995, and in following years Philips sold its entire shareholding in the company.

While the desertion of it parent companies seemed to indicate the semiconductor field was not a growth industry, ASML's superior engineering has allowed it to weather periodic downturns quite well. The next industrywide slump came in the mid-1990s, as progresses in chip production again outpaced demand. Very possibly, ASML was saved by its PAS-2500 invention, a chip-processing machine ideal for outsourcing. However, when chip sales started to slow once more in 2004, ASML began looking for new sources of profit; another breakthrough was the introduction of the TWINSCAN system, a system were wafers are measured and exposed in parallel, which gives an enormous boost to the productivity of the systems and is as yet unique in its kind.

When worldwide chip sales slumped again soon afterward, ASML increased machinery prices by 60 percent in October 2005, helping to bring in profits. The company also adjusted to its volatile market by adopting a flexible work philosophy in 2006, asking staff to work longer hours during times of high demand, and requiring less of them during downturns in chip-making's somewhat predictable cycle.

Expanding East ... and West

In 2006, 83 percent of ASML's customers were based in Asia, but its historical headquarters and suppliers were predominantly European. Citing the need for Europe's largest chip-maker to have a global profile, and no doubt because the region was already home to its main competitors Nikon and Canon, ASML told the press in March 2006 that it would begin expanding its research and development operations in Asia. The company predicted that the formation of a research center in Asia, including training in customer support and manufacturing, would take between three and seven years.

Befitting a company its size, ASML had also been looking to expand west into the American market. In 2000, it announced its intent to acquire the Californian firm SVGL (Silicon Valley Group Lithography) for $1.6 billion, instantly arousing security concerns from a lobbying group, the U.S. Business and Industry Council. In May 2001, ASML succeeded in acquiring SVGL, selling off Tinsley Labs in the process to satisfy security concerns. In 2003, the company also sold off SVGL's track businesses and thermal product lines, choosing to focus specifically on semiconductors and lithography, its traditional strength.

ASML then acquired California-based Brion Technologies for $270 million in December 2006. The buyout gave ASML another holding in computerized semiconductor design and the advanced lithography R&D sector. According to ASML, Brion, founded in 2002, is a leader in the area of design and manufacturing solutions, and the company expected the acquisition to increase profits—and ASML's lead over Canon and Nikon—in the future. *BusinessWeek* agrees that ASML has an edge over its competitors, ranking it ahead of Canon and Nikon on its InfoTech 100 in 2007 (they were ranked No. 40, No. 51 and No. 86, respectively).

Lookin' good

The end of fiscal 2006 brought good news to ASML—it had increased its market share to 63 percent, with record sales of €3,597 million ($4,749 million), up 42

Visit Vault at **www.vault.com** for insider company profiles, expert advice, career message boards, expert resume reviews, the Vault Job Board and more.

VAULT CAREER LIBRARY   **99**

percent from 2005. In addition, the firm's net income more than doubled, going from €311 million ($408 million) in 2005 to €625 million ($820 million) in 2006. However, the boom in sales worried some analysts, who predicted that, based on decreases in cell phone, TV and computer chip sales, 2007 could bring a semiconductor market hurt by oversupply. ASML thought otherwise: the company predicted that sales of chip components, especially flash memory drive pieces and PC chips, would increase over the course of 2007.

GETTING HIRED

Make your mark at ASML

ASML's careers site (www.asml.com/careers) provides links to video interviews with employees, FAQs and a virtual tour of an ASML cleanroom. The page also provides links to local HR personnel ("Career Contacts"), global events ("Events"), benefits and career information ("Working at ASML"), and open positions ("Vacancies"). The vacancies page allows prospective employees to search by region (the U.S., Europe or Asia), country, functional area, and education and training background.

In order to realize the growth ambitions, ASML is looking to hire hundreds of technicians for R&D, production and customer service. In particular the company is looking for experienced (more than five years) physicists, software specialists, mechanical and electronic engineers, who will support its entrepreneurial spirit and international ambitions, and who have the skills to collaborate in multidisciplinary teams. Employees can grow into management functions, but they can also become highly valued fellows with deep knowledge about specific technologies.

Functional areas (aka departments) include administration, communications, facility management, finance and accounting, general management, HR, IT, legal, manufacturing, marketing and sales, planning and logistics, procurement, quality, R&D, service and maintenance, and training and publications. Each job listing page contains links at the bottom to e-mail, save or apply for the position. Prospective interns can visit ASML's intern site (at www.asml.com/asmldotcom /show.do?ctx=1313) for an internship search engine and information on how to apply. Finally, ASML hires can also submit general application for consideration by region.

Atmel Corporation

2325 Orchard Parkway
San Jose, CA 95131
Phone: (408) 441-0311
Fax: (408) 436-4200
www.atmel.com

LOCATIONS

San Jose, CA (HQ)
Anaheim Hills, CA • Annandale, NJ
• Austin, TX • Belleview, WA •
Berkeley, CA • Boston, MA •
Boynton Beach, FL • Chicago, IL •
Cleveland, OH • Colorado Springs,
CO • Columbia, MD • Detroit, MI •
Durham, NC • El Dorado Hills, CA •
Houston, TX • Lake Oswego, OR •
Lakewood, CO • Minneapolis, MN •
Nashua, NH • Plano, TX • Raleigh,
NC • Scottsburg, IN • Totowa, NJ •
Westlake Village, CA

39 locations worldwide.

DEPARTMENTS

Administrative Support/Clerical •
Audit • Design • Engineering •
Facilities • Finance • HR • Investor
Relations • IT • Legal •
Manufacturing/Operations •
Materials/Purchasing •
Sales/Marketing • Technician

THE STATS

Employer Type: Public Company
Stock Symbol: ATML
Stock Exchange: Nasdaq
Chairman: David Sugishita
President & CEO: Steven Laub
2006 Employees: 7,992
2006 Revenue ($mil.): $1,671

KEY COMPETITORS

Infineon
Freescale
Microchip
Renesas
Samsung
STMicroelectronics

EMPLOYMENT CONTACT

www.atmel.com/careers

Visit Vault at **www.vault.com** for insider company profiles, expert advice,
career message boards, expert resume reviews, the Vault Job Board and more.

VAULT CAREER LIBRARY   **101**

THE SCOOP

Chips ahoy!

Atmel makes semiconductors for a wide range of uses: its products include nonvolatile memory devices such as flash and ROM chips, which are used to retain memory when a device is turned off, application-specific integrated circuits and programmable logic chips. Atmel's offerings can be found in consumer electronics, wireless communication devices and industrial and military applications. Its main focus is divided among four units: nonvolatile memory devices, microcontrollers, radio frequency products, programmable logic devices and application-specific integrated circuits (ASICs). Within those units, the company's products run the gamut from industrial image sensors to digital camera and TV processors to chips for high-security "smart" credit cards.

Based in San Jose, California, Atmel sells its products to over 60,000 customers in over 60 countries throughout North America, Europe and Asia, and runs 39 design centers worldwide. Today, it is a global leader in the semiconductor business, and has recently become one of just a handful of companies capable of integrating dense nonvolatile memory, logic and analog functions into a single chip. Atmel also has one of the largest libraries of intellectual property in the industry, holding more than 1,300 analog and digital patents.

Formed in a flash

George Perlegos, a former design engineer for Intel, founded Atmel in 1984. He took the firm public in 1991, and it has since grown into the world's largest manufacturer of parallel and serial EPROMs (erasable programmable read-only memory chips) and EEPROMs (add "electrically" to what came before). In 1993, Atmel acquired Concurrent Logic, which makes user-programmable chips.

The company opened a Colorado Springs manufacturing facility in 1994 and began building a manufacturing plant near its San Jose headquarters one year later. Also in 1995, Atmel penned an agreement with Paradigm Technology to license its SRAM (static random-access memory) technology used in multimedia chips. By 1996, nonvolatile memory was in high demand, and Atmel's revenue and profits soared— the top line nearly doubled to just over $1 billion—as it was one of the major suppliers of flash chips.

Temic, then tough times

Atmel purchased Temic Semiconductor, a European chipmaker previously owned by Daimler Benz, in March 1998. Operating as an independent entity, the acquisition gave Atmel expertise in radio frequency circuitry design, microcontrollers and a silicon germanium process technology. In April 1999, Atmel purchased Motorola's Smart Information Transfer semiconductor business to strengthen its phone data storage position. The company debuted its programmable system-on-a-chip, the world's first, in October 1999. Now indispensable in the consumer electronics industry, this technology puts all the varied functions of a gadget (say, audio, image and memory functions, in the case of a cell phone) onto one microchip. By 2000, Atmel was bursting with good news; revenue stood at $2 billion, far above the previous year's $1.3 billion, and net income, at $266 million, greatly outshone 1999's $53 million.

The good news didn't last long, though. An overall slump in the economy, partly due to the September 11 terrorist attacks, prompted Atmel to eliminate 2,500 jobs—more than a quarter of its staff—to stay afloat in what was, according to the company, "the most difficult business climate our industry has ever faced." The cuts saved $600 million annually and would enable Atmel to post an operating profit of $415 million in 2002. Atmel ended volume manufacturing at two chip-making factories and indefinitely postponed plans to launch a new eight-inch wafer fabrication plant in the U.K.

Soldering on

Like its chip-making brethren, Atmel endured after the tech meltdown. Fiscal 2001 revenue came in at $1.5 billion, 25 percent less than the year before. Year-end restructuring, to the tune of $481 million, sent income into the red zone, and the company recorded a loss of $331 million. In the meantime, Atmel kept its nose to the grindstone and continued to roll out new products and alliances. In December 2001, the company announced the opening of a mixed-signal design center in Germany, focused on automotive applications.

In January 2002, Atmel released a security chip featuring encryption technology, aimed at providing secure storage of sensitive government and commercial information. Next, Atmel gained licensing rights to silicon chip technology from Tantivy Communications, a privately held wireless data communications maker in Melbourne, Florida, allowing it to produce silicon chips for wireless Internet access.

The cycle keeps on spinning

Though restructuring continued, profits continued to fall despite the company's best efforts—2002 brought a net loss of $641 million. However, by the end of 2003, Atmel was pulling in better-than-expected revenue, signaling what CEO Perlegos referred to as "the upturn for the next semiconductor cycle." Indeed, year-end income for 2003 was up more than $500 million, although that still meant a loss of $117 million. By the close of 2004, loss had shrunk to $2.4 million, and profitability seemed within reach (revenue for 2004 increased to $1.6 billion, up 24 percent from 2003).

The big player for Atmel that year was its ASIC division, which accounted for more than a third of the company's revenue. The increasing need for security (and efficiency in maintaining it) brought heightened demand in that division's Smart Card technology, which has beefed up security for debit cards, eased travel time in contactless mass transit cards and burrowed into difficult-to-counterfeit electronic passports. Atmel, assured by its shrinking losses that it would be back in black in no time, was able to reduce its debt by $48 million with cash generated throughout 2004, while also investing $241 million for new equipment.

There's a hole in Atmel's bucket

But 2005 presented still more challenges to the company's bottom line. Unfavorable exchange rates took a $6 million toll on income, a drain felt all the more keenly because of the company's inability to stop it. Since 86 percent of the company's revenue comes from foreign markets (more than half of its buyers are Asian), Atmel is at the mercy of foreign currencies. Struggling to cut costs, Atmel took a loss on the sale of its Nantes, France, facility in December 2005 and later that month, finally abandoned its plans for a Colorado Springs wafer fabrication plant (construction for that facility started in 2002 only to be halted later that year).

Then the year-end financial report came in with more bad news. Sales in the nonvolatile memory segment fell 12 percent over 2005 due to competitive pricing, while microcontroller sales fell another 6 percent as a result of declining demand from military and aerospace organizations. Despite the positive indicators of 2004, Atmel reported a scant $26 million increase in revenue overall for 2005, and a depressing $32 million loss.

He said, they said

The year 2006 proved to be Atmel's most tumultuous yet. Like so many of its tech industry cohorts, the company was forced to launch an internal investigation into past granting of stock options that July. As a result of this investigation, Perlegos and his brother, company executive vice president Gust, were fired in August by the board after it concluded that the brothers had footed $406,000 in plane ticket bills to the company for members of the Perlegos brood. Steven Laub, former CEO of semiconductor firm Silicon Image and a member of Atmel's board since February 2006, was appointed by unanimous vote as president and CEO.

But Perlegos fought back with a lawsuit, claiming that his termination was not valid because he had called a special board meeting (slated for October 5th) before he was removed as chairman, to fire the directors who had fired him. The case was tried in Delaware, and a judge ruled in favor of the board in February 2007. Perlegos appealed and was granted the right to hold a special stockholders meeting in May 2007, where he made a bid to replace five of the six current board members with people of his choosing. But Perlegos's plan didn't fly with shareholders, who voted in favor of the new management team (overwhelmingly so, says the company press release).

The numbers are in, at last

The finally Perlegos-less Atmel wasn't in the clear yet. Because of all the hoopla surrounding the Perlegos controversy, 2006 year-end financial filings were delayed, putting Atmel in hot water with the SEC and Nasdaq. When the company finally came into compliance and filed its annual report, in June 2007, Atmel found it had had a year worth celebrating. Revenue grew seven percent to $1.67 billion, and income crawled into the profit-zone again, with the company recording a take of $14 million after the previous year's loss of $33 million. Year-end totals were boosted by Atmel's sale of its operations in Grenoble, France, for $140 million in July 2006.

GETTING HIRED

At work at Atmel

Atmel welcomes resume from potential employees at every stage of their career, from fresh-faced college grad to seasoned professional. To this end, the company's career web site (www.atmel.com/careers) divides its job pool into three sets—

professional positions (accessible on the main career page), entry-level openings and internships (both found in the "college recruiting" section of the site). These are all searchable by department and location. To aid the jobseeker in narrowing down a possible work location, a directory of skill sets lets you know what kinds of work is being done at each Atmel facility. You can also skip all this and see a full list of all opportunities. To apply, you will be asked to fill out a brief application form before e-mailing your resume to recruiter@atmel.com.

If the folks at Atmel are interested, they'll contact you for an interview. According to one insider, interviews are generally "relaxed and informal," and consist of one-on-one interviews of 30 minutes to one hour in length. If you are applying for a technical position, "you are sure to encounter a set of technical questions," which will vary according to the level of the job. "Expect to meet with the hiring manager, three or four engineers and an HR representative." Atmel aspirants are advised to come in with "a basic understanding of what the company does." In addition, the company's "major concern when interviewing people is how well we think they will fit in to the group."

What could fitting in to the group mean for you? Atmel employees enjoy a slew of health benefits—including medical, dental and vision coverage along with a prescription drug plan. Financial benefits include 401(k) with company matching (up to $500 annually) and an employee stock purchase plan. Tuition reimbursement and in-house training opportunities help employees keep their edge, while paid vacation, accrued immediately upon hire, allows them to take the edge off.

Autodesk, Inc.

111 McInnis Parkway
San Rafael, CA 94903
Phone: (415) 507-5000
Fax: (415) 507-5100
www.autodesk.com

LOCATIONS

San Rafael, CA (HQ)
Alpharetta, GA • Greenwood Village,
CO • Itasca, IL • Ithaca, NY •
Manchester, NH • McLean, VA • New
York, NY • Novi, MI • Petaluma, CA •
Princeton, NJ • San Francisco, CA •
Tualatin, OR • Waltham, MA

54 locations in 20 countries
worldwide.

DEPARTMENTS

Administrative Services • Business
Development • Consulting Services
• Corporate Communications/PR •
Customer Service • Facilities •
Finance & Accounting • HR • IT •
Internet/Web • Legal • Localization
• Marketing • Operations • Product
Design • Product Management •
Product Support • Program
Management • Project Management
• Quality Assurance • Sales •
Software Development • Technical
Publications

THE STATS

Employer Type: Public Company
Stock Symbol: ADSK
Stock Exchange: Nasdaq
Chairman: Carol Bartz
President & CEO: Carl Bass
2007 Employees: 5,169
2007 Revenue ($mil.): $1,839

KEY COMPETITORS

Dassault Systemes
Parametric Technology
UGS

EMPLOYMENT CONTACT

www.autodesk.com/jobs

Visit Vault at **www.vault.com** for insider company profiles, expert advice,
career message boards, expert resume reviews, the Vault Job Board and more.

VAULT CAREER LIBRARY   **107**

THE SCOOP

3-D glasses not required

Ever stopped to look at those ultra-sharp 3-D billboards standing at the site of soon-to-be condo developments, depicting a towering high-rise replete with swanky restaurants and beautiful residents that will soon take the place of the vacant lot before you? That hopeful vista is most likely brought to you by Autodesk—the firm creates the 2-D and 3-D design software that allows architects, engineers and their clients to fully visualize and analyze complex projects before any ground is broken. Although Autodesk got its start in the computer drafting field, the company portfolio has expanded in recent years to include special effects software for the entertainment industry, mapping software for civil engineers and product design software used in automotive manufacturing. Out of its headquarters in San Rafael, California, Autodesk delivers its design and imaging software in over 160 countries to more than eight million users. The company ranked No. 960 on the Fortune 1000 in 2006 and snagged second place in the computer software industry on *Fortune*'s 2007 list of America's Most Admired Companies.

No more Mr. Softee

A dozen engineers, led by CEO John Walker, founded Autodesk in 1982 (before he chose Autodesk, Walker puzzled over such possible company monikers as Future-Ware, Desktop Software and, in jest, Mr. Softee). The firm succeeded early with the introduction of the first computer-aided drafting (CAD) program—AutoCAD—written not for mainframes but for PCs.

The company then rode the PC boom, going public in 1985 and acquiring interior decorating software developer Creative Imaging Technologies a year later. Walker relinquished his CEO position to Alvar Green in 1989, who held down the fort until Carol Bartz was appointed chairwoman, president and CEO in 1992. In her executive role, Bartz blazed a trail for women business leaders in the technology industry, leading Autodesk for the next 14 years.

King of the 3-D hill

Much of AutoCAD's success through its early years, like that of Microsoft, came from the dominance of a particular file format. While Microsoft capitalized on DOS to transform itself into an industry titan, AutoCAD saved its files in a format it doesn't share with competitors. Thus, if an architect designs a building using

AutoCAD, the engineer who builds the structure must also use AutoCAD to open and read the plans. Revising this format slightly every time it issues a new release of the software, Autodesk prevents other companies from cracking the code.

Rounding itself out

But CEO Bartz, who was wisely wary of pinning all of the firm's hopes on one program, led the company toward diversification. Accordingly, throughout the 1990s Autodesk began applying its technology to new areas and acquiring other software firms to round out its offerings. In 1997 the company released its home improvement software line, Picture This Home, and acquired rival computer-aided drafting software developer Softdesk for $90 million. In 1999, Autodesk launched its Inventor software (for use in mechanical design) and paid $520 million for Discreet Logic, a Canadian special effects firm that helped create, among other films, James Cameron epic *Titanic*.

Slimming down

Revenue and profit grew steadily during the 1990s, but with the new century came a host of new problems. In fiscal year 2000, net revenue fell 6 percent due to surprisingly low sales of AutoCAD software; in response the company cut its workforce by 10 percent. Autodesk poured still more energy into diversifying its portfolio in the following years, splurging on a seemingly unbridled acquisition binge that brought 17 companies into the fold in only six years. But revenue slid again in 2003, this time by 13 percent, to $824 million. Execs blamed a slowing economy, also stating that the company's focus on long-term development that year resulted in fewer product releases. Sales of the flagship AutoCAD line fell to $69 million in 2003, down 70 percent from the product's $230 million sales in 2002. The company, reeling, let go of 18 percent of its workforce in November 2003.

The much-needed turnaround in company fortunes finally came in fiscal 2004. After opening a development center in Shanghai, China, that year, the company enjoyed 9 percent profit growth in its Asian markets—strong enough on its own, but pale in comparison to the whopping 28 percent growth in Autodesk's Europe, Middle East and Africa segment. By 2005, the company had exceeded $1 billion in revenue for the first time, a 30 percent increase from 2004, with profit nearly doubling to $222 million.

Bringing home the gold

In January 2006, Autodesk added an Oscar to its trophy collection with the $182-million-dollar acquisition of 3-D animation and modeling firm Alias. The private Canadian company was honored by the Academy in 2003 for its Maya software, used to add some special effects flavor to such films as *Spiderman*, *Harry Potter* and *The Lord of the Rings*.

Three years later, Autodesk's Discreet Lustre software provided the digital coloring of Oscar-winning films in the visual effects, documentary feature and foreign film categories. (Part of this prowess in digital coloring was due to the acquisition of digital post-production and coloration science firm Colorfront in June 2005.) However, Autodesk's media and entertainment segment still sits in the shadow of the mainstay design segment, accounting for only 11 percent of overall revenue.

Another good year ...

Also in January 2006, CEO Bartz passed her crown down to former COO Carl Bass, who served with Autodesk for 10 years before gaining the prime leadership position. Bartz stayed on as chairwoman of the company, stating that for the future, her focus will be on gaining a foothold for the company in the growing markets of China, India and Eastern Europe. Revenue for the year increased 20 percent to break a new record at $1.5 billion. The 60 percent growth in sales of 3-D software was especially promising, as the company had struggled in recent years to convert its staunchly 2-D customer base to embrace the third dimension. Still going strong after all these years, sales of AutoCAD increased 23 percent and comprised 44 percent of the company's overall revenue in 2006.

The company branched out into some new frontiers in May 2006. In conjunction with the unveiling of its new sustainability-centric web site (www.autodesk.com/green), Autodesk sponsored a PBS series on sustainable design. The firm encouraged eco-friendly engineering with, of course, the aid of Autodesk's many design software tools, positioning itself to reap from the current "green building" trend. And its not just environmentalists who should be thanking Autodesk, the company is bringing families together as well: also in May, Disney tapped Autodesk's LocationLogic mapping technology to help parents find their stray little ones in Disney Mobile's Family Locator wireless GPS system.

... with a couple of snags

Later that year, in October, Autodesk announced that—like so many other tech firms—it had discovered improperly dated stock option grants to former top executives in the company's skeleton closet, resulting in a $4 million cut in company revenue to cover legal expenses. A few months later, in February 2007, the company said that it would have to restate past profits by $45 million to correct the errors.

Meanwhile, Autodesk delayed filing its fiscal 2007 (February 2006 through January 2007) results to the SEC. Pressure came from Nasdaq in the form of a noncompliance letter threatening to delist Autodesk if it didn't file its annual report. When the results finally came out in June 2007—three months late—Autodesk was pleased to announce that revenue had increased more than $300 million over 2006, to $1.8 billion. Profit slid by $44 million, but remained healthy at $289 million. On the heels of this release came the announcement that Autodesk would acquire Opticore AB, a Swedish software firm that helps users create interactive 3-D graphics and presentations. Opticore's product line falls well within the bounds of the fare already offered by Autodesk, but executives expect that Opticore's advanced technology and interactive features will improve Autodesk's 3-D prototyping tools.

GETTING HIRED

Go ahead, be a CAD

Selected as one of *Fortune*'s 100 Best Companies to Work for in 2006, Autodesk wants to put its logo on your business cards—it says so right on its careers site (www.autodesk.com/jobs). From there you can search for your dream job at Autodesk by job type, location or industry, and also apply online; just post your resume, please, no cover letters here. A handy on-site resume builder helps you make the best possible impression. Upon submitting your resume you will be considered for not only the job at hand but also for similar openings in the future.

Autodesk offers about 100 summer internship positions for full-time students, preferably those studying mechanical engineering, civil engineering, computer science or similar technical majors. Openings are posted throughout the spring season and are available mostly at the company's San Rafael headquarters, although opportunities do pop up elsewhere. A full listing of slots can be found by selecting "intern" from the job function search category through the company's main job board.

Passing the buck

What makes a desk at Autodesk so desirable? Well, the company offers its employees such niceties as a 401(k) and an employee stock purchase plan in addition to a hearty smorgasbord of health insurance coverage. There is vacation time and a generous holiday calendar (including the week off from Christmas Eve through New Year's Day), and the folks at Autodesk also enjoy one paid six-week sabbatical for every four years of full-time employment. The company's "Autobucks" program awards cash and better-than-cash bonuses (theater tickets, dinners and the like) to reward individuals or teams.

Benchmark Electronics, Inc.

3000 Technology Drive
Angleton, TX 77515
Phone: (979) 849-6550
Fax: (979) 848-5270
www.bench.com

LOCATIONS

Angleton, TX (HQ)
Austin, TX
Beaverton, OR
Dunseith, ND
Hudson, NH
Huntsville, AL
Redmond, WA
Rochester, MN
San Jose, CA
Winona, MN

24 locations in 9 countries
worldwide.

DEPARTMENTS

Administrative
HR
Manufacturing
Technical

THE STATS

Employer Type: Public Company
Stock Symbol: BHE
Stock Exchange: NYSE
Chairman: Donald Nigbor
President & CEO: Cary Fu
2006 Employees: 9,548
2006 Revenue ($mil.): $2,907

KEY COMPETITORS

Flextronics International
Jabil Circuit
Sanmina-SCI
Solectron

EMPLOYMENT CONTACT

www.bench.com/viewer/employment.
asp

THE SCOOP

They make the parts that make the whole world run

Benchmark Electronics aims to set the standard for electronics manufacturing. Though the company keeps a low profile in the high-flying tech industry, its products are ubiquitous; Benchmark makes the parts that make computers, medical devices, testing products and telecommunications equipment. Under the blanket term "electronics manufacturing services," Benchmark designs and produces circuit boards for original equipment manufacturers (OEMs) like computer firm Sun Microsystems—which also happens to be its biggest customer, comprising 39 percent of Benchmark's revenue in 2006.

From its headquarters in Angleton, Texas, Benchmark oversees a wide-ranging global business, with 24 manufacturing facilities as far flung as the Netherlands, Brazil and Thailand; it pulls in more than $1 billion in sales a year from Asia and Europe. Ranked No. 664 on the Fortune 1000, Benchmark is also a bit of a teacher's pet, having won numerous supplier awards from its loyal customers over the past few years, including recognition from EMC Corporation, Guidant and Medtronic.

Pacemaker forefather

G. Russell Chambers, then the CEO of Intermedics, a medical implant manufacturer, formed Electronics Inc. in 1979. Intermedics had shot to great profitability in 1976 with the invention of the first modern pacemaker, but Chambers soon began exhibiting erratic business habits: he spared no expense in R&D for future lines of pacemakers and financed it by raising pacemaker costs, which were paid for by the U.S. government, under Medicare's cost-plus reimbursement policy at the time.

By 1982, federal authorities were investigating Chambers and Intermedics for allegedly providing bribes and kickbacks to pacemaker manufacturers. (Chambers also owned a hunting lodge and fishing boat that he leased to the company under suspicious circumstances.) Although the company was never found guilty after investigations by five different federal authorities, it had defaulted on a $100 million bank loan by 1984, and it soon sold off Electronics Inc. and ousted Chambers as its CEO.

Setting a new pace for Benchmark

Donald Nigbor, Steven Barton and Cary Fu, former Intermedics executives who had formed the company Electronic Investors Corp. (EIC), were waiting to snatch up Electronics Inc. when it went up for sale in 1986. EIC and Electronics Inc. merged in 1988;

the new firm changed its name to Benchmark Electronics, and the company went public in 1990. The name change represented a shift away from the medical technology market, and it didn't hurt to cut loose any association with Intermedics. Nigbor, Barton and Fu predicted that OEMs would soon start outsourcing the manufacturing of tech products to contract equipment manufacturers (CEMs), and wanted to get Benchmark in on the action.

The 1990s would prove Benchmark's business strategy correct, and the firm grew modestly, while looking for acquisition possibilities to strengthen its technology manufacturing capabilities. In 1995, the company enjoyed $97 million in revenue, with a $6 million profit, then boldly purchased fellow electronics component manufacturer EMD of Winona, Minnesota, for $51 million the next year. The acquisition more than doubled the company's size and gave it a new foothold in the Midwestern and Northeastern markets. Another swell in the ranks followed in 1998, when Benchmark acquired the commercial electronics segment of Lockheed Martin for $70 million. Fu and Nigbor are still with the company, Fu as its president and CEO and Nigbor as its chairman.

Still bigger

More acquisitions followed and the company grew ever larger, becoming the sixth-largest publicly held CEM provider on the planet with the purchase of Avex Electronics in August 1999. The purchase of Avex brought with it 14 facilities in eight countries abroad, giving Benchmark entrée into international markets. By the end of 2000, Benchmark had achieved $1.7 billion in sales, more than five times what it drew in 1997. The company predicted that the upward trend would continue indefinitely and opened manufacturing facilities in Huntsville, Alabama, in 2000 and Singapore the next year.

Head east

However, the company suffered a heavy dose of reality as a result of the economic downturn that followed September 11th. The entire tech industry was pulled into a slump, meaning less demand for Benchmark's electronic components and a $428-million decrease in sales, with a year-end loss of $54 million. The company subsequently restructured, closing two plants in Tennessee and Massachusetts, along with a whopping 25 percent reduction in the workforce. It also strengthened its international presence with the purchase of the U.K. and Thailand CEM firm ACT Corp. in 2002. That same year, it broke ground on a manufacturing plant in Suzhou, China, which started production in 2003.

Visit Vault at **www.vault.com** for insider company profiles, expert advice, career message boards, expert resume reviews, the Vault Job Board and more.

VAULT CAREER LIBRARY   115

Most of its eggs in three baskets

The efforts paid off, as revenue bounced back to $1.6 billion in 2002. Just over half of this impressive figure came from one customer—Sun Microsystems—which accounted for $843 million in sales, nearly three times the sales to that company the year before. Other big contracts came from EMC and Medtronic; combined with Sun, the trio comprised 70 percent of Benchmark's revenue in 2002.

Despite the year's promising returns, Benchmark's heavy dependence on such a small handful of corporations worried analysts and executives alike. So, the company continued attempts to widen its customer base—by 2004, Sun only accounted for a little less than a third of Benchmark's sales.

Wish on a Pemstar

In another bid to broaden Benchmark's list of buyers, the company acquired Minnesota-based contract electronics manufacturer Pemstar for about $300 million in January 2007. The purchase brought in IBM, Applied Materials and Motorola as customers, in addition to 11 manufacturing facilities and 3,500 employees. But it also brought some risk; although Pemstar's revenue has steadily risen to its 2006 peak of $871 million, the company has not been profitable since 2001. Benchmark also took on the corporate equivalent of emotional baggage—Pemstar's $89 million debt.

However, after bouncing back from its own slump, Benchmark may be able to cut costs effectively and bring its new business back into the black. The company is taking on Pemstar from a relatively secure position—Benchmark's profit in 2006 rose to $111 million on revenue of $2.9 billion, an increase of 23 percent over 2005's $2.2 billion. The company has already absorbed much of the costs of its purchase; by March 2007, it had paid back $66 million of Pemstar's debt, and it reported quarterly earnings in July of $25.9 million, only $1.6 million down from the same period in 2006. Also in July, the company announced the beginning of a share repurchase program of up to $125 million of company stock.

GETTING HIRED

Set some benchmarks of your own

Benchmark Electronics takes a minimalist approach to job listings—what you see is what you get, and what you get is a list of available openings, sorted by location; these can be found at the company's careers web site, www.bench.com/viewer/employment.asp. Benchmark lists new positions as they open up, so aspiring employees are encouraged to check back often to catch the latest opportunities. Upon finding a role you'd like, you may fill out a brief application form and post your resume online.

Visit Vault at **www.vault.com** for insider company profiles, expert advice, career message boards, expert resume reviews, the Vault Job Board and more.

VAULT CAREER LIBRARY   117

Black Box Corporation

1000 Park Drive
Lawrence, PA 15055
Phone: (724) 746-5500
Fax: (724) 746-0746
www.blackbox.com

LOCATIONS

Lawrence, PA (HQ)
Atlanta, GA • Brooklyn Park, MN •
Chicago, IL • Cincinnati, OH •
Cleveland, OH • Dallas, TX •
Denver, CO • Fayetteville, AR • Ft.
Meyers, FL • Houston, TX •
Johnson City, TN • Lincoln, NE •
Los Angeles, CA • Maitland, FL •
Minnetonka, MN • Murfreesboro,
TN • Nashville, TN • Orlando, FL •
Phoenix, AZ • Rancho Cucamonga,
CA • Redwood City, CA • San
Antonio, TX • Seattle, WA • West
Des Moines, IA

DEPARTMENTS

Administrative • Customer Service •
Finance & Accounting • HR • IT •
Legal • Quality Control • Sales &
Marketing • Technical

THE STATS

Employer Type: Public Company
Stock Symbol: BBOX
Stock Exchange: Nasdaq
Chairman: Thomas G. Greig
Interim President & CEO: Terry
 Blakemore
2007 Employees: 3,400
2007 Revenue ($mil.): $1,016

KEY COMPETITORS

CDW
CompuCom Systems
Ingram Micro

EMPLOYMENT CONTACT

www.blackbox.com/About_Us/Careers
.aspx

Visit Vault at **www.vault.com** for insider company profiles, expert advice,
career message boards, expert resume reviews, the Vault Job Board and more.

VAULT CAREER LIBRARY   119

THE SCOOP

Inside the box

The little black box that got this company on its feet is a device that connects office network technology like phones, computers and printers; it is now just one out of a multitude of little black boxes, little blue boxes, long black cables and all manner of multicolored gadgets offered by the company.

The Black Box Corporation hooks up offices with all the connectivity they need through its data and voice services divisions. The company even goes so far as to custom-build devices when its 118,000-strong product roster doesn't quite cover a customer's needs. To further woo its clients, it supports a dedicated technical hotline that's available 24 hours a day, 365 days a year. Calls to the hotline are answered within 20 seconds—guaranteed! It's that dedication to customer service that keeps 175,000 clients in 141 countries coming back to Black Box for all their connectivity hardware and service needs. And it's that customer loyalty that has made Black Box the world's largest dedicated network infrastructure provider.

Mail-order moguls

Eugene Yost and Dick Raub founded the company in 1976 as Expandor Inc., offering its first catalog in 1977. The humble publication had a scant six pages, graced by only nine products, but went on to generate $170,000 in sales. By 1982, the company was synonymous with its popular catalog, and "Expandor" was wisely dropped in favor of "Black Box." Yost and Raub sold the company to Micom Systems in 1983 for $19 million.

Third time is the charm

After the sale, Black Box and Micom operated as subsidiaries of the private holding company MB Communications Inc., but the new owners didn't mess with success, as Black Box grew through the 1980s, developing a strong customer base. By 1990 the company was showing $25 million in profit, outshining its sister subsidiary Micom, whose poor performance made MB a bad bet for a public offering (in fact, MB made two attempts at a public offering, in 1989 and 1990, but both were canceled due to investor reluctance). MB finally made a successful public offering in 1992, and Black Box was able to shed its dead-weight sister company when MB spun off the floundering Micom two years later. Then, MB Communications learned the same lesson that Yost and Raub had with "Expandor," and changed its name to Black Box Corporation in the year of its IPO.

Shopaholics

Expansion followed in the 1990s—Black Box opened a subsidiary in Brazil in 1994, with another center to open in Mexico in mid-1995. Despite offering its catalog online and on then-vogue CD-ROM by 1996, the paper catalog remained the main sales-generator for this direct marketing firm. By 1997, Black Box offered more than 7,000 products from subsidiaries in 15 countries to customers in more than 77 nations. Sales that year clocked in at $232 million, 47 percent of which came from markets outside of North America.

An acquisition spree, which started in 1998, fueled further growth—Black Box purchased 40 companies in just over two years. This allowed Black Box to expand its service to customers who were no longer satisfied with chatting to a technician over the phone; now Black Box could send out a technician for on-site installations or fix-ups from its growing multitude of service providers. After the acquisitions, the catalog swelled to 90,000 products and revenue rocketed to $826 million in 2001, more than double that of 1999.

Low tide and high aims

The economic downtown that followed September 11th hurt the tech industry as a whole and, despite its diverse product offering, Black Box couldn't keep growing amid the deflating market, and revenue began to fall. By 2004 sales had shrunk to $520 million, with income falling as well, to $75 million. Although Black Box had traditionally let its acquired companies continue running pretty much unchanged, restructuring was in order to keep the company in the black. It trimmed staff down to 3,700 in 2002, from a peak of 4,800 the year prior. An October 2003 Securities and Exchange Commission probe into insider trading—concerning whether members of the company brass had unloaded their stock before announcing bad news—didn't help matters. The company settled the resulting class-action lawsuit in October 2004 for the relatively small amount of $2 million.

What did help matters was the January 2005 acquisition of telecommunications service company Norstan for $95 million. The purchase increased Black Box's stake in telephone equipment and installation and set Black Box back into a growth groove, as 2005 sales increased 3 percent to $535 million. Black Box acquired eight new

Visit Vault at **www.vault.com** for insider company profiles, expert advice, career message boards, expert resume reviews, the Vault Job Board and more.

V/\ULT CAREER LIBRARY

121

firms in 2006, including Nu-Vision Technologies and the U.S. operations of NextiraOne, which sent revenue for that year up a robust 35 percent, to $721 million.

The acquisitions significantly boosted Black Box's presence in voice systems services, as sales in that division nearly tripled in one year to $310 million. Indeed, voice services represented 43 percent of Black Box's revenue that year, a dramatic increase from 2004, when it only garnered 13 percent. Citing these gains, CEO Fred Young ambitiously believes that 2007 will prove to be the year Black Box sails past $1 billion in revenue.

Suit optional

Although Black Box's financial future is definitely rosier, shareholders and federal investigators have found cause for alarm in the possible illegal backdating of stock options. A lawsuit launched by one upstart investor in November 2006 alleged that 14 executives, including CEO Young, conspired to date employee stock options to an especially low price in 1996, allowing them to later sell the stock for high profit.

The suit spurred another SEC probe and an internal review in February 2007, which in turn delayed annual financial statements and put Black Box in hot water with Nasdaq. Guilty or not, the company may luck out, according to some industry analysts: while New York and California state prosecutors have jumped on backdating scandals, the U.S. attorneys in Pennsylvania, where Black Box is headquartered, are not expected to press charges.

It got old for Young

Meanwhile, the internal audit determined in March 2007 that Black Box will have to pay $63 million (a figure later adjusted to $70 million) to correct the improper accounting. Within two months of this announcement, CEO Young resigned from his post after serving a nine-year stint with Black Box. Although press releases on the matter gave no indication of why Young would want to leave, speculation abounded regarding his involvement with the company's stock option misbehavior. Erstwhile-senior vice president, Terry Blakemore, stepped into Young's shoes while Black Box seeks out a new, permanent leader.

Heedless of whatever shakeups are furrowing brows at the top office, business goes on at Black Box. The company acquired private Florida firm ADS Telecom, a provider of voice and data communication services in the southeastern swath of the U.S., in February 2007. With its diverse client list, including companies of the commercial, financial, health care and governmental persuasion, ADS brought annual revenue of about $14 million into Black Box's coffers. The company's fiscal 2007 statements, released in late June, confirm the picture of a profitable company spending too much money on internal problems. Revenue was up to $1 billion from the 2006 total of $721 million, but profits improved by the smallest of margins, to $37.5 million from $37.3 million in 2006.

GETTING HIRED

Get inside the box

Black Box Corporation is looking for "versatile, highly motivated and forward-thinking candidates" to augment its 3,300-strong workforce, and these types of folks are encouraged to browse Black Box's open positions at the company's careers site (www.blackbox.com/About_Us/Careers.aspx). Here you will find a list of job openings sorted by location, although those seeking employment in the voice services division of Black Box—which makes up 40 percent of its business—are directed to a separate directory of positions. Applicants to non-voice services departments can submit their resume by fax, e-mail or regular delivery directly to the hiring manager. Voice services positions, on the other hand, are handled via an online form, with space for applicants to post their resume.

BMC Software, Inc.

2101 City West Boulevard
Houston, TX 77042
Phone: (713) 918-8800
Fax: (713) 918-8000
www.bmc.com

LOCATIONS

Houston, TX (HQ)
American Fork, UT • Atlanta, GA •
Austin, TX • Cary, NC • Charlotte,
NC • Chicago, IL • Clearwater, FL •
Columbia, MD • Dallas, TX •
Detroit, MI • Englewood, CO •
Irvine, CA • McLean, VA •
Minneapolis, MN • New York, NY •
Parsippany, NJ • Plano, TX •
Pleasanton, CA • Santa Clara, CA •
Seattle, WA • Sunnyvale, CA •
Waltham, MA • West Trenton, NJ

78 locations in 36 countries
worldwide.

DEPARTMENTS

Administration • Business Planning •
Facilities • Finance & Accounting •
HR Management • IT • Legal •
Marketing • Product Development •
Professional Services • Sales

THE STATS

Employer Type: Public Company
Stock Symbol: BMC
Stock Exchange: NYSE
Chairman: B. Garland Cupp
President & CEO: Robert E.
  Beauchamp
2007 Employees: 6,000
2007 Revenue ($mil.): $1,580

KEY COMPETITORS

CA
Hewlett-Packard
IBM

EMPLOYMENT CONTACT

www.bmc.com/careers

THE SCOOP

Management is their middle name

As one of the world's largest independent software vendors, BMC Software makes its money helping large corporations manage their operations and information through applications, databases and other types of infrastructural software. The firm's product offerings are divided into four primary categories of "management" service: impact, IT service, application and IT operations, database and infrastructure. Combined, these divisions deliver a little bit of magic called "business service management," or, the implementation of day-to-day business processes through technology for the greatest possible efficiency.

Headquartered in Houston, Texas, BMC runs with the big boys, drawing its customer base mainly from Global 2000 companies—the kind that need comprehensive database systems to connect and manage their sprawling and diverse operations. Indeed, BMC estimates that 95 percent of the Forbes Global 100 and 80 percent of Fortune 500 companies use BMC's software to manage their enterprises.

The main event

BMC grew out of a trio of computer programmers from the Houston area in the 1970s, Scott Boulett, John Moores and Dan Cloer (the B, M and C of the company's name, respectively). The business was incorporated in 1980, and initially focused its energies on developing software tools for the mainframe computers used by large corporations. But with the rise of the PC, BMC's business changed to incorporate desktops into the mix (although handling massive and complex business organizations with mainframe mammoths in their closets has remained its forte). Under the direction of CEO Max Watson, who assumed that top role in 1987 after founder Moores scaled back his involvement with the company, BMC went public on Nasdaq in 1988 to the tune of $9 per share.

Acquisition and aggressive direct marketing (often of the tele-persuasion) led the company to grow through the 1990s. Chairman Moores fully retired in 1992, leaving the company totally in the hands of CEO Watson (Moores would go on to buy the San Diego Padres baseball team). BMC purchased PATROL Software two years later, bringing that firm's network surveillance software into its portfolio, which marked a departure from its former reliance on mainframe tech. More acquisitions followed to round out its non-mainframe offerings, including DataTools in 1997, BGS Systems in 1998 and New Dimension Software in 1999.

Mainframe, meet Internet

As the Internet grew, BMC hustled to position itself as the go-to firm for e-business management. The launch of its MAINVIEW software in October 1999 connected stable and dependable mainframes to the versatile and ever-changing World Wide Web, much to the joy of the credit card companies, airlines and utilities companies who benefited from BMC's innovation.

The company grew exponentially in this period, and revenue had sailed past $1 billion by the end of 1999, more than twice as much as the company pulled in just three years before. The headcount at BMC also grew, sitting at 5,500 upon the closing of the New Dimensions purchase, a threefold increase over 1996. The company would grow to over 7,300 employees by 2001.

Setbacks

With the company doing so well, CEO Watson decided it was time for him to pass the job onto someone else. Senior vice president of product management, Robert Beauchamp, filled Watson's shoes just before, as he would later put it, "the industry changed suddenly and just knocked our legs out from under us." Although the company was decreasing its reliance on sales of its mainframe management software, it didn't move fast enough to avoid the crippling blow suffered across the entire tech industry in the early years of the new century. Its stock peaked at $85 per share in 2000, only to plummet to $11 per share one year later. Similarly, the revenue that had soared up to $1.7 billion in 2000 fell nearly $500 million in two years, taking profit with it—the company reported a loss of $184 million in 2002. In response, BMC eliminated 1,000 employees from its workforce over the course of one year; the total number of employed fell to 6,300 by mid-2002.

Slow and steady wins

The company quickly recognized the need to change, and introduced a new business strategy—business service management—in April 2003. Perhaps a more conceptual than tangible change, the new shift to connecting IT resources to business management processes was spurred by the acquisitions of service-management software companies Remedy (in November 2002) and IT Masters (in March 2003). Later acquisitions—including that of Marimba in 2004 and Calendra and

OpenNetwork in 2005—allowed the company to broaden its software line from specific programs to an entire suite of management tools.

With these subtle shifts in direction, BMC has downplayed its once-encompassing focus on mainframe technology, embracing up-and-coming technologies. This strategy seemed to work, as revenue slowly rose over the next few years, from $1.32 billion in 2003 to $1.46 billion in 2006. Profit in 2006 clocked in at $102 million, the highest BMC had seen in six years.

Diversify or die

Portfolio diversification continued with the May 2006 acquisition of Israel-based Identify Software, which boosted BMC's foray into transaction management software. It also gave BMC a stronger foothold in foreign markets, and revenue for fiscal 2007—which grew slightly over 2006, to just over $1.5 billion—was split pretty much down the middle between U.S. and international buyers. Profit for the year more than doubled to $215 million.

BMC celebrated the good news with more shopping, picking up ProactiveNet, Inc. in June 2007; previously a private firm, the new acquisition produces analysis software that generates early warnings of potential problems in business operations. These functions will be integrated into the business service management products already offered by BMC. Acquisitions like this are keeping expectations high for the company. Despite the low rise of the company's growth trend, analysts put BMC in a strong position for continued profit, thanks to its cost-cutting efforts and investment in new technologies. Furthermore, its relatively narrow range of products— compared to its Goliath competitors HP and IBM—allows it to respond with agility to the needs of its customer base.

Ask what BMC can do for your country

As BMC's focus changes, so does its customer base. The U.S. government has made increasing use of its new offerings as that bulky behemoth of the people attempts to update its technology and streamline its operations. The Department of Defense contracted BMC for its management software in February 2007, shelling out $8.5 million for BMC's services.

The company had contracted to other fingers of Uncle Sam's stern fist in recent months, including The Missile Defense Agency, The Defense Logistics Agency and The Military Sealift Command. BMC made headway into the government's civilian offices following the completion of its defense obligations, winning $15 million in contracts by June 2007 from such groups as The Environmental Protection Agency, The U.S. Postal Service, and The Center for Medicare and Medicaid Services.

GETTING HIRED

Is it in your blood?

Through BMC's careers site (www.bmc.com/careers) you can explore just what it takes to join the business service management team. In its "Leadership DNA" section, the company outlines the core competencies that should already be built into the very fiber of successful applicants, including such attributes as the drive for innovation, the ability to communicate directly and honestly, and a yen for customer advocacy. These are the ideals that make up BMC's company culture, and job seekers who fit the bill are welcome to search company's openings. Searchable by function, location and position type, BMC's job database allows you to upload your resume and cover letter upon finding a position you'd like to fill.

Once you're in the company, BMC will supply you with comprehensive health coverage, including a prescription drug plan and a 401(k) with company matching to get you going on that long march toward retirement. Employees may also purchase company stock at a discounted rate. Tuition assistance is available for those still looking to learn, provided your study relates to the job at hand.

Business Objects S.A.

157-159, rue Anatole France
92309 Levallois-Perret Cedex
France
Phone: (33-1) 4125-2121
Fax: (33-1) 4125-3100
www.businessobjects.com

Business Objects Americas
North American Corporate
Headquarters
3030 Orchard Parkway
San Jose, CA 95134
Phone: (408) 953-6000
Fax: (408) 953-6001

LOCATIONS

**Levallois-Perret, France
(Corporate HQ)**
San Jose, CA (North American HQ)
Atlanta, GA • Burlington, MA •
Centennial, CO • Edina, MN • Edison,
NJ • Houston, TX • Irvine, CA •
Irving, TX • Madison, WI • McLean,
VA • Miami, FL • New York, NY •
Novi, MI • Phoenix, AZ • Portland,
OR • Rosemont, IL • San Jose, CA •
San Francisco, CA • Scottsdale, AZ •
Seattle, WA • Troy, MI

46 locations in 29 countries
worldwide.

THE STATS

Employer Type: Public Company
Stock Symbol: BOBJ
Stock Exchange: Nasdaq
Chairman: Bernard Liautaud
CEO: John Schwarz
2006 Employees: 5,428
2006 Revenue ($mil.): $1,254

DEPARTMENTS

Administration/Support Services •
Customer Advocacy • Customer
Support • Educational Services •
Executive • Finance • HR • IT •
Internet/Web • Legal • Marketing •
Presales • Product Management •
Professional Services • Project
Management • Purchasing • Sales •
Sales Operations • Software
Development/Engineering

KEY COMPETITORS

Cognos
Oracle

EMPLOYMENT CONTACT

www.businessobjects.com/company/
careers

Visit Vault at **www.vault.com** for insider company profiles, expert advice,
career message boards, expert resume reviews, the Vault Job Board and more.

VAULT CAREER LIBRARY

129

THE SCOOP

Who's got BO? Most of the Fortune 500, for starters ...

From its world headquarters just outside of the City of Lights and from its U.S. locus in San Jose, California, Business Objects is a leading software company concerned with business intelligence—that is, the organization of a company's multitude of data into meaningful information. Business Objects facilitates and fosters that intelligence—like a tutor, of sorts.

It has three main departments: enterprise performance management, information discovery and delivery, and enterprise information management. Aside from being a jumble of similar-sounding words, these three components basically collect data from different sources within company networks and organize it into something that makes sense to a company's managers, analysts and customers. And 42,000 customers in 80 countries agree—it works! Business Objects serves 82 percent of Fortune 500 companies with its special brand of data analysis and organization, including firms in the communications, financial services, health care, manufacturing and technology industries.

Start up with a bang

Bernard Liautaud, who had worked for Oracle's French subsidiary upon graduating from Stanford, founded the company in Paris in 1990; Liautaud conceived Business Objects as a response to the growing need for databases that could share information with each other, as more and more companies' statistics came to be stored on computers. This brand of interdepartmental organization quickly became profitable, and in 1994 Business Objects became the first European company to be listed on Nasdaq. Revenue doubled annually from 1993 to 1995, leaping from $11 million to $48 million. By 1998 the company was pulling in more than $100 million in sales and had made its entrée onto the information superhighway with its WebIntelligence platform, which enabled, over the Internet, the same query, reporting and analysis that made the Business Objects program a success.

If you can't beat 'em, buy 'em

Growth continued for Business Objects into the 21st century, prompting the company to open its headquarters for U.S. operations in San Jose, California, in 2001.

Acquisitions followed in 2002—Blue Edge Software brought more web-based capabilities to the company, and the purchase of Acta Technology amped up Business Objects' analytic applications. BO caught the acquisition fever and reeled in a big buy in 2003 with the purchase of fellow business intelligence software firm Crystal Decisions for $1.2 billion, which led in part to that year's record-high revenue of $560 million.

In the meantime, the company had prevailed in a patent lawsuit against rival firm Cognos in May 2002, winning $24 million in damages. The Crystal Decisions deal had placed Business Objects ahead of Cognos, and, together with the lawsuit, it ignited fierce, head-to-head and toe-to-toe competition between the firms for years to come.

Bye bye Bernard

In September 2005, Bernard Liautaud stepped down from his CEO role, passing the torch to the former president of Symantec, John Schwarz. But Liautaud stayed on as chairman, and was on hand to celebrate the company's landmark achievement that year—$1 billion in revenue, nearly double that of 2003. Part of this growth came from the acquisitions of the firms SRC, Medience and Infommersion that year, with four more to follow in 2006, including FirstLogic and ALG. The purchases are part of Business Objects' larger plan to outrun the competitor(s) snapping at its heels— namely, Cognos, which has historically shown greater profit, despite bringing in smaller revenue.

Mind the gap

When software big boy Oracle announced its intention in October 2006 to enter the business intelligence game through acquisition, all eyes turned to Business Objects, leader of the sector, as the top draft pick. Although Oracle ended up purchasing its rival Hyperion in March 2007, the industry gossip didn't subside, as analysts immediately began speculating that Cognos was next in line to get snapped up.

In the meantime, Business Objects continued in its onwards-and-upwards trend in 2006, reaping $1.2 billion in revenue. Profits, however, fell nearly 19 percent compared to 2005, to $75 million. Paradoxically, these falling profits might save BO's independence, for, although it is the top dog in business intelligence, Cognos' higher profitability margin might make it the more attractive takeover candidate.

Just one more, pretty please?

Business Objects made another acquisition in April 2007. This time, the must-have morsel was Cartesis, a privately held enterprise performance management software firm. Business Objects shelled out $300 million for Cartesis, which brings products with a finance-management bent into the company's lineup, and which will accordingly be marketed toward chief financial officer-types.

In the same month, an earlier acquisition (and its lawsuit baggage) resulted in a slap-on-the-wrist for Business Objects. When it bought Acta Technologies in 2002, Acta was already involved in a patent lawsuit with Informatica. The four-year-old case regarded a specific feature on the Acta Technologies's ActaWorks software (now sold by Business Objects as Data Integrator) and was decided in favor of Informatica in April 2007; Business Objects was ordered to pony up $25 million. By May, Business Objects had released a new version of Data Integrator, minus the offending feature, which had to do with the software's data warehousing process.

GETTING HIRED

Contribute to BO

As a firm at the leading edge of a very competitive industry, Business Objects is interested in getting and keeping top talent. Accordingly, the company treats its employees well, offering a slew of health care benefits in addition to vacation time and floating holidays. Those looking out for the future will appreciate the savings plan with company matching and the employee stock purchase program. Through the company's Business Objects Learning and Development (BOLD) program and TechED, workers can keep abreast of industry developments and get the necessary training to rise through the levels of the company. Thanks to the global reach of Business Objects, the travel-savvy can look forward to permanent or temporary relocation opportunities.

Oh, Canada!

Business Objects is most emphatically recruiting the college set, albeit in a decidedly Canadian fashion. Students (local or international) of computer science, software engineering, engineering physics, math and plain-old physics are invited to apply for internships and co-op opportunities through Business Objects' Canadian recruiting program. Applicants will go through a four-part interview process, consisting of a written technical evaluation, a meeting with a recruiter, a technical interview and, finally, a meeting with the manager. Interviewees are encouraged to come prepared by researching the company history and product portfolio.

For experienced hires and recent graduates with degrees in computer science, software engineering, math or IT, the way to get a foot in the door may be found by perusing the company's careers site at www.businessobjects.com/company/careers. Employment opportunities are sorted on that page by location and job category; job seekers ready to apply may upload their resume to submit online. A short application follows, with the option to be notified when similar positions open up in the future.

Visit Vault at **www.vault.com** for insider company profiles, expert advice, career message boards, expert resume reviews, the Vault Job Board and more.

VAULT CAREER LIBRARY   **133**

CA, Inc.

One CA Plaza
Islandia, NY 11749
Phone: (631) 342-6000
Fax: (631) 342-6800
www.ca.com

LOCATIONS

Islandia, NY (HQ)
Atlanta, GA • Bellevue, WA •
Bloomington, MN • Boulder, CO •
Charlotte, NC • Chesterfield, MO •
Columbus, OH • East Windsor, CT •
Ellicott City, MD • Englewood, CO •
Framingham, MA • Franklin, TN •
Herndon, VA • Houston, TX •
Independence, OH • Jacksonville, FL
• Leawood, KS • Lisle, IL • Los
Angeles, CA • Mason, OH • New
York, NY • Palo Alto, CA • Petaluma,
CA • Pittsburgh, PA • Plano, TX •
Portsmouth, NH • Princeton, NJ •
Redwood City, CA • San Francisco,
CA • Scottsdale, AZ • Tampa, FL

234 locations in 46 countries
worldwide.

DEPARTMENTS

Administration • Child Care •
Corporate Communications •
Development • Finance • HR •
Learning Office • Legal • Marketing
• MIS/Corporate Information
Systems • Sales • Technical
Support • Technology Services

THE STATS

Employer Type: Public Company
Stock Symbol: CA
Stock Exchange: NYSE
Chairman: William E. McCracken
President & CEO: John Swainson
2007 Employees: 14,500
2007 Revenue ($mil.): $3,943

KEY COMPETITORS

BMC
EMC
IBM
Symantec

EMPLOYMENT CONTACT

www.ca.com/career

THE SCOOP

Think you can manage?

CA (or, the firm known as Computer Associates until 2006) is a supplier of systems management, information management and business management software products for hardware platforms. As one of the world's largest software companies, CA does business all over the globe, with branches and subsidiaries in 46 nations abroad. The company, which has been a major player on the software scene for nearly three decades, ranks No. 547 on the Fortune 1000.

Billion-dollar baby

Computer Associates has its roots in a Swiss company of the same name, formed 1970, which employed Queens College graduate Charles Wang in 1976 as the head of a new American branch. Wang started with four employees and one product, CA-SORT, a data organizing tool that was tailored to IBM computer systems. Serving existing software rather than innovating, CA's future success would follow the same philosophy. The company became the first to offer what is now known as multi-platform software, which is compatible with multiple computer systems.

By 1980, Wang's division had grown large enough to buy out its Swiss partners. The next year, the company filed an IPO valued at over $3.2 million, and used the money raised to continue adding to its stable of customer-friendly software, often at the expense of smaller, struggling companies. The largest of many notable acquisitions came in 1987, when CA purchased its chief rival UCCEL in a swap of stock then valued at $780 million. In 1989, as a result of its large growth, CA earned the lofty title of first software company to reach $1 billion in revenue.

Market growth

During the 1990s, CA expanded into the Far East, Africa, and Latin America. The firm promoted Sanjay Kumar, a protégé of Wang's, to president in 1994, and he spearheaded an acquisition philosophy away from older systems, and towards network software, particularly of the "enterprise storage" type. CA acquired Cheyenne Software, Inc. in 1996, sold today under the BrightStor line of products. At the time, the purchase made CA the only provider of end-to-end storage management solutions, a business approach revolutionized by Wal-Mart.

Visit Vault at **www.vault.com** for insider company profiles, expert advice, career message boards, expert resume reviews, the Vault Job Board and more.

VAULT CAREER LIBRARY   135

The company then failed in its 1998 hostile takeover attempt on Computer Sciences Corp., despite offering a hefty all-cash bid of $9.8 billion. The deal would have created a combined company with over 50,000 employees and annual revenue of $11 billion. Instead, CA turned to other acquisitions, picking up Platinum Technology International for $3.5 billion in May 1999. Overall in fourth-quarter results posted that month, CA's revenue increased 11 percent to $1.6 billion. CA continued its aggressive pursuit of acquisitions the next year, buying Sterling Software in 2000 for a deal worth $4 billion in stock, then the largest acquisition in software industry history.

If the suit fits ...

The new century brought an unhealthy dose of unrest to the company. Mr. Kumar had been promoted to CEO in 2000, followed closely by an attempted coup at the August 2001 shareholder meeting, which was staged by Sterling Software's co-founder, Sam Wyly, for reasons of "corporate mismanagement." The next month, the U.S. Justice Department filed a civil antitrust lawsuit, stemming from actions the company took regarding its 1999 acquisition of Platinum Technology. DoJ officials claimed CA violated a premerger waiting period requirement and various price-fixing laws, but the company escaped with relatively little harm, settling the matter for $638,000 a year later.

However, the company did struggle with the economic downturn affecting the entire tech industry, laying off 900 workers In October 2001. Also, legal issues continued to emerge, as the SEC opened an inquiry in early 2002 related to CA's methods of recording sales of software and maintenance fees; the agency was particularly interested in learning if CA invented an extra $500 million in revenue to activate stock incentives for top officials. Later the same year, Wyly attempted another coup, to no avail, although Wang subsequently resigned his chairmanship, the CA board naming Kumar as his replacement. Wang has since become known for his unorthodox business practices as owner of the New York Islanders hockey team, most notably when he signed a goaltender to a record-length 15-year contract.

I think I can, I think I can, I think I can ...

CA attempted to continue business as usual during the litigation. In 2002, the company was rocked by a $1 billion gouge out of its revenue, which fell from $4 billion the year before to $2.9 billion, and reported a net loss of over $1 billion.

Included in these losses was a $10 million settlement payment to major shareholder Sam Wyly, he of the frequent coup attempts.

CA returned to its principal growth strategy and showed that it could turn things around in a hurry, acquiring Netreon Inc. in February 2003. CA also renewed its commitment to management software, releasing The Security Command Center, which offered advanced technology to help customers manage security from a central point, and Managing On-Demand Strategy—both products helping cement its position as an expert in the field.

Hello, goodbye

The ongoing criminal inquiry into accounting practices brought about a major upheaval in corporate structure in 2004. Many new faces came into the CA fold, as Lewis S. Ranieri became chairman of the board, while Jeff Clarke, former CFO of Compaq, signed on as COO and CFO, and Kenneth D. Cron, a former CEO and chairman at Vivendi Universal, took over as interim CEO, later replaced permanently by John Swainson, a 26-year industry veteran, in November 2004.

In September, just as it was entering into a deferred prosecution agreement with the SEC, committing a $225 million fund for the reimbursement of shareholders, the company announced it was cutting 800 jobs as part of a restructuring plan, expected to result in annual savings of $70 million. Apparently, neither event impinged much upon CA's operating procedures, as it picked up Netegrity Inc., another provider of security products, for $340 million the next month.

In October 2004, CA paid the federal government $2.2 billion for accounting fraud. Meanwhile, former CEO Sanjay Kumar, who was indicted in 2004 on charges of accounting fraud and obstruction, went to trial. Despite the turmoil, CA kept at its comeback, reaping $3.3 billion in sales for 2004, with year-end loss shrinking to $89 million. In 2005, the company launched its largest product release ever, trucking out 85 products and 26 updates to older software lines. Perhaps tellingly, the company unveiled a new logo and slogan the same year, urging the industry and the public to "Believe Again."

Dealing with fallout

2006 brought yet more shake-ups to the war-battered company. Controversially, CA awarded swelling commissions to sales representatives despite stagnant revenue returns, prompting several executives to leave. Also, after twice-postponing reporting financial results to the SEC, CA announced in June that an internal review found irregularities in the employee stock option system, which could result in costly corrections.

Throughout these troubles, CA again restructured its organization with an eye to cutting costs. In March 2006, in the far away land of India, CA began stocking up on staff to fill out its support and development functions abroad. The company brass announced an investment in overseas employees that would double the Indian workforce to 2,500 in two years, at which time it would surpass that of the company's headquarters in Islandia, New York. After profits slid later in the year, CA set free 1,700 employees, or 10 percent of its workforce, in August 2006 (half of those cuts comprised U.S. workers). The past two years had already seen a reduction of company ranks by 1,600.

Richards and Kumar go to prison

Meanwhile, Sanjay Kumar, CA's former CEO, was sentenced to 12 years in prison and fined $8 million—pocket change for someone who netted a $330 million bonus in 1998, though the jail time's another story—after pleading guilty to charges of accounting fraud and inflating profits during 1999 and 2000. Kumar was accused of plumping sales figures with a "35-day month," lying to federal investigators and bribing a witness to keep mum. A few weeks later, Stephen Richards, former head of sales at CA, was sentenced to seven years in prison for his role in colluding with Kumar to inflate revenue figures.

Tacked on to the paltry $8 million fine is the harder-hitting $800 million in restitution Kumar agreed to pay investors in April 2007 for the losses his fraud created. Although it will be difficult to actually collect the entire sum, CA has already contributed $255 million to this restitution fund, Kumar will pay $52 million in 2007, and the government will be able to garnish up to 20 percent of his wages when he's back on the job in 12 years.

What now? Buy more stuff

Through all the ups and downs, business did—by some miracle—continue at CA. The company was gearing up for a turnaround through acquisition in 2005, and four more firms found themselves engulfed by the behemoth software seller. In June 2005 it purchased Concord Communications, a service management software firm, for $359 million, with IT management software company Niku Corporation to follow a month later, for $345 million. Archiving software developer iLumin joined the party the following October to the tune of $48 million, and Wily Technology brought up the fiscal rear, coming aboard in March 2006 for $374 million. The new additions helped boost year-end revenue by nearly $200 million, to $3.8 billion. Profit was back in black for the second consecutive year, clocking in at $156 million.

However, the high turnover continues at the highest levels of the firm, as CA announced in June 2007 that William E. McCracken would succeed Lewis S. Ranieri as chairman, the latter remaining with the company in a directorial role. Ranieri cited the seeming culmination of CA's legal battle as a major consideration, stating that with the "prosecution agreement successfully completed and the company moving in the right direction, it is both fitting and appropriate that I step down as chairman." Whether McCracken's tenure outlasts his predecessors remains to be seen.

GETTING HIRED

Send 'em your resume

Applicants can use the online job search on CA's career web site (www.ca.com/careers) to track down open positions. Employment is available the world over, in Europe, Middle East, Africa, the Americas and Asia Pacific. Job seekers can create and submit a résumé to the company through the site's online resume builder, or upload it via PDF, Microsoft Word, or plain text.

Benefits for CAers include 401(k) with company match, an employee stock purchase program and a pretax flexible expense plan. In addition to the usual health coverage, CA sees to the health of its worker bees with on-site fitness centers and even looks out for its worker bees' birds, dogs, cats and such, with optional pet insurance. Parents cheer for the child development centers, where trained staff will turn your crayon-eating tot into a Mozart-appreciating genius while you work, and "baby bonds" award workers for incubating or adopting the next generation of CAers.

In addition, CA offers internship opportunities to current college students in a variety of departments. These positions can be found through the main job search page, under the "internship" job category.

OUR SURVEY SAYS

Meritocracy

Though it tends to fight dirty on the corporate playing field, on the inside, CA is "a true meritocracy," where "everyone has a chance to go as far as they want to." One source on the technical side reports, "I have found my raises accurately track what I did in the previous year," and "my responsibilities have increased as I have asked for them to be." Further, they advise that "it's important to be very proactive." One respondent criticized CA as a "very low performance culture. The bar is set so low you can trip over it."

Moneymakers: in

Insiders say the company is very open to women and minorities, which is common for the industry. *Fortune* listed the company as one of America's 50 Best Companies for Minorities in its July 2001 issue. One contact adds that, "the overall attitude is positive towards advancement of anyone who can generate revenue." If they can't, however, they're out. Some insiders—salespeople, at least—say "there is a medium to high turnover rate" for those who do not meet the company's quotas. Some, however, feel that the situation is "very relationship-driven," bordering on "incestuous. Either you're part of the in crowd or not."

Bagels for nothing and your kicks for free

CA offers an extensive array of perks, including free breakfast, tuition reimbursement, fully paid medical benefits and a "discretionary distribution"— where the company matches up to 8 percent of your base salary in a 401(k) plan. CA maintains "first rate" corporate fitness facilities and "fantastic" on-site child care centers in its offices all over the world. However, one contact notes that "Many perks have been removed. The HQ gym used to be free; now employees are charged for membership." The dress code in the corporate offices is pretty straightforward: "If you meet clients, you wear full business attire; if you don't, you wear business casual." The lucky employees in development and tech support outside the New York headquarters say they wear jeans and T-shirts on a regular basis.

Dissenting opinions

Computer Associates was formerly ranked high on "top places to work" surveys in publications around the country. Recent surveys suggest a trend toward staff dissatisfaction. "Poor integration and communication with employees. Software designed by committees, and overburdened process takes forever, so nothing gets done," wails one respondent. "There is a very top-heavy corporate culture that seems to be very out of touch with the remainder of the company," adds an engineer. It's hardly any surprise that his co-worker adds, "Employee morale … is not very good at all." "Opportunities for advancement are slim," interjects an engineer. Diversity seems also to be an issue. "Women and African-Americans are overwhelmingly underrepresented in the company," notes a source. It's not all gloom and doom, however. "There are high quality employees in the company," says one programmer.

CACI International, Inc.

1100 North Glebe Road
Arlington, VA 22201
Phone: (703) 841-7800
Fax: (703) 841-7882
www.caci.com

THE STATS

Employer Type: Public Company
Stock Symbol: CAI
Stock Exchange: NYSE
Chairman: Dr. J.P. London
President & CEO: Paul M. Cofoni
2006 Employees: 10,000
2006 Revenue ($mil.): $1,755

DEPARTMENTS

Administrative & Support Services
Defense & Intelligence
Technology

KEY COMPETITORS

Computer Science
General Dynamics
SAIC

EMPLOYMENT CONTACT

www.caci.com/job.shtml

LOCATIONS

Arlington, VA (HQ)
Albuquerque, NM • Alexandria, VA •
Ashburn, VA • Aurora, CO • Austin, TX
• Baltimore, MD • Beavercreek, OH •
Bloomington, IN • Boulder City, NV •
Bremerton, WA • Chantilly, VA •
Charleston, SC • Charlottesville, VA •
Chesapeake, VA • Colonial Heights, VA
• Columbia, SC • Columbus, GA •
Dahlgren, VA • Dayton, OH • Draper,
UT • Eatontown, NJ • Elkridge, MD •
Fairborn, OH • Fairfax, VA • Falls
Church, VA • Ft. Walton Beach, FL •
Gaithersburg, MD • Garden Ridge, TX •
Hanover, MD • Hapeville, GA •
Herndon, VA • Honolulu, HI •
Huntsville, AL • Indian Head, MD •
Jacksonville, FL • La Jolla, CA •
Landover, MD • Langhorne, PA •
Lanham, MD • Layton, UT • Lexington
Park, MD • Manassas, VA •
Mechanicsburg, PA • Middletown, RI •
Millington, TN • Montgomery, AL •
New Orleans, LA • Norfolk, VA •
O'Fallon, IL • Odenton, MD • Oklahoma
City, OK • Olathe, KS • Omaha, NE •
Orlando, FL • Oxnard, CA • Panama
City, FL • Pensacola, FL • Pittsburgh,
PA • Portsmouth, NH • Richmond, VA •
Ridgecrest, CA • Rockville, MD • Rome,
NY • San Antonio, TX • San Diego, CA
• San Francisco, CA • Scottsdale, AZ •
Shalimar, FL • Sierra Vista, AZ •
Smyrna, GA • Southern Pines, NC •
Springfield, VA • Swansea, IL • Tinton,
Falls, NJ • Tukwila, WA • Valley Lee,
MD • Vienna, VA • Virginia Beach, VA
• Warner Robins, GA • Washington, DC

THE SCOOP

This CACI's eagle-eyed

Under the guiding motto "ever vigilant," CACI looks out for the technological well-being of the U.S. armed forces abroad, while also keeping an eye on network management for a few large corporations, primarily in the U.K. The company offers IT and communication services to mostly American government agencies, and operates in four specific arenas: systems integration, managed network services, knowledge management and engineering services.

CACI, with a broad range of expertise in things computer, is one of Uncle Sam's primary go-to tech firms. It works especially hard for the Department of Defense, which accounted for 73 percent of company revenue in 2006. When it's not helping our military wage the war on terror, CACI also enjoys gathering kudos as a leader among its peers, including a No. 932 ranking in the Fortune 1000, with an extra nod from *Fortune* as the seventh-largest information technology services company in the U.S.

A child of the Sixties

So, a businessman and a programming genius are sitting on a park bench, and the businessman says ... No, it's not a setup for the kind of joke your uncle might tell, it's the birth of a successful tech firm. Originally called the California Analysis Center Incorporated, this now-gigantic technology firm started out very small. In fact, company legend has it that founders Harry Markowitz and Herb Karr used to do business from a park bench during the company's infancy in the early 1960s.

Those are especially humble beginnings for a firm that would be going public and making more than $1 million in revenue by 1968. That may sound like small potatoes now, but in 1968, $1 million constituted a much larger potato, especially when served alongside lucrative contracts with the U.S. Navy, IBM and the U.S. Commerce Department. The business that rocketed Markowitz and Karr from the park bench to the corner office so quickly was SIMSCRIPT, a software language the entrepreneurs commercialized and developed into their own line of simulation software, with applications especially useful to the government. It could be used to build computer models of complex functions ranging from airport traffic maps to tactical military plans.

Paging Dr. London

CACI soon expanded, opening offices in Washington, D.C., and New York City in 1967 to keep up with the brisk business. It chose the Hague as its European headquarters in 1974, and also opened regional offices in London, Dublin, Milan and Bermuda. In 1979, CACI released its nifty ACORN program to commercial clients, a demographic information system that analyzed neighborhoods based on consuming trends, so that purveyors of stuff could geographically pinpoint their buying public.

Consumer goods aside, CACI's real bread-and-butter was (and is) the federal government. The company developed its relationship with the federal government throughout the 1980s, winning contracts to develop data management systems for the Department of Justice and the U.S. Navy. By 1983 the company had topped $100 million in revenue; Dr. John London was named president and CEO of CACI the next year, having worked for the firm since 1972. Dr. London would later be elected chairman of the board after the death of founder Herb Karr in 1990.

Networking is everything in business

With the change of leadership came a change of tactic; the company shifted from providing small-scale software packages to tackling larger, broader networking projects. CACI achieved the switch by snapping up smaller firms throughout the 1990s, including American Legal Systems Corp. (1992), Automated Sciences Group (1995), IMS Technologies (1995) and MapData (1999). As the company swelled in size, so did its contracts—in 1996 CACI won a Department of Justice job worth $375 million, and company revenue increased past $300 million for the first time by 1998.

Hey, big spender!

From here the company grew exponentially, riding the coattails of government investment in defense technology following September 11th. CACI acquired several companies after the turn of the century to gain further access to specialized and competitive government contracts, including Premier Technology Group, a Department of Defense professional services provider, in March 2003, and MTL Systems, which supplied its engineering expertise to the DoD, in January 2004. By the end of that fiscal year, U.S. government contracts accounted for 93 percent of CACI's revenue; the company earned just over $1 billion dollars from Uncle Sam that year.

Unwanted attention

These facts became big news in April 2004, after the Abu Ghraib prison abuse scandal hit the headlines in newspapers across the country. When CACI was named in an Army report regarding the abusive interrogation practices inflicted on prisoners at Abu Ghraib, a media circus erupted around the increasing use of private contractors in once-military functions. At the time, CACI had 36 employees stationed in Iraq as part of an interrogation services contract it inherited upon the acquisition of Premier Technology Group in 2003; the Army report alleged that one CACI employee was involved in prisoner abuse.

In response to the heightened scrutiny around government contractors, the General Services Administration announced in May 2004 that it would investigate the case, specifically to determine if CACI had broken federal regulations and should be banned from future contracts. The company was cleared of any wrongdoing in that investigation the following July. CACI then launched an internal investigation and determined in August 2004 that there was no evidence implicating its employee in abuse. While CACI came out of the various probes and investigations with clean hands, the entire fiasco made the work of private contractors in Iraq a major concern for the American public, resulting in some negative attention for the erstwhile low-profile CACI.

In bed with the feds

Still, the bad publicity hasn't hurt the company's financial relationship with the U.S. government: in June 2005 CACI won a $73 million contract to provide technical support to the U.S. Navy. This was followed by another Navy job ($188 million) to assist in developing unmanned mine-detection technology. All this, however, paled in comparison to the contract CACI beat out other firms to win in March 2006—its largest ever, the deal is a broadly-based, 10-year arrangement with the U.S. Army worth a massive $19 billion.

In 2006, government contracts made up 94 percent of CACI's $1.7 billion in revenue. Profit of $84 million remained healthy, as well, thanks in part to the acquisition of three companies that year: ISS, NSR and AlphaInsight. Together, these new firms accounted for more than half of 2006's growth in DoD contracts, bringing in an extra $54 million.

Ever vigilant, even in the dark

Big government contracts continued to roll in for CACI throughout 2007. Under a $24 million contract with the U.S. Pacific Air Forces command, awarded in January, CACI will provide mission planning and control services to increase efficiency for the flyboys. The following month, the company signed a contract for $39 million to assist the army with developing and testing night vision surveillance technology. CACI's longstanding relationship with the U.S. Navy was further fortified with a contract to set up databases to assist the Navy in its worldwide inventory and supply chain operations; the contract was awarded in April, and will bring another $48 million into company coffers.

Oh Captain, my chairman

After leading the CACI crew for more than 20 years, CEO Jack London stepped down in July 2007, passing the role on to Paul Cofoni, who has served as president of CACI's U.S. operations since 2005. London will keep a place at the company round table as chairman, a post he's held since 1990. Although an announcement of the executive musical chairs came only a couple of weeks before the annual shareholders' meeting, where Cofoni officially took the lead, London said the transfer of power had been planned since Cofoni joined the company.

Together, the gentlemen plan to take things up a notch for CACI, leading it from its position as a really big company to a super-big company. To that end, CACI is in the process of shifting its strategy from gathering a lot of little contracts to catching the big $100 million-plus kahunas passed out by the government, with whom CACI has a long and friendly history.

GETTING HIRED

Support the company that supports our troops

Because of its close relationship with the DoD, CACI goes where the military goes. For the globe-trotter, that can mean job opportunities in the far-flung corners of the world. For the local-yokel, that also means plenty of jobs within the U.S. proper. All of the positions at CACI's many locations can be found through the company's jobs site, at www.caci.com/job.shtml. Here open positions are searchable by type or location. A more cerebral "concept search" allows you to post your resume or enter keywords and find matching opportunities. Seekers are encouraged to build a profile on the employment site by entering some basic information and uploading a resume; then CACI can contact you when a position that could fit you opens up. If this is all a bit too impersonal, CACI also appears at several professional career fairs throughout the year. A full list can be found on its career site.

For those initiated into the CACI ranks, life is good. The company offers a 401(k) program with company matching, comprehensive health coverage, and paid vacation and holidays. Tuition reimbursement is available through the company, as are legal services and financial planning. For the puppy, kitty, birdie and fishy lovers, CACI offers a pet care discount plan (pet rocks are, unfortunately, not covered). The company also provides training to keep its employees abreast of the latest technology and to promote advancement through the company ranks.

Canon Inc.

30-2, Shimomaruko 3-chome
Ohta-ku, Tokyo 146-8501
Japan
Phone: (81-3) 3758-2111
Fax: (81-3) 5482-5135
www.canon.com

Canon USA Incorporated
One Canon Plaza
Lake Success, NY 11042-1198
Phone: (516) 328-5000
Fax: (516) 328-4809
www.usa.canon.com

LOCATIONS

Tokyo, Japan (World HQ)
Lake Success, NY (US HQ)
Burlington, NJ • Chesapeake, VA •
Conshohocken, PA • Gardena, CA •
Gloucester, VA • Irvine, CA • Mount
Laurel, NJ • Miami, FL • Newport
News, VA • Rockville, MD • Salt
Lake City, UT • Schaumburg, IL

DEPARTMENTS

Accounting • Administration •
Clerical • Finance • HR • IT • Legal
• Logistics • Manufacturing •
Marketing • Product Support • Sales

THE STATS

Employer Type: Public Company
Stock Symbol: CAJ
Stock Exchange: NYSE
President & CEO: Yoroku Adachi
2006 Employees: 118,499
2006 Revenue ($mil.): $34,916

KEY COMPETITORS

Hewlett-Packard
Ricoh
Sony
Xerox

EMPLOYMENT CONTACT

www.usa.canon.com/templatedata/
Careers/ciwcrcareer.html

THE SCOOP

Canon, a captivating company

As one of the world's top producers of cameras, Canon urges you to take a picture; it will last longer. Additionally, as one of the world's leading manufacturers of printers and fax machines, Canon urges you to make lots and lots of copies of that picture, for posterity. The company specializes in all things visual, and is best known for its business machines (copiers, printers, fax machines, et al.) and its cameras (digital, video, LCD projectors and the like). Through its optics arm, it also produces medical X-ray machines and the steppers and mask aligners used in semiconductor production.

Canon's headquarters and much of its production facilities remain in Japan, but the Canon USA slice of the pie accounts for about a third of the parent company's sales. And don't let the name fool you; Canon USA is not just limited to the States. This division of the company operates through more than 30 offices across North, South and Central America and the Caribbean, and employs more than 10,000 people in North America. *Business Week* ranked Canon at No. 51 on its InfoTech 100 in 2007 (behind rival Hewlett-Packard who came in at No. 35, but ahead of Sony and Xerox, who did not make the list).

A photogenic star is born

Canon got its start in 1937 as Precision Optical Industry Co. Ltd. in Meguro, Japan, a research laboratory producing cameras at a rate of 10 per month. The company changed its name to Canon in 1947, a transliteration of the name of the Kwanon camera that founder Takeshi Mitarai (a gynecologist) and his technicians developed in 1933, before the company was incorporated. The Kwanon was Japan's first 35mm camera. As the company accrued capital from investors and experience, more staff was added and Canon began rolling out new camera models.

Though World War II had slowed production and overseas sales, by 1955 Canon had opened a branch on Fifth Avenue in New York City, its first step in becoming a global company. Sales activity increased through aggressive marketing campaigns, and in 1966, Canon officially christened its subdivision in New York as Canon USA, the HQ moving from Manhattan to Queens. In 1971 it moved again, to Lake Success, where the company currently resides, although plans were announced in June 2007 to build a new headquarters about 20 miles up the road, in another Long Island locale, Melville, New York.

Not just cameras anymore

Canon's portfolio expanded in the 1980s, and it took on big competitors for market share. The company's electronic typewriter snatched 11 percent of that market out from under IBM in 1982, while the company bested Xerox just a year later with the release of its significantly cheaper but equally competent laser printer. Canon also formed a valuable friendship with Apple Computers during this time; the former began marketing the latter's products in Japan in 1984 and also jointly developed software for the Japanese market.

Getting smart in the 1990s

In part because of a general recession in Japan, Canon's sales slowed down in the 1990s. To cushion itself against future local economic woes, Canon poured more money into globalization and research and development. In 1994, Canon teamed up with one-time competitor IBM to produce PowerPC laptop computers. Two years later, the company opened facilities in Miami and the Philippines to extend its reach into Latin America and the Pacific Rim. It opened subsidiaries in China and Finland in 1997, and a software development center in India the following year.

Going, going, gone digital

Canon's renewed commitment to R&D bore fruit in 1996, when the company introduced the PowerShot 600, its first digital camera. The product originally emerged as a computer periphery through a subsidiary company, but digital photography soon became an "it" technology, and Canon reorganized its camera division. The company's camera arm had previously had success with traditional 35mm cameras, but now it had responsibility for the design, development, manufacture and marketing of digital cameras. The division did Canon proud—its digital cameras were honored in 2004 by a wide variety of publications, from PC World to Popular Science, from Outside to American Photo. Canon also creates its own lenses, image sensors, image processors, color-rendering software and imaging technologies to keep ahead of its competitors (the company is consistently one of the top holders of technology patents in America).

Adachi assumes control

In April 2005, Yoroku Adachi took over as Canon USA's president and CEO. A former president of Canon Canada, Canon China and Canon Hong Kong, Adachi has also served as chairman for Canon Singapore. CEO Kinya Uchida, whom Adachi was replacing, returned to Tokyo to take on an advisory role at the parent company's headquarters.

Although the company had been showing a steadily (if somewhat sluggishly) growing profit since the turn of the century, the tempo picked up significantly after 2004. By 2005, profit was double that of just three years earlier—this was aided in part by the acquisitions that year of ANELVA Corporation (in September) and NEC Machinery Corporation (in October), both of which pumped up Canon's manufacturing muscles.

Still life with new products

Canon rolled out a bunch of new products in 2006, including 22 digital cameras, printers, and projectors at the Photo Marketing Association trade show alone that February. Two months later, the company debuted its new line of high-def lenses for professional broadcast use at the National Association of Broadcasters trade show. In August Canon released the world's smallest and lightest high-def video camcorder for the space-challenged consumer. Weighing just under a pound and only two inches by four inches in size, the Canon HV10 HDV camcorder takes up less space in your pocket than the cash needed to purchase it—the estimated selling price upon its release was $1,299. More innovations followed in 2007, including a trio of DVD camcorders, two high-end but low-price photo inkjet printers and nine new portable high-definition camera lenses.

King of the copier hill

Canon, which constantly vies for the top of the heap in its competitive consumer electronics markets, triumphantly snagged the No. 1 U.S. market share in copiers for 2006, making that the 24th time in 25 years that the company led all other brands in the copier market. Overall, sales for 2006 increased 10 percent over 2005, reaching the mind-boggling figure of $34.9 billion by year's end. Besides reaping great profits, the company continued to invest in its own innovation, spending over $2.5 billion in research and development that year. The company has the goal of holding onto the No. 1 position worldwide in all its core businesses, and trusts its internal resources to get it there.

He SED, she SED

One market sector where Canon is not No. 1 is in flat-screen TVs, and Canon ran into a roadblock in early 2007 with its planned October rollout of new flat-panel televisions, using the latest SED (surface-conduction electron-emitter display) technology. Nano-Proprietary had sued Canon in 2005, after Canon licensed SED display technology from the firm, on the grounds that that the contracted license didn't extend to another flat-screen venture, which Canon had entered into with Toshiba. Canon bought out Toshiba's stake in the venture in January 2007 in an effort to backpedal out of the sticky situation. The case was dismissed and Canon was found guiltless in May 2007, but then Canon appealed an earlier decision in an effort to get back the $5.5 million licensing fee it had paid Nano-Proprietary. As a result, that May Canon also indefinitely suspended the product's October release.

The delay in launching the televisions is unfortunate for Canon, which is hoping SED will give it a stake in the flat-panel TV market. Although SED screens have a quicker response time than LCD (liquid-crystal display) or plasma TVs, and use much less power than traditional cathode-ray-tube models, they are still relatively unknown in the U.S., where LCD technology currently reigns supreme.

Global (heart)warming

Canon prides itself on corporate responsibility. At the heart of Canon's values is the concept of kyosei, most succinctly defined (by the company) as "living and working together for the common good." To this end, Canon is involved in many social welfare and humanitarian efforts, including its Canon4Kids program, which helps locate missing children across the U.S. Canon also puts its sponsorship to good use; for each birdie, eagle and hole-in-one Briny Baird and Michelle McGann make during PGA golf tournament play, Canon donates money to the National Center for Missing and Exploited Children. Both duffers also place photographs of local missing children on their golf bags during tournaments. At Yankee Stadium in New York City, as part of Canon4Kids, parents are offered wristbands with seating location information for their children to wear, in order to help assist security personnel in aiding lost children.

Kyosei applies to the common good of the environment as well. Through its Factor 2 initiative, Canon has set specific goals to reduce its carbon dioxide emissions in inverse proportion to company sales. Canon USA also funds a national parks scholars program to encourage doctoral research in support of wildlife and sponsors the award-winning public television series *NATURE*. Finally, in 2006 the company launched its first 100 percent recycled copier and printer paper, and also started using biodegradable plastic in its packaging materials.

GETTING HIRED

Extra points for using Canon's printers to print your resume

Employment information can be found on Canon USA's web site at www.usa.canon.com. Applicants can search for open positions by location or by job type, and then apply for a spot online by uploading their resume or by building one on the career site. There is also a wealth of information about the company's American locations, its benefit programs and its diversity on the Web. First, the benefits: medical, dental and vision insurance coverage, flexible spending account, a 401(k) plan with company match and profit sharing. Canon will also reimburse those thirsting for knowledge for up to 90 percent of their tuition expenses, if the course of study is relevant to the company.

Canon also offers a summer internship program for college upperclassmen and recent grads. The program runs from June to August, and is a full-time, paid internship, offering experience with business, marketing and technical issues, as well as the opportunity "to gain hands-on work experience by participating in significant projects." Potential interns can also apply online for positions in a number of different areas, including marketing, IT, business administration, strategic planning, finance, economics, environmental science, industrial engineering, logistics, HR, legal, corporate communications and statistics. Canon stresses it is looking for internship candidates with "strong communication skills, technical and/or computer skills, plus the ability to work independently and as part of a team." The company prefers college students who have completed their junior year.

Cisco Systems, Inc.

170 West Tasman Drive
San Jose, CA 95134
Phone: (408) 525-3777
Fax: (408) 526-4100
www.cisco.com

LOCATIONS

San Jose, CA (HQ)
Boxborough, MA
Lawrenceville, GA
Research Triangle Park, NC
Richardson, TX

144 locations throughout the US,
194 locations worldwide.

DEPARTMENTS

Administrative • Chief Technology
Office • Cross-Functional •
Customer Advocacy • Engineering •
Executive • Finance • HR • IT •
Legal • Manufacturing • Marketing •
Sales • Workplace Resources

THE STATS

Employer Type: Public Company
Stock Symbol: CSCO
Stock Exchange: Nasdaq
Chairman & CEO: John T. Chambers
2006 Employees: 38,416
2006 Revenue ($mil.): $28,484

KEY COMPETITORS

Avaya
Juniper Networks
Nortel Networks

EMPLOYMENT CONTACT

www.cisco.com/web/about/ac40/abo
ut_cisco_careers_home.html

THE SCOOP

Huge and loving it

One of the brightest stars of the 1990s tech bubble, Cisco Systems was one of the most valuable companies in the world at its peak, with a market capitalization exceeding both Microsoft and GE. Its fortunes have since slid from those lofty heights, but Cisco is not exactly a corporate charity case; in its fiscal year 2006, Cisco's sales were approximately $28 billion, and the company currently holds its own at No. 77 on the Fortune 500. *BusinessWeek*, also appreciative, ranked the company at No. 20 on its 2007 edition of the InfoTech 100.

Today, Cisco is a top manufacturer of networking hardware such as routers, switches and hubs—all devices constituting what is often called the "backbone" of the Internet. Switches and hubs connect computers to each other, creating local area networks, or LANs. Routers, meanwhile, connect one network to another—for example, from a LAN to a wide-area network (or WAN), such as the Internet. Cisco also develops and sells the software used to manage these networks. Its primary client base consists of large corporations, educational institutions, government agencies and telecom service providers, all of which deploy large and complex networks. But with its acquisition of Linksys in March 2003, Cisco dove into the small business and home networking markets as well.

The company has also been one of the pioneers of voice over-Internet protocol (VoIP) technology. VoIP uses the same technology that enables computers to communicate with each other, but it applies to phone calls—enabling phone calls to bypass traditional phone lines. The company's goal is to eliminate the need for separate voice, data and video infrastructure, replacing these with a single multipurpose network.

Cisco has United States R&D facilities at its HQ in San Jose, California, as well as in Massachusetts, Texas, Georgia and North Carolina. Internationally, the company has major research facilities in India, China and Israel. It also looks outside its own researchers for innovation, having acquired over 108 companies since its inception.

Cisco vs. Stanford

The history of Cisco Systems is part of Silicon Valley lore, infused with drama, conflict and a bit of myth. Its name even harkens to the major city in the area: San Francisco. In the early 1980s, married couple Leonard Bosack and Sandy Lerner worked as part of a team of computer engineers at Stanford University who were looking for a way to connect the separate and incompatible computer networks scattered around campus. Their research resulted in an early prototype of the multi-protocol router—and Bosack and Lerner founded Cisco out of their home in 1984 to commercialize this technology. They continued to work at Stanford, however, refining their product largely with the university's time and resources, which inevitably led to a contentious battle over ownership rights. Stanford relented in 1987 and agreed to license the technology to Cisco in exchange for $19,300 in cash, $150,000 in future royalties and equipment discounts.

Cisco Systems took off, and its founders eventually decided to get help in growing the company, attracting the interest of local investment firm Sequoia Capital. Investors chose John Morgridge, another Stanford-ite and former executive at Honeywell, to lead the company as CEO, effectively becoming Bosack and Lerner's boss. Relations among the three were tense from the start. Nevertheless, the company went public in 1990 and became something of a Wall Street darling, attracting attention as the newest hot tech stock. But Bosack and Lerner had become disillusioned with the increasingly corporate atmosphere of the company they'd founded. That August, Cisco's seven VPs met with Morgridge to demand their resignations—they quit and sold their stake that same day, netting about $170 million.

Boom then burst

In January 1995, John Chambers succeeded Morgridge as president and CEO of Cisco, while the latter retained the chairmanship. Chambers was one of the most vocal and visible proponents of what was dubbed the "new economy." Nimble Cisco was where all of corporate America was headed, and Chambers dazzled Wall Street analysts by releasing finalized financial data within 24 hours of each quarter's close.

But Cisco was caught flat-footed when the tech bubble burst in 2000, as orders suddenly dried up and the company was left with a mountain of unsold and aging equipment. In the second quarter of 2001 (no matter the 24-hour financials), Cisco's revenue dropped for the first time in its history, and Chambers was forced to write off $2.5 billion worth of inventory (a near-scandalous development on Wall Street at the time). The company's stock also took a dive, from a high near $80 a share in March 2000 to a low of around $12 in September 2001. In an effort to right the ship, Chambers instituted aggressive restructuring, laying off nearly 10,000 employees; also, in a largely symbolic move, he slashed his own salary to $1 per year until the company recovered.

Target: Cisco

But Cisco was not out of the woods yet. Competitors such as Dell and China's Huawei Technologies are seeking to eat into Cisco's market share by introducing low-cost networking equipment and software. Just as commodification effectively slashed the prices of PCs in the 1980s and 1990s, a similar process in the networking equipment business could spell disaster for Cisco.

The company has defended its turf, though, terminating its partnership with Dell and bringing a patent-infringement lawsuit against Huawei. After some legal wrangling, the companies settled in September 2004 on terms seen as favorable to Huawei. Though the Chinese router company will amend its user interface and some other code, Huawei is free to continue doing business in the lucrative U.S. market, and could establish itself as a major global player and threat to Cisco.

Expanding abroad, securing at home

In more international news, Cisco has shelled out some big bucks to build its empire abroad. In October 2005, the company said it would spend $1.1 billion in India over the next three years—the company's largest international investment for that time period, CEO Chambers told investors. Nearly three quarters of that hefty sum is earmarked for research and development at Cisco's global research center in Bangalore. Furthermore, in December 2006, the company announced that, by 2011, it would triple its workforce in India to 6,000. The company is also looking at opening a manufacturing plant in India.

Dealing with disasters

September 11th and Hurricane Katrina provided the impetus for a more recent Cisco innovation, when the company recognized the need for reliable communications between first responders to national emergencies. They set out to do something about this, designing IP-based gadgetry that would solve existing "voice communications interoperability," thereby enabling communications over different varieties of devices. Called IPICS (IP Interoperability and Collaboration System), the product is one of the first to integrate various different push-to-talk radio systems together with widely positioned voice, video and data networking. Cisco considers the ability of IPICS to allow preexisting communications networks to interoperate—sort of a recycling of old technology with the aid of new IT—as one of its most promising recent developments, saving money for its clients by eliminating the need to entirely reinstall already deployed systems.

Cisco turned another snarly problem into a lucrative opportunity in February 2007, and all the fuss was over a six-letter word. CEO and media darling Steve Jobs had just unveiled Apple's new iPhone gizmo to much fanfare, until Cisco crashed the party with the announcement that it had owned the trademark on the name "iPhone" since 2000 and was using it for a line of Internet-connected telephones. Things got a little ugly—words like "silly" and "malicious" were bandied about by both sides with lawsuits in tow—but in the end a mutually beneficial deal was struck. Apple would get to use the name, and Cisco products would get interoperability with Apple's highly-coveted music phone.

It just keeps getting bigger

Already a sprawling business, Cisco is diversifying its product line. Swerving into consumer-oriented technology, Cisco acquired the No. 2 cable box manufacturer, Scientific-Atlanta, for $7 billion in February 2006. The deal was especially noteworthy for a company that is already known for its acquisitorial enthusiasm, since this was the second-biggest purchase in Cisco's history (behind that of home and small office networking company Linksys two years earlier).

The new addition to the Cisco family has so far held up its end of the bargain. Net sales for 2006 increased 15 percent and broke a company record with the staggering figure of $28.5 billion. At the same time, income for the year dropped a notch, falling from $5.7 billion in 2005 to $5.6 billion a year later. This minor slip may be due to the massive effort involved in integrating 8,000 employees from Scientific-Atlanta into Cisco, with another 3,500 new employees hired in 2006 to boot.

Buy, baby, buy—Cisco inferno!

More large deals—and, accordingly, more diversification—came in 2007, when Cisco agreed to purchase IronPort Systems, a security software firm, for $830 million and Tivella, a digital signage software company, in January 2007, with the acquisitions of 'Net-tech firms Reactivity and Five Across to follow in February. In March, Cisco went for the triple-threat, buying file storage software maker NeoPath, network processor-provider SpansLogic, and finally WebEx, an online conferencing firm, for $3.2 billion. Video surveillance software maker BroadWare Technologies rounded out Cisco's springtime feeding frenzy, joining the company in May 2007.

GETTING HIRED

See ya at Cisco!

The employment section of Cisco's site, www.cisco.com/web/about/ac40/univ/index.html, has a database of available jobs around the world, searchable by country, town, keyword and job function. Employment-seekers can submit their resume for a specific position, saving their job profile in the meantime for future ease of application. Cisco has special listings for recent graduates, and especially encourages those with degrees in sales, services, engineering, marketing, finance and operations to apply. Those who haven't graduated yet can work as an intern in the summer or during the semester as a co-op. Either way, the company promises a worthwhile immersion in the company environment and—better still—it pays. In 2007, Cisco landed at the No. 11 spot on *Fortune*'s list of the Top 100 Best Places to Work. The company's strong stock-options plan, which gives a majority of employees the opportunity to be company shareholders as well, received special mention from the magazine.

Ready for your interview? One source reports that "Interviews for regular employees [are] typically one-on-one sessions with department managers or directors." He reports going through five rounds of interviews with department staff on two separate days, around 30 minutes each."

OUR SURVEY SAYS

Cisco's always on

Once hires get on board, they get to experience the Cisco culture. The company offers laptops and Treos, but they come at a price. The "work from home lifestyle means many Cisco employees work around the clock—we are always online," says one sleep-deprived employee. "The hours are flexible, with the primary criteria being to get the job done, whatever/whenever it takes," adds a source. This, of course, leads to days "easily much longer than standard working hours." The good news about being constantly accessible means that the "culture is one of empowerment, with individuals encouraged to take ownership. In general, Cisco treats its employees as mature adults," according to her co-worker. "Voicemails are also a huge part of the company culture," adds a respondent posted in Asia. The dress

Visit Vault at **www.vault.com** for insider company profiles, expert advice, career message boards, expert resume reviews, the Vault Job Board and more.

VAULT CAREER LIBRARY   159

code is as relaxed as the working hours, with one contact noting that "There is no dress code. Dress is expected to be appropriate to the situation."

There is some turmoil in the ranks, however. "No job is secure as everyone typically needs to go through interviews all over again to secure their roles during departmental shake-up. Cisco is tight with headcounts and will be difficult for contractors and temps to get converted," says a source. Another points out that the company has a "fairly good track record on hiring minorities at the Cisco corporate office (other than African-American), and fairly poor results with hiring and retaining women. One source from sales adds that "opportunities for advancement [are] good if you can network and meet the right people."

Compuware Corporation

One Campus Martius
Detroit, MI 48226
Phone: (313) 227-7300
Fax: (313) 227-7555
www.compuware.com

LOCATIONS

Detroit (HQ)
Addison, TX • Appleton, WI •
Atlanta, GA • Baltimore, MD •
Bellevue, WA • Blue Bell, PA •
Brentwood, TN • Cambridge, MA •
Cedar Rapids, MI • Charlotte, NC •
Cincinnati, OH • Columbus, OH •
Denver, CO • Des Moines, IA •
Durham, NC • East Rutherford, NJ •
Eden Prairie, MN • Englewood, CO •
Fort Lauderdale, FL • Fort Myers, FL •
Framingham, MA • Grand Rapids, MI
• Houston, TX • Independence, OH •
Irvine, CA • Lake Mary, FL • Lansing,
MI • Long Beach, CA • Madison, WI •
McLean, VA • Merrimack, NH •
Milwaukee, WI • Oakland, CA •
Overland Park, KS • Phoenix, AZ •
San Diego, CA • Schaumburg, IL •
Tampa, FL • Woodland Hills, MI

THE STATS

Employer Type: Public Company
Stock Symbol: CPWR
Stock Exchange: Nasdaq
Chairman & CEO: Peter Karmanos Jr.
2006 Employees: 7,539
2006 Revenue ($mil.): $1,213

DEPARTMENTS

Accounting/Finance • Business &
Systems Analysts • Corporate
Administration • Data Resource
Management • E-Business & Web
Development • Facilities •
HR/Recruiting • Legal • Marketing •
Network & Systems Administration •
Project Management • Sales & Sales
Support • Software Development •
Software Testing • Systems
Programmers • Technical
Communications • Technical Customer
Service

KEY COMPETITORS

BMC Software
Borland Software
CA, Inc.
IBM

EMPLOYMENT CONTACT

www.compuware.com/careers

Visit Vault at **www.vault.com** for insider company profiles, expert advice,
career message boards, expert resume reviews, the Vault Job Board and more.

VAULT CAREER LIBRARY   161

THE SCOOP

Motor City/Shmotor City

Detroit isn't just about cars anymore. One of its largest corporate citizens, Compuware Corporation, has been providing software and information technology services to businesses around the globe for more than three decades. Now serving the world's leading IT companies, including more than 90 percent of the Fortune 100, the company is a leading example of Detroit's role in the technology industry. Compuware typically offers IT governance and management, application development, quality assurance, service management and support services.

Expanding through acquisition

In Southfield, Michigan, in 1973, Peter Karmanos Jr., Thomas Thewes and Allen Cutting established the Compuware Corporation with the goal of providing clients with professional technical services to allow them in turn to focus on their own core businesses, or, in their own words, to "help people do things with computers."

Within five years, Compuware introduced its first software product, Abend-AID, designed to detect bugs and suggest corrective action in corporate mainframe systems, and opened an office on the East Coast to service the Washington, D.C. and Baltimore area. Compuware's next big product was released in 1983, called File-AID. It was the first in a line of data management products from the company, and it allowed IBM-compatible computers to easily organize information. File-AID products compatible with other types of computers would soon follow.

Going shopping in the 1990s

After acquiring some European companies, Compuware moved its headquarters from Southfield to Farmington Hills, Michigan, in 1987. The additions abroad accompanied a long string of purchases stateside that dominated Compuware's business dealings in the 1990s. Over the course of the decade, the company picked up a slew of tech firms, including Centura Software, EcoSystems and UNIFACE; and Hiperstation and Coronet soon after. The subsequent purchases of Direct Technology Limited and DRD Promark, Inc., served as the foundation for Compuware's QACenter, a key software for testing product quality. Finally, Compuware expanded into the Southeastern and Western U.S. with the addition of Data Processing Resources Corporation in 1999. The numerous deals helped Compuware augment its

Windows/NT software business, increase its number of employees and technicians, and expand into new markets. Along the way, Compuware sold shares to the public, first in 1992 and again in 1994, and topped $1 billion in revenue in 1998.

Not all the transactions were smooth sailing, though. In 1999, Compuware attempted to buy Viasoft, Inc., which specialized in trouble-shooting Y2K computer issues, for $164 million. The Justice Department requested additional information in August and then announced it would block the merger in October, bringing an antitrust case against the two companies which argued that Compuware would attain a dominant position over the de-bugging/trouble-shooting industry. After only a few months of litigation, Compuware and Viasoft reneged on their agreement.

Innovation and recognition

Undaunted, Compuware continued to grow through acquisition, picking up e-business service firms BlairLake and Nomex, which Compuware then combined to form Compuware Digital Development Centers in 2000. The company put new technological advancements and product launches at its forefront, releasing OptimalView, an enterprise portal; DevPartner Java Edition 3.0; File-AID/Data Solutions 3.3; Vantage 9, a program to improve application performance; and CARS, the Compuware Application Reliability Solution.

The hard work and innovation did not go unnoticed: Compuware picked up Network Computing's Product of the Year for its Application Expert in 2001. More accolades followed in 2003, when the company won the Yphise Award for the Best Application Performance Management Software for its Vantage line, and *Visual Studio* magazine's Readers' Choice Award for DevPartner Studio and TrackRecord products.

Tough times in the 2000s

Despite the accolades, Compuware hit a stumbling block with the turn of the century. Revenue rose impressively up to the 2000 peak of $2.2 billion as Y2K-fearing companies threw money at Compuware to protect them against an imagined technological breakdown. But on the heels of that success came the pop heard 'round the world—that is, the bursting of the tech bubble—which dropped sales by 22 percent to $1.7 billion for 2002. That year, the company posted a net loss of $245 million. Compuware eliminated 1,600 jobs through 2002 and slashed prices to compete with free or cheaper software from IBM.

Taking on a giant

In fact, IBM proved to be a particularly painful thorn in the company's side that year, as Compuware sued the computer industry's Goliath in May for software piracy. The suit alleged that IBM copied Compuware software products and resold them under the IBM name. IBM denied the charges (naturally), claiming it had simply used public information and its own expertise to create products that lowered prices in a market formerly dominated by Compuware. The litigation drained Compuware's coffers for three years, costing the firm $12 million in 2002, $34 million in 2003 and $45 million in 2004, but paid off with a March 2005 settlement, in which IBM agreed to license and purchase $400 million dollars-worth of Compuware software and services over the next four years.

Community ties

In the meantime, corporate headquarters moved to Detroit in 2003, bringing over 4,000 workers to the area, and cementing the longstanding relationship between Compuware and the Motor City (the software firm had been providing IT services to Detroit Public Schools since 2001, and by 2003 had saved the district more than $3 million). The company continued to make nice with the city in March 2004, when it purchased Covisint, a company specializing in infrastructure technology for the automotive industry.

More acquisitions were in the works, as it Compuware scooped up IT governance provider Changepoint Corporation in May 2004, with the purchase of performance monitoring tech firm Adlex, Inc. to follow a year later. But the company still turned in disappointing financial results, as revenue slid further—to $1.2 billion in 2005. Thanks to restructuring, however, profit was on the mend, increasing 53 percent from 2004's $49 million to $76 million.

Latest developments

In March 2006, Compuware bought a treat for its Covisint division, acquiring privately-held ProviderLink to support Covisint's expansion into the health care industry. With the goal of spreading technology to health organizations still based on paper, Compuware married Covisint's web-based communication applications to ProviderLink's expertise in online health care tools. Another company, SteelTrace, joined Compuware the following April, to beef up the firm's application development management business.

As the company optimistically moves to diversify its product offerings and grow through acquisition, it's still a little early for company brass (and stockholders, for that matter) to issue sighs of relief. Although profit for fiscal 2006 doubled to $142 million, revenue for fiscal 2006 dropped $25 million from 2005. Fiscal 2007 revenue, however, increased slightly to $1.21 billion; profit increased about 10 percent to $158 million. CEO Karmanos admitted in a March 2007 e-mail to employees that the company "must change the way it does business." If Compuware doesn't clean up its act soon, Karmanos warned, "an outside entity will use the tremendous amount of capital currently available in the marketplace to take over the company."

GETTING HIRED

Detroit and Compuware, sitting in a tree ...

The company posts job openings online (at www.compuware.com/careers) and has set up a search engine through which interested applicants can find opportunities from Australia to the United Kingdom, and several places in between. Applicants can upload their resume to the career web site and will be contacted if Compuware likes what it sees. The company also hosts recruiting events; a schedule of these is available through the web site.

Detroit loves Compuware and has increasingly exhibited its approval of the company. In 2006 Compuware was named to *Metropolitan Detroit*'s 101 Best and Brightest Companies to Work For list, sponsored by the Michigan Business and Professional Association; and readers of the local *Hour Detroit* voted it the very best place to work in the city.

Compuware also touts its commitment to diversity in the workplace, and was recognized for backing up that pledge last year. The National Black Data Processing Associates cited Compuware as a top 10 company for blacks in technology, and *DiversityInc* named Compuware to its list of Top 50 Companies for Diversity in June 2006. So what is everybody crowing about? In addition to a full run of health care benefits, Compuware provides an employee stock purchase plan and 401(k) with company match. Veterinary insurance takes care of Fido and a generous vacation and sick day program allows workers to stay home while Fido recovers. Extensive training and certification opportunities keep employees' minds razor sharp, while health club membership discounts keep their bodies taut.

Visit Vault at **www.vault.com** for insider company profiles, expert advice, career message boards, expert resume reviews, the Vault Job Board and more.

VAULT CAREER LIBRARY   **165**

Cypress Semiconductor Corporation

198 Champion Court
San Jose, CA 95134
Phone: (408) 943-2600
Fax: (408) 943-4730
www.cypress.com

LOCATIONS

San Jose, CA (HQ)
Austin, TX • Beaverton, OR •
Bloomington, MN • Boise, ID •
Burlington, VT • Cambridge, MA •
Colorado Springs, CO • Dallas, TX •
Lexington, KY • Lynnwood, WA •
Minneapolis, MN • Moscow, ID •
Nashua, NH • Phoenix, AZ •
Portland, OR • Round Rock, TX •
San Diego, CA • Seattle, WA •
Starkville, MS

112 locations in 20 countries
worldwide.

DEPARTMENTS

Administrative & Support Services •
Advertising/Marketing/Public
Relations • Computers & Software •
Engineering • Executive
Management • HR/Recruiting •
Installation, Maintenance & Repair •
Legal • Manufacturing & Production
• Product Management

THE STATS

Employer Type: Public Company
Stock Symbol: CY
Stock Exchange: NYSE
Chairman: Eric A. Benhamou
President & CEO: T.J. Rodgers
2006 Employees: 5,800
2006 Revenue ($mil.): $1,091

KEY COMPETITORS

Integrated Device Technology
Samsung Electronics
Texas Instruments

EMPLOYMENT CONTACT

www.cypress.com/careers

Visit Vault at **www.vault.com** for insider company profiles, expert advice,
career message boards, expert resume reviews, the Vault Job Board and more.

VAULT CAREER LIBRARY   167

THE SCOOP

A tree grows in San Jose

Cypress Semiconductor is a manufacturer of microchips for consumer electronics and communications devices. The company's product line also includes memory, both wired and wireless USB chips, mixed signal chips that can process both digital and analog data, and programmable systems-on-a-chip (PSoC), especially programmable clocks. Cypress also owns a majority stake in SunPower, a manufacturer of solar panels.

Networking pays off

In 1979, T.J. Rodgers was looking to leave the large Silicon Valley firm Advanced Micro Devices and start up his own semiconductor company nearby, but he couldn't work out the financing. Conveniently, that year he met Stanley Fingerhood, a venture capitalist who hooked him up with some friends, who had money to invest. Rodgers estimated that he needed $40 million, and Fingerhood led him to Ben Rosen, chairman of Compaq at the time. Rosen, in turn, introduced Rodgers to his partner, the improbably named L.J. Seven, who, upon receiving Rodgers' business plan (seven handwritten pages), found it worthy of investment. Seven gave Rodgers the startup capital, and in 1982, three years after Rodgers' first attempt, Cypress Semiconductor was born.

Cypress went public in 1986, after its first year of profitability. Sales initially took off as the computer market boomed, and revenue had reached nearly $300 million by 1991. But the following year, Cypress was battered by the slings and arrows of the volatile (some might say outrageous) semiconductor industry, as competitors made inroads into its markets, and the company found itself in the red. Thanks to a swift restructuring, though, it was once again turning a profit the next year.

Acquiring time!

The Internet age dawned in earnest in the late 1990s, and Cypress responded by acquiring several companies. At the start of 1999, it bought IC Works, a producer of integrated-circuit timing equipment, for $130 million; CEO Rodgers said the purchase would decrease Cypress' dependence on memory chips and expand its holdings in the wireless RF business. In July of the same year, Cyprus bought Arcus Technology, a privately held provider of equipment data for telecommunications, for

$20 million. A year later, Cypress picked up Silicon Light Machines, designer of optical networking chips, for $166 million, adding to its optical networking technology cache. Next on the list was International Microcircuits, maker of timing circuits, bought in January 2001 for $125 million. Recent years have been trying for Cypress, as it has endured a series of convulsive restructurings in an attempt to return to profitability; the company posted losses in 2002, 2003 and 2005.

Regrowth

2006 was a good year for Cypress: the company took in revenue of just over $1 billion, a 23 percent increase over 2005, and posted a profit of $39 million. That same year, it shed two divisions to refine its product line (not to mention perk up its bottom line). The PC clock division was sold to Spectra Linear for $8 million, and Cypress sold its network search engines unit to NetLogic Microsystems a few months later.

Sun, power, power, light

Much of Cypress' growth during 2006 can be attributed to strong results from its solar panel subsidiary SunPower. But Cypress has owned the company since 2000, sticking with it through years of unprofitability, and the story of its success is an odd one. Richard Swainson, an old Stanford schoolmate of Cypress founder Rodgers, had founded SunPower with an interest in engineering efficient solar-power cells. The two chums ran into each other in a bay area coffee shop right as the tech bubble had burst, and Swainson admitted that he would soon have to lay off 40 people, half of his entire workforce. On an impulse, Rodgers wrote a $750,000 check, buying his buddy's company. He then spent years convincing his board that buying a friend's floundering, non-semiconductor-related company had been a wise business decision.

Under Cypress, SunPower now manufactures unusually efficient solar panels that transform light into electrical energy at a rate of around 21 percent (most other solar panels operate in the 15 percent range). SunPower's solar cells were used in an experimental solar-powered plane developed by NASA (which, regrettably, crashed into the Pacific in 2003). In late 2006, SunPower acquired PowerLight, a company that specializes in installing solar panel systems for the residential, commercial and utility markets, for $265 million. Rodgers can be forgiven for feeling vindicated: he announced record quarterly revenue in July 2007, citing the "continued strong growth of (Cypress') SunPower division" as a main reason.

Visit Vault at **www.vault.com** for insider company profiles, expert advice, career message boards, expert resume reviews, the Vault Job Board and more.

VAULT CAREER LIBRARY   **169**

GETTING HIRED

Put down some roots at Cypress

Cypress' careers page, www.cypress.com/careers, provides information on job openings, benefits and student opportunities. Job postings are divided into openings in the U.S. and in the Philippines, and are searchable by location, keyword and category. To apply, job seekers must first create an online profile. The company offers a flexible benefits package, larded with incentives like a bonus for the publication of articles and a stock purchase plan. Students can check out Cypress' pages directed at recent graduates, where they can view a PowerPoint slideshow concerning the company's rotational program.

OUR SURVEY SAYS

Mind your P's and Q's

Sources report that Cypress has a very unique corporate culture. One describes it as "a highly data-driven company where 'the right data' wins." His co-worker adds, "Driving towards root causes of issues is valued above all. This ethic is exercised using precision question and answering techniques that can seem curt, invasive and accusatory, but … really aren't. Usually the questioner just wants to arrive at the 'real' answer without the hand waving and window dressing. PQ&A, as it's called, is mandatory training for all employees."

Not everyone at the company is fond of PQ&A, however. It "leads to a constant air of tension throughout the company," says one source. Another sums it up: "corporate culture: combative, risk averse, tense, don't question authority." "Some people thrive in this environment, others do not; you will not last long at Cypress if you have a thin skin," observes one hire. Another member of his department adds, "I would recommend Cypress if you enjoy the culture outlined above," but if it's not your cup of tea, expect to wind up like this guy: "I can only speak for my morale: subterranean." Some, however, say that the culture makes the company "a very stimulating place to work."

If the culture isn't to everyone's taste, the compensation and benefits receive few complaints. "Compensation is good," sums up one guy. "Benefits are a 'cafeteria' system, with some excellent choices," adds his co-worker. "Significant bonuses for writing technical articles and filing patents," observes another. "The company's employee stock purchase plan is an absolute godsend, and has made many people in the company very wealthy," notes a source. If promotions are more to your tastes, one respondent adds, "The company never stands in the way of an employee's advancement or career goals. People are free to move around various job functions, as long as they demonstrate an aptitude for their new roles."

Visit Vault at **www.vault.com** for insider company profiles, expert advice, career message boards, expert resume reviews, the Vault Job Board and more.

VAULT CAREER LIBRARY   **171**

Dell Inc.

1 Dell Way
Round Rock, TX 78682
Phone: (512) 338-4400
Fax: (512) 283-6161
www.dell.com

LOCATIONS

Round Rock, TX (HQ)
Austin, TX • Lebanon, TN •
Nashville, TN • Oklahoma City, OK
• Roseburg, OR • Twin Falls, ID •
Waco, TX • West Chester, OH •
Winston-Salem, NC

Locations in 50 countries
worldwide.

DEPARTMENTS

Call Center • Communications •
Corporate Strategy • Customer
Support • Demand Supply •
Desktop Portable • Engineering •
Engineering Operations •
Engineering Technician • Facilities •
Finance • General Manager • HR •
IT • Legal • Logistics •
Manufacturing • Manufacturing
Engineering • Manufacturing
Support • Marketing • Materials
Product Distribution • Procurement
• Product Development • Sales
Inside • Sales Operations • Sales
Outside • Sales Retail • Security •
Server Storage • Services • Services
Deployment • Support Services •
Technical Support • Training

THE STATS

Employer Type: Public Company
Stock Symbol: DELL
Stock Exchange: Nasdaq
Chairman, President & CEO: Michael S.
Dell
2006 Employees: 66,100
2006 Revenue ($mil.): $55,908

KEY COMPETITORS

Hewlett-Packard
IBM
Sun Microsystems

EMPLOYMENT CONTACT

www.dell.com/content/topics/global.
aspx/corp/careers

THE SCOOP

You're getting a Dell!

Dell is one of the leading providers of PCs, notebook computers and printers for the consumer, commercial and government market. Its product line also includes LCD TVs, network switches and servers and data storage. Dell is currently moving away from its direct-to-consumer sales model into the higher-margin realm of services, and offers IT outsourcing, training and installation.

A born entrepreneur

Some people are just born entrepreneurs. Precocious Texan Michael Dell showed strong signs of business acumen at the tender age of 13, when his mail-order stamp business earned him $2,000 in a couple of months. In high school, he sold subscriptions of the *Houston Post* with enough zeal and success to buy himself a BMW at 17. Then Dell famously began to sell IBM components out of his dorm room while a University of Texas undergraduate in the early 1980s. Dell's mail-order business was efficient and user-friendly, tailoring its offerings to specific customer demands and undercutting retail prices by 10 to 15 percent.

By 1984 his company was grossing about $80,000 a month, and Dell dropped out of college the same year to start making and selling IBM clones under the name PCs Limited. Dell Computer Corporation had 100 employees in 1985, and earned $34 million during the fiscal year. Very soon, the company had gone international and public, its first overseas branch in the U.K. arriving in 1987, with a $32.4 million public offering the following year. Michael Dell was showered with praise, earning three consecutive Entrepreneur of the Year award from the Association of Collegiate Entrepreneurs. In 1992, Dell was named by *Fortune* as one of the 500-largest American companies.

Dell became the world's leading direct seller of computers in the 1990s, as computers moved from being a niche product to becoming a household essential up there with the refrigerator. Using its founder's mail-order catalog sales philosophy, Dell took computer orders from its customers and then assembled the parts, saving costs on manufacturing and retailing into the bargain.

This revolutionary "direct sales model" moved onto the Internet in the 1990s, and only increased in effectiveness. Sales took off, nearing $32 billion in 2001. The subsequent burst of the tech bubble hardly caused executives to bat an eye. Though profits dropped, and the company laid off 4 percent of its workers, the company remained in the black. In 2004, Dell anointed President Kevin Rollins as its new CEO, but the company's forward progress came to a screeching halt soon after, in 2006.

Burned computers and cooked books

Ironically, the first blow to the company's prestige came from its good friend the World Wide Web. Online footage of a Dell laptop bursting into flames began to circulate with ever increasing popularity, and Dell announced the recall of 4.1 million laptop computer batteries in August 2006—the biggest safety recall in consumer electronics industry history. The online video was apparently accurate, as the company admitted that, in fact, its lithium-ion batteries (made by Sony) could occasionally burst into flames.

Dell had documented six computers that had overheated or caught fire as a result of the defective batteries, though no injuries or deaths were attached to any of the incidents. The batteries were installed in some 2.7 million laptops sold in the U.S. and another 1.4 million sold abroad—about 18 percent of the company's total notebook output between April 2004 and July 2006. A *New York Times* article questioned the numbers released by the company, though, citing pictures released by a disgruntled former Dell employee of almost 100 melted laptops returned by customers between 2002 and 2004.

The following month, Dell revealed that federal prosecutors had contacted it for information in an expanding accounting investigation, forcing the company to cancel a meeting with analysts and postpone a stock buyback strategy. As part of the federal investigation, the U.S. Attorney for the Southern District of New York subpoenaed documents related to the company's financial reporting from 2002 to the present, preventing Dell from filing its second-quarter report. Ominously, Dell had previously scrubbed a meeting in April with industry financial analysts—not people who like to see signs of weakness.

The feds were quite right about Dell's financial reporting, as the company's own audit committee turned up a chaotic accounting situation and evidence of fiscal impropriety in March 2007. Dell had only reported preliminary figures for the last two quarters, and the company said its 2006 annual report would be further delayed while executives determined if any prior-period results were affected by this mistake. At the time, the U.S. Department of Justice had not yet concluded its own investigation of the company, and Nasdaq was even threatening to delist it.

For whom the Dell tolls

In November 2006, Dell made an official announcement of what industry insiders had suspected and investors had dreaded: the company was under investigation by the SEC. Dell's release revealed very little other than the existence of the inquiry, but it sent Wall Street into a tizzy. The company delayed its third quarter earnings by a month, citing compliance with the investigators' requests, and canceled another analysts' meeting.

The company said the 3Q results would be forthcoming at the end of November 2006, but that there would be no conference Q&A call with company execs. The other shoe dropped in August 2007, when Dell admitted to falsifying quarterly returns from 2003 to 2006, and will restate its earnings by $50 to $150 million. *The New York Times* reported that many industry analysts are actually relieved to see the investigation resolved, and called it a "turning point" for the company.

The difference is Dell

Earlier in 2007, Dell rounded an arguably more significant turning point—Rollins, previously CEO and right-hand man to Michael Dell, was ousted from the big seat, and the founder went back in. The shake-up stirred much speculation, as sales at the company had slowed, and industry criticisms of poorly designed computers and disastrous customer service were getting louder and louder, not to mention the company's financial issues with the SEC and the Justice Department.

The board believes the company is better off with Dell in the top job, although perhaps the move is a superficial one for the sake of Wall Street, as Rollins had always emphasized his close working relationship with the founder. Dell, though, has harkened back to his old dorm room days. As if to restore the company's luster; he was quoted by *The New York Times* as saying, "It feels like 1984 and I am starting over again. Only this time I have a little more capital."

Pulling the purse strings

Dell's increased input on day-to-day operations of the company meant new confidence for the company—and no bonuses for its workers. After a review of the company's performance in February 2007, Dell announced belt-tightening companywide to make up for losses in recent years, including a suspension of bonuses for fiscal 2007.

In a call for more efficiency, an internal memo declared that lack of profit was unacceptable and that employees should focus efforts on streamlining operations to push sales. Dell has followed his own advice by reducing the number of executives reporting to him from 22 to 12. How many of these changes were voluntary is unclear, though, as six executives quit the company from December 2006 to February 2007, ahead of this restructuring.

Tryin' to throw their arms around the world

Since 2006, Dell has made it abundantly clear that it sees international growth as the cure to its ills. As other executives left the company in droves, it created new posts in its international affairs division. A new position was created in late 2006 called "head of global services," and the lucky guy to fill it was Steve Schuckenbrock, formerly of the outsourcing pioneer Electronic Data Systems (Ross Perot's old company). His responsibilities include sprucing up and expanding Dell's IT service offerings, including outsourcing and supply chain automation. At the time, these services were badly in need of an overhaul, as 64 percent of Dell's revenue came from North and South America, and lackadaisical, outsourced customer service was a persistent complaint from Dell's customers.

Mere months later, Dell hired another executive (from outside the company) to oversee its supply chain operations, creating another position from scratch in the process. In February 2007 Dell hired former Solectron CEO Michael Cannon as the firm's first president of global operations (Schuckenbrock's title was shifted to "senior vice president of global operations"). Cannon's time at Solectron gave him experience in manufacturing services, and the company hopes to use his supply chain expertise to improve Dell's international component manufacturing, purchasing and distribution operations.

These changes have received mixed reviews, as many pundits view more outsourcing as a problem and not a solution, and suggest that Dell open more stores locally to deal with customer service issues. Michael Dell has hinted that he agrees with these criticisms, at one point publicly wondering whether to shift away from the company's standby direct sales model.

Dude, it's not 1999 anymore!

Rival Hewlett-Packard beat out Dell for the No. 1 spot in the PC industry in the fourth quarter of 2006, and maintained its lead through spring 2007, prompting Michael Dell to roll out an amibitious plan to perk up business in April 2007. Dell's four-point plan, reported in the pages of *The Wall Street Journal*, involves making the company's products less confusing and threatening for consumers; moving into emerging economies like India and Brazil; flattening and revising the management structure of the company; and providing cost-effective consulting services.

Dell's brand might have simply become less attractive, as laptops/notebooks, a particular weakness of Dell's, are increasingly in vogue—industry analysts expect they will surpass desktops in U.S. market overall sales by 2008. Dell falters with laptops because they are more difficult to customize than the traditional Dell desktop, and are more expensive to manufacture. So far, though, the company's emphasis on international growth appears to be working, as international shipments exceeded American shipments in the fourth quarter of 2006, a company first.

Dell also might be shifting away from its trademark sales model, as it has been (gasp!) opening its own retail stores and making alliances with major outlets to feature their computers. The first step came in May 2007, when the company announced a deal to sell PCs through Wal-Mart, which was followed by similar deals in the U.K., Japan and Russia. In the quarter ended August 3, 2007, sales were up 5 percent, and more Dell laptops and desktops will probably soon be headed to a store near you.

GETTING HIRED

It's not just a career, it's a Dell!

The careers page on Dell's web site, www.dell.com/content/topics/global.aspx/corp/careers, provides information on positions available at Dell the world over. The Dell careers site for the U.S. provides information on career paths in sales, customer service, IT, engineering, manufacturing and business. Job seekers, once they have selected their career path, can search open positions by function, location and keyword. To apply, they must first fill out a profile as well as an assessment questionnaire. To make applications stand out, Dell advises job seekers to put as many keywords and abbreviations—and as much industry jargon as possible—in their resumes.

In addition, Dell posts a schedule of recruiting events on its web site, and potential hires are invited to visit Dell's representatives. The company also has an employee referral program, so job seekers are strongly advised to work their contacts to see if they know anyone at Dell.

Benefits at Dell in the U.S. include health, vision and dental insurance, flexible spending accounts for health care or dependent care costs, and a health-improvement scheme, with programs aimed at healthy pregnancies and managing certain conditions, that rewards employees with money when they reach certain goals. Dell also offers a 401(k) with company match and a discount stock purchase plan. Nice perks include adoption assistance, discounts on company merchandise, employee referral program, discounts on gym memberships and time off between Christmas and New Year's.

Dell offers a wealth of opportunities for students and recent graduates at the college and MBA levels. Dell's internships are offered to undergraduate students with at least one semester remaining. Internships are offered to students pursuing a major relevant to the company's business need. Dell provides interns with company housing (including housekeeper) reimbursement for travel to the internship site, discount on Dell products and a discount health club membership. Dell also offers internships to students pursuing an MBA in the areas of marketing, HR, operations, program management, strategy, IT, finance, logistics and procurement.

OUR SURVEY SAYS

Dude, you've got your foot in the door!

One source in sales gives the following advice: "It's nearly impossible to get into Dell by simply sending in your resume from the street. Dell has this sort of 'buy and try' system worked out with a temp agency known as Spherion. [It] screens and tests candidates and then hires them on as contract workers for a set 'per hour' pay rate, and if you work out well and do your job, you will get a permanent job offer from Dell after about four to six months of contracting with Spherion."

Most interviewees report two rounds of interviews. "Both the first and second rounds were behavioral," notes one insider. His co-worker adds that he was asked "traditional interview questions with qualitative focus. There is much less focus on problem-solving. For some positions, such as corporate strategy, there is limited use of the case method." "Questions revolved around stories for leadership, teamwork, dealing with ambiguity, multitasking, project management, conflict management, which other companies was I considering, why Dell, walk me through your resume, innovation, and a big portion of the time for asking questions to the interviewer, around 25 minutes," reports an insider.

Another hire cautions, "HR is known for delays and slow turn around on hiring. Get the e-mail and/or phone contact information for the hiring manager and keep in touch."

GETTING AHEAD

Responses about Dell's culture were mixed. "No matter what people say, Dell can be a great place to work," says a staunch defender. "The culture and work/life balance has improved in Dell IT over the past two years," observes a member of that department. Expect, however, once hired that "Dell is your life ... work from 8 a.m. to 5 p.m. but if you leave at 5 p.m. people will talk. 6 p.m. is the minimum." "Dell is a very aggressive environment. [Hundreds] of e-mails a day and lots of conference calls," notes a respondent.

His colleague adds that "promotions and raises can be very hard to get." "Networking is very important to progress, it matters a lot who knows you rather than what you know," adds a member of the sales team. "There are lots of social activities and learning opportunities but never enough time to take advantage," says a co-worker. However, there are "diversity networking groups ... which are very active and help you a lot in networking." And they appear to have many members. "You can find people from all nationalities, experience background, ethnic groups, [and] genders ... in large numbers," adds a source. Her co-worker adds, it's a "High-stress environment. Lots of lunch-and-learns to keep you up to speed." "You will learn more than you'd ever imagine," notes an awestruck intern.

Diebold, Incorporated

5995 Mayfair Road
North Canton, OH 44720
Phone: (330) 490-4000
Fax: (330) 490-3794
www.diebold.com

LOCATIONS

North Canton, OH (HQ)
Atlanta, GA • Chicago, IL • Coppell,
TX • Dallas, TX • Green, OH •
Hebron, OH • Marietta, GA •
Newark, OH • Plainsboro, NJ •
Princeton, NJ • Schaumburg, IL •
Uniontown, OH

Additional operations in Argentina,
Australia, Austria, Barbados,
Belgium, Brazil, Canada, Chile,
China, Colombia, Czech Republic,
Ecuador, France, Germany, Greece,
Hong Kong, Hungary, India,
Indonesia, Italy, Malaysia, Mexico,
Namibia, Netherlands, New Zealand,
Panama, Paraguay, Peru,
Philippines, Portugal, Poland,
Romania, Russia, Singapore, South
Africa, Spain, Switzerland, Taiwan,
Thailand, Turkey, the United Arab
Emirates, the UK. Uruguay,
Venezuela and Vietnam.

THE STATS

Employer Type: Public Company
Stock Symbol: DBD
Stock Exchange: NYSE
Chairman: John N. Lauer
President & CEO: Thomas W.
Swidarski
2006 Employees: 15,451
2006 Revenue ($mil.): $2,906

KEY COMPETITORS

NCR
Wincor Nixdorf International
IBM

EMPLOYMENT CONTACT

www.diebold.com/careers

THE SCOOP

Diebold to user: do it yourself

Diebold is a leading manufacturer of self-service kiosks, which range in function from ATMs to automatic check-in and checkout kiosks to electronic voting machines. The company also manufactures and installs safes and security systems for banks, stores and office buildings, as well as creating custom hardware and software for its customers. Finally, in addition to guarding countless reams of currency in bank safes, Diebold security systems protect the Constitution, Bill of Rights and Declaration of Independence.

Play it safe

Charles Diebold founded his eponymous company in Ohio in 1859 as a manufacturer of safes and vaults. After the Great Chicago Fire in 1871, Diebold safes became known for their ability to protect their contents in the face of catastrophe. By the 1880s, Diebold had manufactured safes for major banks in the U.S. and in Mexico, and a decade later the company announced the development of safe doors strong enough to withstand a direct blast of TNT. During the 1920s, Diebold specialized in large safes, manufacturing the world's largest strong room in 1921, outdoing themselves in 1964 when it cast the doors for the vault of the Federal Reserve Bank in Cleveland.

Business suffered a temporary decline during the Depression, so the company relied on its bulletproof reputation to win armor plating contracts as the U.S. began to gear up for the World War II. Following the armistice, Diebold returned to its safe business and bought a company that specialized in alarm systems for buildings in 1947.

The ATM-ic age

In 1964, just as computers were beginning to affect Diebold's products, the company went public. After the automatic teller machine (ATM) was invented in 1967, Diebold created its own version of it, releasing one in 1970, along with a high-tech (for that year, anyhow) security system, whose cameras could be controlled by a computer. Diebold was soon dueling with rival ATM manufacturers IBM and NCR for dominance in the fledgling ATM market. The company also sought out other markets for its self-serve kiosks beginning in 1986, such as pay-at-the-pump gas pumps and self-service ticket kiosks for theaters, movies and trains.

Content for the moment with its efforts to automate the lives of people in its home markets, in 1993 Diebold embarked on a venture to sell ATMs in China, thereby gaining market share right before that country's meteoric economic growth in the ensuing decade. Three years later, Diebold expanded on the ATM concept to create machines that ejected medicines instead of cash, as well as ATMs that could sell postage stamps. By 1996, revenue came in at over $1 billion.

Diebold moved aggressively into the Brazilian market in 2000, when it agreed to supply $100 million in voting machines to the Brazilian government. The state of Georgia followed with a purchase of Diebold machines in time for its 2002 gubernatorial election. In 2003 the governments of San Diego, Calif., also bought Diebold voting machines, as did the state government of Ohio and Maryland. That same year, the company bought two firms that specialized in ATM distribution in Taiwan and India, as it continued its strategy to profit from the growth of economies in Asia.

Rise (and fall) of the machines

But nearly as soon as Diebold's voting machines hit the market, people began discovering flaws in their design, both hardware- and software-related. Whether these flaws were intentional or not, and whether or not the machines were jiggered to hand elections to Diebold-favored candidates, has become much more of a newsworthy topic than the company would surely prefer.

The first batch of wonky Diebold voting machines popped up in the 2002 Georgian election. Diebold gave the state government a machine running one version of the code to inspect for certification purposes. Shortly before the election, Diebold found bugs in the code and patched them, but the state didn't get a chance to vet the code again before the election. Since the patched code was not cleared with Georgia election officials, the possibility exists that it could have tampered with electoral results.

Code red

Things only got worse for the company in 2003 when it accidentally released the source code for its machines on the Internet. Months later, state-appointed programmers in California criticized the kind of touch-screen voting machines that

San Diego had just purchased—and that Georgia had used for its 2002 election—saying that they had gaping security holes large enough that someone could easily rig an election by tampering with the machines or infecting their codes with a computer virus.

The problems were compounded by the fact that these touch-screen machines don't produce countable paper votes the way traditional voting machines do. Diebold machines are designed to store votes on memory cards, which can easily be erased or overwritten. In light of these issues, in 2004 California disallowed the use of Diebold voting machines in its elections. In a 2007 review of the voting machines in the state, the government of California supported its earlier decision to disallow Diebold machines.

Time for an overhaul

Diebold's machines were also used in each of the last two U.S. presidential elections, and a wealth of conspiracy theories abound (especially on the Internet) that link the company's well-documented security struggles with specific voting machine performance in Florida and Ohio in 2000 and 2004, respectively. In late 2005 CEO Warren O'Dell resigned; though a company statement cited "personal reasons" for his departure, O'Dell had a well-documented history as a fund raiser for the Republican Party. In one August 2003 fund-raising letter targeted at the party faithful, O'Dell stated he was "committed to helping Ohio deliver its electoral votes to (President Bush)."

Besides changing its executive lineup, Diebold attempted to shore up its voting machines business in other ways. In June 2004, the company revealed in an SEC filing that it had changed its ethics policy to ban all executives from any political activity except the act of voting. In 2007 it renamed its voting machines division Premier Election Solutions and changed its upper management, giving the division its own executives and board of directors. Also that year, Diebold tried to sell off its voting machines division, but there were no takers, unsurprisingly.

If you can't count votes, try rupees

While things might not exactly be going swimmingly for Diebold in its voting machines division, the company still successfully produces ATMs, and it is moving aggressively into supplying India with them. As of 2007, the company controls about a third of the market, which is largely dominated by rival NCR. The market for ATMs is expected to grow briskly in the country, partly because a larger portion of the population is keeping its money in banks, and partly because credit cards are difficult to obtain and almost prohibitively expensive there. Diebold expects that the number of ATMs in India will grow 140 percent between 2006 and 2010.

India has plenty of room for the market to expand: while the U.S. had about one ATM for every 100 people in 2007, India had only one ATM for every 5,000. Diebold is mostly selling a version of the standard ATM it modified for use in the Brazilian market, which can run on less power than the versions sold in the West. It is also making attempts to move into the market for rural banking, despite the small size of deposits and withdrawals, and spotty connections to power and communications.

GETTING HIRED

Make yourself a career at Diebold

Diebold's careers site, at www.diebold.com/careers/default.htm, provides information on job openings at the company. Jobs are searchable by location, title and function. In order to apply to a posting, job seekers must first create a profile.

What can you expect at your Diebold interview? One longtime employee fills us in: "After an initial interview (or possibly two), the candidate would take a test to determine personality fit, intellectual fit, etc. If the test is administered, and the candidate does well, there is an offer extended. The candidate must pass a background check and drug test before officially joining the company."

OUR SURVEY SAYS

Moving on up

"Most days, working for Diebold is a good experience," notes one source. He goes on to add that the company is "having a bit of a personality conflict right now, as [it] morphs into more of a high-tech company vs. an equipment company." His co-worker agrees, "morale is very good." Another chimes in, "Good local management … some really smart people, and a staff that is overall, very passionate about doing the right thing for the customer." "Dress code is usually business casual unless otherwise specified," says a respondent. "Hours are flexible from 7 a.m. until 7 p.m.," notes another.

One thing that distinguishes Diebold is its emphasis on allowing employees to move up the ranks. "Jobs are posted internally, and you have to be with the company for a year before changing positions. Internal promotion is very good," says one worker who's been around a while. "[The] company believes in internal grooming and growth for associates, but also has interest in new and diverse talented individuals," his associate agrees.

Visit Vault at **www.vault.com** for insider company profiles, expert advice, career message boards, expert resume reviews, the Vault Job Board and more.

V/\ULT CAREER LIBRARY   **185**

Electronic Data Systems Corporation

5400 Legacy Drive
Plano, TX 75024
Phone: (972) 604-6000
Fax: (972) 605-6033
www.eds.com

DEPARTMENTS

Administrative Support •
Communications • Consulting •
Corporate Operations • Customer
Business Services • Data Processing
Operations • EDS Advanced
Solutions • Engineering/
Manufacturing • Financial • Global
Field Services • HR • Insurance •
Legal • Manager • Marketing •
Sales • Service Delivery • Systems
Engineer • Technical Delivery

THE STATS

Employer Type: Public Company
Stock Symbol: EDS
Stock Exchange: NYSE
Chairman & CEO: Michael H. Jordan
2006 Employees: 131,000
2006 Revenue ($mil.): $21,268

KEY COMPETITORS

Computer Sciences Corp.
Deloitte Consulting
IBM

EMPLOYMENT CONTACT

www.edscareers.com

THE SCOOP

Got data? They can help

EDS is the original outsourcer—it handles all manner of pesky IT tasks for its customers, managing everything from data centers to desktops, security to networks. Other services include the creation of specialized software and of finance and customer relations services. The company offers HR outsourcing through a joint venture called ExcellerateHRO. EDS serves companies in the government, defense, retail, agriculture, media, manufacturing, transport, energy and finance markets (to name a few). EDS is listed as No. 111 on the 2007 Fortune 500.

Outsourcing pioneers

EDS was founded in Texas by Ross Perot—yes, that Ross Perot—in 1962. Its early business strategy involved selling computing time on mainframes to other companies needing to crunch data. EDS resold computing time on a life insurance company's computer during off-hours, until it bought its own computer in 1965.   State bureaucracies were soon swamped with more data to crunch, as the Medicare and Medicaid health insurance programs came into being that same year—now that it owned a computer, EDS was happy to step in and start crunching numbers. By 1969, the company had won contracts to process data for the Medicare and Medicaid programs in Texas, California and Kansas, and parlayed this expertise into a program for the commercial insurance industry in 1972.

Four years later, EDS started handling airline ticket transactions, and by 1980 counted over 100 airlines among its customers. In 1984, General Motors acquired EDS in a deal valued at $2.5 billion. In 1995, EDS spent $600 million for A.T. Kearney, a management consulting firm, and EDS was spun off the following year. The acquisition of Structural Dynamics Research Corp. in 2001 expanded EDS' offerings in the realms of software development, engineering and design.   The ExcellateHRO outsourcing program for HR began in 2005.

Still going strong

In 2006, EDS took in revenue of $22 billion, an increase of 8 percent year-on-year. Profits for the year were $470 million, an increase of over 200 percent from the year before. In 2006, EDS signed deals for contracts worth $26.5 billion, and anticipates revenue in the neighborhood of $22 billion by the end of 2007.

Also that year, EDS sold off A.T. Kearney, its consulting arm, to 170 members of the company for an undisclosed amount. EDS also divested Kearney's executive search arm to another group of investors. Meanwhile, the company acquired a significant stake in the Indian firm Mphasis, provider of outsourcing services and call centers for banks and health insurers, for $380 million, and then promptly folded it into its Indian division a few months later.

Hiring in Delhi, firing in Dallas

The acquisition of Mphasis is only part of a larger plot afoot at EDS. In a plan that began in 2004, EDS is laying off workers in areas with high labor costs while hiring workers in lower-cost regions like India, Eastern Europe and Latin America. In fact, even though EDS laid off about 5,000 workers in the U.S. in 2006, it hired about the same number in India (not counting the 12,000 workers there it acquired through its purchase of Mphasis). By 2008, when the plan is complete, EDS will fire 20,000 people in the U.S., and have accumulate nearly 50,000 employees overseas.

These low-cost employees are critical for maintaining the bottom line, as competition is quickly driving down the margins on outsourcing. Having offices scattered around the world provides EDS with another cost-saving benefit as well—it can provide 24-hour service to clients without having to employ pricier late-shift employees.

GETTING HIRED

Source a job at EDS

EDS' careers site, at www.edscareers.com, provides the hopeful EDS-er with information on career opportunities the world over. (A site specific to U.S. jobs can be found at www.edscareers.com/us.) In order to submit a resume for a job posting or as a general inquiry, job seekers must first create a profile. Positions in the U.S. include project management, account management, accounting, sales, HR, communications and legal staff.

Benefits offered by EDS vary by location. In the U.S., they include college savings plans, heath, disability and life insurance, flexible spending accounts and retirement savings programs. Nice perks include a vehicle purchase plan and employee discounts. EDS also offers its employees over 5,000 courses to boost their skills in everything from leadership and sales to technical pursuits.

Here's what to expect of your interview at EDS: One insider reports he went through "two round of interviews ... [The interviewers] asked weak technical questions." " Very positive experience," says one hire.

OUR SURVEY SAYS

Conservative culture, flexible hours

EDS is known for its "very formal" corporate ethos, but a source reports enjoying the "reportedly conservative culture." Don't brace yourself for black-tie Mondays, however. "The dress code is basically business casual. Occasionally I need to wear suits given my job. When the company wins big deals, though, we get dress down days and can even wear jeans," notes an insider. "People are terrific, I love the corporate culture ... and I see tons of room for advancement. How's that for hitting everything in one breath?" gushes one member of the finance department. "We work hard and when we have a chance we play hard, and have a good time doing it all," adds a guy from sales. On the other hand, outsourcing is making many of the technology people nervous. "Employee morale is very low," says a techie. "Turnover is increasing and company is hiring more contractors," adds another contact. "I like my coworkers and the type of work I do," says a colleague, diplomatically.

Despite the "formal" culture, hours at EDS are pretty flexible. "[I'm] trusted to do my job without constant supervision and guidance," says one staffer. "I can make my hours work for me," points out a contact in finance. Their co-worker notes, "I report to a supervisor in another state."

"For the most part, as long as we get our work done, the hours we work matter less," adds a contact. "We work on average 45 to 50 hours per week but occasionally work 70+," says a source. "I work 50 to 55 hours. [Work] tends to stay at a constant, elevated level," notes his colleague. That hard work will get you places, however. "Given the size of the company and the amount of change, there is PLENTY of room for advancement," says a particularly emphatic source.

EMC Corporation

176 South Street
Hopkinton, MA 01748
Phone: (508) 435-1000
Fax: (508) 497-6912
www.emc.com

LOCATIONS

Plano, TX (HQ)
Atlanta, GA • Austin, TX • Baltimore,
MD • Boston, MA • Charleston, SC •
Charlotte, NC • Chicago, IL •
Cleveland, OH • Dallas, TX • Denver,
CO • Des Moines, IA • Detroit, MI •
El Paso, TX • El Segundo, CA • Fort
Worth, TX • Honolulu, HI • Houston,
TX • Indianapolis, IN • Jacksonville,
FL • Kansas City, MO • Little Rock,
AR • Los Angeles, CA • Memphis, TN
• Miami, FL • Minneapolis, MN •
Nashville, TN • New York City, NY •
Oklahoma City, OK • Omaha, NE •
Philadelphia, PA • Phoenix, AZ •
Portland, OR • Rochester, NY • Salt
Lake City, UT • San Antonio, TX •
San Diego, CA • San Francisco, CA •
Seattle, WA • Washington, DC

THE STATS

Employer Type: Public Company
Stock Symbol: EMC
Stock Exchange: NYSE
Chairman & CEO: Joseph M. Tucci
2006 Employees: 31,100
2006 Revenue ($mil.): $11,155

DEPARTMENTS

Administrative Services & Operations •
Corporate Security • Customer Service
& Support • Education Services &
Training • Engineering—Hardware •
Engineering—Software • Finance &
Accounting • Global Facilities • Global
Real Estate • HR • IT • Legal •
Manufacturing • Marketing &
Communications • Product
Management • Professional Services •
Sales • Six Sigma • Technical Support
• Technology Solutions Group

KEY COMPETITORS

Hewlett-Packard
IBM
Sun Microsystems

EMPLOYMENT CONTACT

www.emc.com/hr

THE SCOOP

Information lives!

A world leader in information storage and management, EMC provides information lifecycle management (ILM) services, as well as servers, switches and routers for data storage. EMC's products are based on RAID (redundant array of independent disks) arrangements, which use inexpensive, easily replaceable components to store large quantities of data. These systems are used by clients as diverse as banks, Internet service providers, airlines, government entities and schools and universities. The company also provides network attached storage (NAS) servers and a line of software.

Remembrance of things—past and present

Richard Egan and Roger Marino, two friends from their college days together as engineering students at Northeastern University, founded EMC in 1979 (they were the E and M in EMC). They wanted to manufacture circuit boards, but their only way into business was through distributing boards of a different kind—a friend had designed particle-board desks for computer workstations, and wanted the two to sell them on the East Coast.

The desk business provided the nascent EMC with some seed money, and in 1981 Egan used his connections at Intel, where he was a former manager, to start selling memory for desktop computers there. In 1986, EMC went public, and added disc drives to its product line the following year. In 1989, the company started selling data servers based on RAID formats, which were less expensive, less-failure prone and far more scalable than other storage systems then available. They proved to be a foundation for the company's success—by 1995, EMC's RAID systems were outselling IBM's data storage.

Five years later, EMC controlled nearly 35 percent of the data storage market. From there, the company looked into acquisitions that would complement its core products, purchasing two companies in 2003 that specialized in information lifecycle management, a concentrated approach to data storage that separates information along specific and variable guidelines. The following year, EMC acquired a company that specialized in data backup and retrieving data from failed systems.

Visit Vault at **www.vault.com** for insider company profiles, expert advice, career message boards, expert resume reviews, the Vault Job Board and more.

V/\ULT CAREER LIBRARY   **191**

Everyone needs memory

The acquisitions continued in 2006 for EMC. It bought up seven companies that year, with areas of expertise ranging from software for content management, data de-duplication, customer relationship management and data protection. The most notable acquisition was that of RSA in September for $2.1 billion. RSA's information security services, including identity verification, encryption and secure transfer of data, will be a vital addition to EMC's line of offerings in the age of the Patriot Act, as security becomes an increasingly higher priority for businesses.

Data, data everywhere

In 2006, EMC took in $11 billion in revenue and made $1.2 billion in profits; revenue was up 15 percent over 2005. EMC is seeing a rise in demand for its services, as legislation like the Sarbanes-Oxley act requires companies to hold onto more data for a longer period of time. In addition, the amount of data that people are producing is increasing dramatically, as music and video content become easier to create and spread—EMC estimates that 988 billion GB of data will be created in 2010.

Catching a wave

In 2007, EMC announced plans to expand its facilities in both India and China. In India, EMC is planning on investing $500 million by 2010 as demand for information management continues to grow on the subcontinent. It is doubling its timetable in China, planning to invest $500 million by 2013 as it establishes a facility for software development there.

And catching waves back home in the States

If the past is anything to go by, EMC's new employees in India should enjoy their new facilities. In the late 1990s, EMC, was flush with cash and not thinking about any oncoming tech bubble bursting, and was planning a new headquarters in Hopkinton, Mass. By law, it was required to have 200,000 to 250,000 gallons of water on hand for fire protection. After some consideration, the company decided not to invest in a water tower for its liquid storage needs, but a swimming pool. And yes, employees are allowed to swim in it—no word yet on how it's serving the company's flame-retardant needs. Employees have a rockin' good time at EMC, too, as the company proudly introduced an in-house band at the 2004 Documentum Developer Conference, called the Rockumentums.

Rocking and Rolling in dough

EMC is in fine corporate shape right now, as it recently recouped on its 2003 purchase of the VMware software for $635 million. The software was an obvious candidate to go public, as its sales increased 83 percent in 2006 to $703.9 million. But EMC was stunned when it raised around $1.1 billion in its August 2007 IPO, the largest technology offering since Google went public in 2004. EMC will remain the principal shareholder, offering approximately 10 percent of VMware's shares on NYSE, under the symbol "VMW." The move should increase EMC's stock prices, which have already gained 43 percent so far in 2007. *BusinessWeek*, writing before the announcement of the IPO, called the VMware acquisition a "home run," and ranked ECM at No. 88 on its 2007 edition of the InfoTech 100.

GETTING HIRED

Get some data for your search

EMC's careers site, at www.emc.com/hr/, provides information on entry-level and internship programs, benefits and, obviously, how to apply for jobs. Job listings are searchable by business, function, location and keyword. To apply, candidates must first fill out a profile. Job seekers must also pass a background and drug test as a condition of employment.

The company offers internships and co-ops for students majoring in business, marketing, finance, computer science, electrical engineering, law and allied fields of study. Applicants must have a 3.0 GPA or above, and will work on projects related to their major. Interns are eligible for employee benefits like discounts, paid time off and training. EMC also offers two entry-level programs, a rotational program in HR, marketing, finance or engineering, designed to groom the next generation of EMC leaders, and an associates program in customer service, sales, and educational and technical services.

Benefits offered by EMC include disability, health and dental insurance, flexible spending accounts, 401(k) with company contribution and discount stock purchase plan.

OUR SURVEY SAYS

Hop on board with EMC

One programmer describes his experience: "Two rounds of interviews, one with the HR recruiter and another with the hiring manager/supervisor." Another source points out he found that the "hiring process was clumsy."

"Anyone interested in working for EMC should definitely try to get a job there," says one hire. Another notes that there is a "disdain for slow decision-making and political BS. Exceptionally strong sales and engineering culture." One less-satisfied insider observes that "there are many more VPs than necessary; far too many management layers."

Hours at the company verge on the lengthy. "There are no 40-hour workweeks," wails one employee. "Fifty to 60 hours is normal at EMC," adds his co-worker. "Absolutely no such thing as work/life balance," moans a programmer. "Weekly hours in the 60s," explains another.

Sources universally praise the company's training. EMC "provide training that is second to none," says one. "Very good training," agrees his co-worker. Other benefits get high marks, too. "Benefits [are] great. Benefits also extended to gay partners," adds an insider. "Tuition reimbursement is good," notes another.

Fiserv, Inc.

255 Fiserv Drive
Brookfield, WI 53045
Phone: (262) 879-5000
Fax: (262) 879-5013
www.fiserv.com

LOCATIONS

Brookfield, WI (HQ)

Atlanta, GA • Austin, TX • Baltimore, MD • Baton Rouge, LA • Boston, MA • Boulder, CO • Bowling Green, KY • Chicago, IL • Cincinnati, OH • Dallas, TX • Denver, CO • Des Moines, IA • Fort Worth, TX • Honolulu, HI • Houston, TX • Las Vegas, NV • Los Angeles, CA • Miami, FL • Milwaukee, WI • Minneapolis, MN • New York, NY • Oakland, CA • Oklahoma City, OK • Omaha, NE • Orlando, FL • Philadelphia, PA • Phoenix, AZ • Pittsburgh, PA • Portland, OR • Richmond, VA • Salt Lake City, UT • San Antonio, TX • San Diego, CA • San Francisco, CA • Scottsdale, AZ • Seattle, WA • St Louis, MO • Tampa, FL • Tulsa, OK

DEPARTMENTS

Accounting & Finance • Administrative & Clerical • Banking • Communications • Creative Art & Media • Customer Service • Education & Training • Engineering • HR • IT • Insurance • Legal & Law Enforcement • Manufacturing & Production • Marketing & Advertising • Medical/Health • Sales & Sales Management • Telecommunications

THE STATS

Employer Type: Public Company
Stock Symbol: FISV
Stock Exchange: Nasdaq
Chairman: Donald F. Dillon
President & CEO: Jeffery W. Yabuki
2006 Employees: 23,000
2006 Revenue ($mil.): $4,4

KEY COMPETITORS

Fidelity National Information Services
Open Solutions, Inc
TSYS

EMPLOYMENT CONTACT

www.fiserv.com/careers.htm

Visit Vault at **www.vault.com** for insider company profiles, expert advice, career message boards, expert resume reviews, the Vault Job Board and more.

VAULT CAREER LIBRARY 195

THE SCOOP

Keeping the banking and insurance worlds going 'round

Fiserv provides more than 18,000 banks, credit unions, thrifts and insurers with services ranging from funds transfer and credit card services to services for the providers of medical, property and life insurance. Fiserv also offers data warehousing and printing services, and specialized software for lenders.

On top of all that, the company supports nonprofits and self-insured businesses with their health benefits, with services that include everything from benefits administration to help for employees managing chronic diseases. The company is also America's largest independent processor of checks. In 2006, Fiserv offered insurance services to four million people and processed 18 billion financial transactions.

Cents and connectivity

Fiserv was created in 1984, when George Dalton and Leslie Muma merged their two data processing companies, a feat they had been trying to accomplish since the late 1970s. Fiserv grew throughout its first decade. As the business of banking transformed throughout the Reagan era, changes in technology and tax laws and other factors such as decreased interstate regulation, all fostered higher competition among banks, as well as a thirst for cheaper, automated data processing, much to Fiserv's delight. Success came so quickly, the company was able to go public in 1986, just two years after its origin.

Another key component to Fiserv's success has been an aggressive acquisition strategy; between 1984 and 2006 Fiserv bought more than 140 other companies—the company's $50 million acquisition of the data processing arm of Citibank in 1991 was a notable coup. As Fiserv grew, it added related services like credit cards, stock transaction processing and data processing for the insurance industry. In 2006, the company bought seven more companies, two of them adding to its stable of auxiliary financial products. In March, it bought the financial training division of Wolters Kluwer, along with some financial software assets. In July, the firm bought the Jerome Group, a provider of customized printed promotional materials for banks and insurers. Then in 2007, Fiserv acquired NetEconomy, the leading international anti-money laundering, anti-fraud and crime management company based in The Hague.

The company's revenue has shown strong growth since 2002, as banks and insurers increasingly outsource their data processing functions. In 2006, Fiserv reported revenue of $4.4 billion, an increase of 12 percent over 2005. (The company's profits, which reached $450 million in 2006, fell by 13 percent from 2005, as they were inflated by a large contract termination fee the previous year)

Fiserv, take two

Fiserv's strategy for its acquisitions was to buy up profitable, well-managed companies in its target industries, and then largely leave them alone, while reaping the profits in order to buy more companies. Clearly, this strategy worked as Fiserv's revenue continued to increase, giving executives and shareholders reason to smile. The downside became obvious in 2006, when Fiserv had a patchwork of 77 more-or-less nonintegrated business units, all of which sourced their own supplies and sometimes competed for business.

In 2006, following his accession as CEO the previous year, Jeffery Yabuki embarked on a plan (cleverly called "Fiserv 2.0") to reorganize the company into a lean, mean profit-making machine. Yabuki's plan calls for integrating the efforts of Fiserv's 77 business units into four groups, including buying of supplies in bulk, allowing customers to have a single point of contact at the company and cross-selling services to boost revenue. Fiserv 2.0 also involves offshoring some of the company's outsourcing to countries with lower labor costs. Yabuki predicts that by 2010, his plan for Fiserv will save the company $100 million—in addition to bringing in hundreds of millions more in revenue.

Keep it clean

Fear not, acquisition fans, Yabuki's strategy doesn't change Fiserv's long history of buying up companies. If anything, the new strategy gave the company an altruistic streak, as it began using its powers to fight crime with the March 2007 acquisition of NetEconomy. NetEconomy is a provider of software to prevent financial misdeeds like identity theft, money laundering, insider trading and fraud, and will make Fiserv's service offerings to banks just that much more alluring.

In August, the company announced its intent to acquire its largest acquisition ever, CheckFree, the nation's leader in electronic bill payment programming. CheckFree is the leading online bill payment system in the U.S., processing more than one billion transactions in a year.

GETTING HIRED

Find yourself at Fiserv

But first, find yourself some information about Fiserv's career options at www.fiserv.com/careers.htm. There, Fiserv helpfully provides a list of open positions (searchable by department, keyword, location and category) as well as information about benefits and company policies. To apply to a position, candidates must create a profile.

Benefits offered by Fiserv vary by business unit, but commonly consist of health and dental insurance, health spending accounts, 401(k) with company contribution, stock purchase plan and educational assistance. Fiserv may conduct credit and background checks as a condition of employment.

Foxconn Electronics Inc.

Number 3-2 Chung-Shan Road
Tu-Cheng, Taipei
Taiwan
Phone: 886-2268-0970
Fax: 886-2268-7176
www.foxconn.com

LOCATIONS

Taipei, Taiwan (HQ)
Austin, TX
El Paso, TX
Fullerton, CA
Harrisburg, PA
Indianapolis, IN
Lexington, KY
Ontario, CA
Raleigh, NC
San Jose, CA
Santa Clara, CA

38 locations in 15 countries
worldwide.

THE STATS

Employer Type: Public Company
Stock Symbol: 2354
Stock Exchange: Taiwan Exchange
Chairman & CEO: Terry Gou
2006 Employees: 200,000
2006 Revenue ($mil.): $10,381

KEY COMPETITORS

Flextronics
Sanmina-SCI
Solectron

EMPLOYMENT CONTACT

www.foxconn.com/location.html

Visit Vault at **www.vault.com** for insider company profiles, expert advice,
career message boards, expert resume reviews, the Vault Job Board and more.

VAULT CAREER LIBRARY 199

THE SCOOP

The source for outsource

Taiwan-based Foxconn is a major player in the consumer electronics industry. The company is a manufacturer of parts for computers and communications devices (mainly cell phones) and other consumer electronics. Its areas of expertise include heat transfer and wireless technology. Currently the company is exploring options in nanotechnology and environmentally-friendly manufacturing. Its customers include Apple, Motorola, Hewlett-Packard, Nokia and Sony.

Knobs and buttons

Foxconn was born as Hon Hai Plastics Corporation in Taiwan in 1974, making plastic television parts, although the company shortly changed its name to Hon Hai Precision Industry. In 1981, the company's founder Terry Gou (a/k/a Kuo Tai-ming) shifted the company's focus to computers and adopted the trade name Foxconn (although Hon Hai Precision Industry is still technically the company's name). Foxconn's first order of business was to start selling plastic connectors for the rapidly growing PC industry. The following year, the company started fabricating wires to go between its connectors. Hon Hai went public in 1991; five years later, it moved into the manufacture of computer cases for such clients as HP and IBM. Although the company attracted criticism for its secrecy and lack of financial transparency, within another five years it was Taiwan's leading private sector computer company.

By 1998, Foxconn was listed on *Business Week*'s list of the top 100 IT companies, and Terry Gou was one of the world's 500-richest people. More recently, Foxconn made a number of acquisitions between 2003 and 2005 to boost its position as a major electronic parts manufacturer, while also acquiring many valuable personnel. It had 3,000 engineers and 100 PhD holders by the late 1990s, who produced more than 5,000 patents by the turn of the 21st century.

Business is up

In 2006, Foxconn had an excellent year, driven by strong demand for its products in the global market. It racked up $10 billion in revenue for the year, an increase of 63 percent over the year previous. Profits increased a healthy, if eye-popping, 86 percent year over year. In 2006, Hon Hai merged with Premier Image, a manufacturer of digital cameras, in a move gauged to improve the company's already vast offerings.

... and more is coming

In early 2007, Apple announced the impending arrival of the iPhone, its all-singing all-dancing wireless-Internet enabled cellphone iPod. The announcement of the device, which is expected to go on sale in mid-2007, engendered demented levels of anticipation among Apple fans. Foxconn has the inside line on all this—literally. Foxconn is the manufacturer of the electronic guts of the iPhone, and is ramping up its production lines in order to make the 10 million devices that Apple expects to sell by the end of 2007.

GETTING HIRED

Out-Fox your career

Prepared to realize your Foxconnian ambitions? The company's careers site provides a list of job opportunities searchable by location. At press time, Foxconn seemed to be searching for financial analysts, accountants and engineers who were fluent in Chinese. To apply, resumes can be sent to the e-mail indicated in the job posting.

Visit Vault at **www.vault.com** for insider company profiles, expert advice, career message boards, expert resume reviews, the Vault Job Board and more.

VAULT CAREER LIBRARY   **201**

Freescale Semiconductor, Inc.

6501 William Cannon Drive West
Austin, TX 78735
Phone: (512) 895-2000
Fax: (512) 895-2652
www.freescale.com

LOCATIONS

Austin, TX (HQ)
Albany, NY • Arlington Heights, IL •
Austin, TX • Boca Raton, FL • Boise,
ID • Bothell, WA • Boynton Beach, FL
• Brookfield, WI • Cedar Rapids, IA •
Chandler, AZ • Chesterfield, MS •
Dayton, OH • Duluth, GA • East
Fishkill, NY • Eden Prairie, MN •
Englewood, CO • Essex Junction, VT
• Fairport, NY • Farmington Hills, MI •
Fort Worth, TX • Hauppauge, NY •
Horsham, PA • Huntsville, AL • Irvine,
CA • Kokomo, IN • Lake Zurich, IL •
Lawrenceville, GA • Libertyville, IL •
Lindon, UT • Miamisburg, OH •
Milpitas, CA • North Andover, MA •
Novi, MI • Phoenix, AZ • Plantation,
FL • Pleasanton, CA • Raleigh, NC •
Richardson, TX • Ridgeland, MS •
Rolling Meadows, IL • San Antonio,
TX • San Diego, CA • San Jose, CA •
Schaumburg, IL • Scottsdale, AZ •
Seguin, TX • Sunnyvale, CA •
Tempe, AZ • Vienna, VA • Woburn,
MA • Yorktown Heights, NY

Locations in more than 30 countries
worldwide.

THE STATS

Employer Type: Private Company
Chairman & CEO: Michel Mayer
2006 Employees: 24,000
2006 Revenue ($mil.): $6,400

DEPARTMENTS

Administrative Support • Biotechnology
• Business Development • Electrical
Engineering • Field Applications
Engineering • Finance • Hardware
Engineering • HR • IC Design
Engineering • IT • Legal •
Manufacturing • Mechanical
Engineering • Program/Project
Management • Programming • Quality
• Research Engineering • RF
Engineering • Sales & Marketing •
Software Engineering • Systems
Engineering

KEY COMPETITORS

NXP
STMicroelectronics
Texas Instruments

EMPLOYMENT CONTACT

www.freescale.com/careers

THE SCOOP

Free your mind, the rest will follow

Freescale is a major player in the microchip arena, providing a catalog of 14,000 chips primarily for the transportation, wireless and networking industries. Freescale chips can be found in products from Sony, Whirlpool, Cisco, Bose, BMW, GM and former parent company Motorola. Freescale is a key supplier of microchips to the automotive and communications industry.

The end of the affair

Freescale existed as the microchip-making arm of Motorola for five decades. In 1965, the company began to develop chips that could process radio signals (a forerunner of the chips in today's cell phones). Later, the division pioneered computer microprocessors, supplying Apple with the chips it would use for the revolutionary Macintosh.

Freescale isn't immune to the cyclical trends that often plague microchip companies; the industry constantly weathers slumping sales and predictions of its demise. Such a slump hit Motorola's semiconductor division from 2000 to 2003, as revenue fell each year, resulting in losses of $4.4 billion. Apparently 50 years was long enough for Motorola to play the faithful spouse, and it spun the division off as a separate company in October 2003. Much to Motorola's chagrin, of course, the wireless and mobile industries immediately caught fire after the sale, and the newly minted Freescale Semiconductor went on to post gains almost as soon as it was cut loose, bringing in revenue of $6.4 billion in 2006. Its success that year didn't come without sacrifices, though, as Freescale has cut 1,000 jobs since 2004.

My, how the tables have turned ...

Freescale was suddenly a hot stock and investors could only stay away from it for long. In the fall of 2006, rumors started flying that Freescale was a possible candidate for a buyout. Two teams of investors, one headed up by Blackstone, the other by KKR, both considered purchasing the increasingly profitable company. In September 2006, Blackstone's bid of $17.6 billion was accepted. Freescale should have sent a thank-you card to (guess who?) Motorola, whose business still accounted for over a quarter of its revenue.

Go East, young man

In 2007, Freescale opened a facility in Noida, India. The company has nearly 1,000 employees there, and has already established a facility in Bangalore. The 250 employees in Noida will be engaged in R&D and in the manufacture of systems on a chip.

GETTING HIRED

Set your career Free

Freescale's careers site, at www.freescale.com/careers, provides information for students and experienced job seekers. Jobs are searchable by department, keyword, location and type; to apply, candidates must first create a profile.

Nice perks offered by the company include sponsorship for professional organizations (if necessary); bonuses for good performance—of the employee and the company; bonuses for patents; on-site fitness centers or subsidies for off-site gym memberships; concierge service; and relocation assistance that includes everything from finding a place to live to shipping things.

The student careers section of Freescale's site provides a précis of the company's internship and rotation programs. To qualify for an internship or co-op position, students must have a GPA of 3.0 or higher and be studying finance, accounting or engineering on a full-time basis. Internships last 12 weeks and interns receive benefits and housing assistance. The company offers two rotation programs, one in engineering and one in sales, for recent graduates majoring in electrical engineering or computer science with backgrounds in wafer manufacture or chips for wireless, radiofrequency and mixed-signal applications, and of course, sales. Freescale posts a schedule of its recruiting junket—stops on the tour occur at such fine institutions as Texas A&M, Rochester Institute of Technology, Carnegie Mellon and Arizona State.

Fujitsu Limited

Shiodome City Center
1-5-2 Higashi-Shimbashi, Minato-ku
Tokyo 105-7123
Japan
Phone: 81-3-6252-2220
Fax: 81-3-6252-2783
www.fujitsu.com

Fujitsu America, Incorporated
1250 East Arques Avenue
Sunnyvale, CA 94085
Phone: (408) 746-6200
Fax: (408)746-6260
www.fai.fujitsu.com

THE STATS

Employer Type: Public Company
Stock Symbol: 6702
Stock Exchange: Tokyo Exchange
Chairman: Naoyuki Akikusa
President: Hiroaki Kurokawa
2006 Employees: 158,491
2006 Revenue ($mil.): $43,222

DEPARTMENTS

Administration • Consulting •
Engineering • Marketing •
Operations • Planning • Sales •
Services

KEY COMPETITORS

IBM • NEC • Toshiba

EMPLOYMENT CONTACT

www.fujitsu.com/global/about/
employment

LOCATIONS

Tokyo, Japan (HQ)
Sunnyvale, CA (U. S. HQ)
Addison, TX • Alpharetta, GA •
Atlanta, GA • Austin, TX • Beaverton,
OR • Bellevue, WA • Bloomfield Hills,
MI • Boise, ID • Burlington, MA •
Chesterfield, MO • Columbia, MD •
Columbus, OH • Denver, CO • Edison,
NJ • El Segundo, CA • Everett, WA •
Fairfield, NJ • Falls Church, VA •
Folsom, CA • Frisco, TX • Greenwood,
CA • Honolulu, HI • Houston, TX •
Hudson, OH • Hunt Valley, MD •
Irvine, CA • Irving, TX • Kimberly, WI
• Louisville, CO • Middleton, WI •
Milwaukee, WI • Minnetonka, MN •
New York, NY • Oak Brook, IL •
Okemos, MI • Phoenix, AZ •
Pittsburgh, PA • Pleasanton, CA •
Plymouth, MI • Portland, OR •
Richardson, TX • San Antonio, TX •
San Diego, CA • San Jose, CA • Santa
Clara, CA • Schaumburg, IL •
Upland, CA • Washington, DC

Locations in 61 countries worldwide.

Visit Vault at **www.vault.com** for insider company profiles, expert advice,
career message boards, expert resume reviews, the Vault Job Board and more.

VAULT CAREER LIBRARY   205

THE SCOOP

One-stop IT solution

Fujitsu is one of Japan's flagship companies, often working hand in hand with the government to further the state of Japanese industry and technological advancement. A major player in many areas of the high-tech industry, it is second in manufacturing computers only to IBM. The company's customers include everyone from governments and Global 500 companies to small businesses and consumers. Fujitsu's offerings range from consulting and IT services (including outsourcing), to software for managing online transactions and manufacturing; of course it also makes and sells computers, servers, point-of-sale systems, and electronic parts like memory chips and capacitors.

Fujitsu's progressive policies in regard to the environment and its employees have earned the company a place on both the Dow Jones Sustainability Index and London's FTSE 4Good index.

Tokyo calling

Fujitsu was founded in the 1930s, an offshoot of the Fuji Electric Company. Fuji itself began that same decade as a joint venture of the Japanese Furukawa Electric Company (founded 1884) and the German Siemens AG (founded 1847). Fuji created Fujitsu in 1935 to oversee its production of telephones and automatic exchange equipment (i.e., switchboards).

The company started producing radio equipment in 1937, but returned to the phone business in 1945, by which point World War II had destroyed roughly half of Japan's 1.1 million telephone connections. Japan's government started contracting with Fujitsu to reinstall the country's telephone system, and in more peaceful times the government later sponsored more diverse research and development. Before long, Fujitsu was making advancements in various fields, especially in computers.

Computer nation

Fujitsu went public in 1949, and subsequently began to manufacture computers during the 1950s, first for the government (the first was delivered in 1951) and later for industry (in 1954). IBM introduced the first transistorized computer in 1959, and its level of advancement sent shockwaves through the computer industry. The Japanese government was quick to respond; in 1961 it negotiated the right to license

IBM patents for local R&D and then restricted imports of foreign computers, supplied local companies with low-cost loans and intervened when necessary to discourage Japanese companies from competing with each other at the cost of technical advancement.

Fujitsu invested heavily in computer R&D throughout the 1960s, largely with government support. The company eventually started to make breakthroughs, first with a system to exchange data between banks electronically in 1974, and then with Japan's first supercomputer, the FACOM VP-100/200, in 1982. In 1988, it introduced an ISDN system, which transmits data over telephone wires, in Singapore.

Fujitsu surfs the Web

Fujitsu moved into the Internet age of the 1990s with a vengeance. In 1994 it came out with InfoWeb, an Internet service for businesses, and followed with INTERTop, a user-friendly and portable way for people to access information and software on the Internet. It was also still releasing supercomputers; the latest installment arriving in 1992. Fujitsu worked to expand sales abroad in the 2000s, and also focused on hot new tech areas such as flash memory, outsourcing and systems-on-a-chip.

Building a consulting army

In fiscal 2006, Fujitsu took in revenue of $43 billion, and profits of $850 million. Revenue was up just over 6 percent compared to 2005, driven by notable performance in the services division outside Japan, which increased nearly 24 percent year over year.

In the face of these numbers, Fujitsu saw an opportunity to expand its consulting arm abroad. It made two acquisitions of American consulting firms in 2006, in order to expand its ability to reap revenue in this area. In February, Fujitsu acquired Greenbrier & Russel, an Illinois-based firm that specializes in enterprise resource planning. This acquisition also strengthens Fujitsu's consulting offerings in the Midwest. Later that month, the company gobbled up Pittsburgh consulting firm Rapidigm. This deal boosts the number of Fujitsu consultants in the U.S. to nearly 5,000—a 75 percent increase.

Also, Fujitsu's research and development remains strong. In 2006, the company opened a joint venture with Advantest to experiment in making microchips: engraving silicon wafers with electrons, which are capable of making finer lines in silicon than the current technology. A few months afterwards, Fujitsu acquired two microchip factories in Japan, which it will use to manufacture flash memory.

French connection

In 2007, Fujitsu made a bid to buy a majority stake in French consulting firm GFI Informatique SA as part of its plans for global expansion. Fujitsu bid $600 million for its stake in the company, but the deal is yet to close.

GETTING HIRED

Connect with Fujitsu

Fujitsu's global careers site, at www.fujitsu.com/global/about/employment/ provides information about opportunities at the company's global locations. In the section that deals with careers in North America, job seekers will find a list of openings at Fujitsu America, Fujitsu Microelectronics America, Fujitsu Computer Systems, Fujitsu Transaction Solutions and Fujitsu Consulting Holdings. Positions are searchable by category, location company subsidiary and keyword. To apply, candidates must first create a profile.

Garmin Ltd.

1200 East 151st Street
Olathe, KS 66062
Phone: (913) 397-8200
Fax: (913) 397-8282
www.garmin.com

LOCATIONS

Olathe, KS (HQ)
Denver, CO • Houston, TX • Los
Angeles, CA • Minneapolis, MN •
Raleigh, NC • Salem, OR • San
Francisco, CA • Tempe, AZ

Southampton • Taiwan •

DEPARTMENTS

Administration
Aircraft Certification
Cartography
Customer Repair
Engineering
Field Service Engineering
Finance
HR
IT
Manufacturing
Marketing
Marketing Communications
Marketing/Sales
Office Services
Operations
Purchasing
Quality Assurance
Warehouse

THE STATS

Employer Type: Public Company
Stock Symbol: GRMN
Stock Exchange: Nasdaq
Chairman & CEO: Min H. Kao
2006 Employees: 4,751
2006 Revenue ($mil.): $1,774

KEY COMPETITORS

Magellan
TomTom
Trimble

EMPLOYMENT CONTACT

www.garmin.com/aboutGarmin/
employment.html

Visit Vault at **www.vault.com** for insider company profiles, expert advice,
career message boards, expert resume reviews, the Vault Job Board and more.

VAULT CAREER LIBRARY   209

THE SCOOP

All who wander are not lost

No matter how far you go, Garmin will tell you where you are. The company is a leader in the market for aftermarket automotive GPS, but its products include GPS receivers for work and play—for every manner of transport, from feet to cars and boats and planes. The company provides a device so consumers can equip nearly everything in their lives that moves—cars, bikes, boats, motorcycles, dogs—with its own GPS receiver. (Unfortunately, the company has yet to make one small enough for a keychain.) Garmin's receivers come with other features, like MP3 players, fish finders, heart rate monitors and Bluetooth connectivity.

Kansas meets Taiwan

Garmin's origins are a mixture of the Midwest and the Far East. In Kansas, an ex-pilot founded King Radio Corporation in 1959 for the manufacturing of plane equipment, mostly navigational in nature (e.g., radios). In 1963, an engineer named Gary Burrell began there.

When the Allied Signal Corporation purchased King Radio in 1983, Burrell met a Taiwanese engineer named Min Kao, who had worked on an experimental technology called the global positioning system (GPS) as a defense contractor. King Radio had pursued such an advance in navigational technology and Burrell excitedly began working with Kao, sensing a breakthrough after 20 years of striving to improve navigational technology.

I told you so!

However, Allied Signal failed to show its charge's enthusiasm for GPS, and Burrell resigned his post in 1989 out of frustration. Within months he was on a plane to Taiwan, as Kao was adamant they should start their own company. Garmin Limited, its name a portmanteau of Burrell and Kao's first names, was incorporated later that year.

The company's first product, released in 1991, was (you guessed it) a GPS device for pilots and boaters. A GPS receiver for drivers followed in 1997, and a cell phone equipped with the system came out a year later. Garmin went public in 2000, shortly after the Department of Defense lifted restrictions on the accuracy of commercial

GPS receivers. Since then, Garmin's revenue has taken off, as the price for GPS devices drops and the technology goes increasingly mainstream.

Revenue's right turn

In 2006, Garmin posted revenue of $1.8 billion with a profit of $514 million. Revenue increased 72 percent over the preceding year, while profits were up 65 percent. The company's automotive division grew by 170 percent, while the fitness division brought in year-over-year growth of 20 percent.

The same year, Garmin acquired Dynastream, manufacturer of wireless monitoring equipment for athletes, for $36 million. Garmin integrated Dynastream's athletic technology into a new generation of GPS devices for runners and bikers that can monitor distance, heart rate, calories burned and other useful bits of telemetry.

GPS for everybody!

The following year, Garmin acquired part of Nautamatic Marine Systems, a manufacturer of an autopilot system for boats, and the company will look to integrate the Nautamatic system with its existing line of nautical GPS solutions. It scored another coup that year when it inked an agreement with Vanguard Car Rental, the parent company of National Car Rental and Alamo Rent-A-Car to put GPS devices in its vehicles. Rental car drivers should appreciate them, as they will usually be driving in alien environments, and the devices will also allow consumers to experience a Garmin system firsthand.

Looking forward, Garmin will no doubt be exploring other opportunities to spread its navigational know-how far and wide, staying true to its King Radio roots. The company remains true to its roots in other ways: Garmin's headquarters are located in the same Kansan city as King Radio's were, 44 years after Gary Burrell started working there.

GETTING HIRED

Put your career in orbit with Garmin

Garmin's careers site (www.garmin.com/aboutGarmin/employment.html) provides information about job opportunities and benefits at the company. Jobs are offered in the areas of administration, aircraft certification, sales and marketing, engineering, corporate communications, accounting, cartography, customer repair, MIS, HR, IT, manufacturing, operations and quality assurance. Positions are listed by function at https://my.garmin.com/iRecruitment/joblist.htm. To apply, candidates must fill out a web form.

Benefits at Garmin include health and life insurance, 401(k), paid vacations and holidays, tuition reimbursements and discounts on company merchandise. Questions or comments can be sent to jobs@garmin.com.

General Dynamics Corporation

2941 Fairview Park Drive
Suite 100
Falls Church, VA 22042
Phone: (703) 876-3000
Fax: (703) 876-3125
www.generaldynamics.com

LOCATIONS

Falls Church, VA (HQ)
Ann Arbor, MI • Anniston, AL •
Appleton, WI • Bath, ME •
Bloomington, MN • Brunswick, GA •
Brunswick, ME • Burlington, VT •
Camden, AR • Charlotte, NC • Dallas,
TX • Garland, TX • Gilbert, AZ •
Groton, CT • Kilgore, TX • Las Vegas,
NV • Lima, OH • Long Beach, CA •
Marion, IL • Marion, VA •
McLeansville, NC • Minneapolis, MN •
Needham, MA • Newton, NC •
Pittsfield, MA • Quonset Point, RI •
Red Lion, PA • Saco, ME • San
Diego, CA • Santa Clara, CA •
Savannah, GA • Scottsdale, AZ •
Scranton, PA • St. Marks, FL • St.
Petersburg, FL • Sterling Heights, MI •
Taunton, MA • West Palm Beach, FL
• Westfield, MA • Westminster, MD •
Woodbridge, VA • Ypsilanti, MI

17 locations throughout Europe.

DEPARTMENTS

Advanced Information Systems •
ATP • C4S • Electric Boat •
Gulfstream • Information
Technology • Land Systems •
Ordnance & Tactical Systems

THE STATS

Employer Type: Public Company
Stock Symbol: GD
Stock Exchange: NYSE
Chairman & CEO: Nicholas D. Chabraja
2006 Employees: 81,000
2006 Revenue ($mil.): $24,063

KEY COMPETITORS

Lockheed Martin
Northrop Grumman
Raytheon

EMPLOYMENT CONTACT

www.generaldynamics.com/
employment

Visit Vault at **www.vault.com** for insider company profiles, expert advice,
career message boards, expert resume reviews, the Vault Job Board and more.

VAULT CAREER LIBRARY   213

THE SCOOP

One if by air, two if by sea

General Dynamics Corporation is a force to be reckoned within the armaments market. It started manufacturing submarines and surface ships as The Electric Boat Company over 100 years ago, and started diversifying its business platform during the Cold War. By now, it is an all-purpose defense contractor.

The company focuses on four specific business areas: aerospace, combat systems, marine systems, and information systems and technology. These divisions account for, respectively, business aviation, expeditionary combat systems, submarines and surface ships (still manufactured under the name Electric Boat), and specialized IT products for defense and law-enforcement customers. GD is looking to its IT department in particular for future growth.

Engines of war

General Dynamics traces its roots back to 1899, when John Holland started The Electric Boat Company. Electric Boat sold submarines to the adversarial parties during the Russo-Japanese War, but primarily devoted itself to cargo ships thereafter, despite receiving orders for submarines from England and the U.S. during World War I. The company nearly went bankrupt during the interwar period of the 1920s and 1930s, but as ship and submarine production boomed during World War II, its fortunes revived. The company has strived to learn from the experience ever since, diversifying its products and services so that it won't overly rely on war machines for its prosperity.

Diversification began in earnest as soon as the war ended. John Jay Hopkins, CEO at the time, purchased Canadair in 1946 and Convair in 1953, both manufacturers of business and military aircraft. The company adopted the General Dynamics brand name in 1952, to reflect its focus on contracting beyond the scope of boats and submarines. With David S. Lewis Jr. at the helm from 1971 to 1985, the company's revenue quadrupled, allowing it to further diversify with the acquisition of Chrysler's combat division in 1982.

Cold War hangover

The immediate post-Cold War era was a tough time for GD, as declining defense budgets and cutbacks in government R&D spending forced the company to sell off all of its divisions except for Electric Boat and its land combat manufacturing in 1995. The company moved decisively the same year to shore up its boat division by acquiring Bath Iron Works. National Steel and Shipbuilding Co. joined the fold three years later.

On the rebound and looking for contracting love

Neither the divestment of its auxiliary businesses nor the consolidation of its shipbuilding signaled a retreat for General Dynamics, however. 1997 began a period of furious acquisitions, as GD attempted to revitalize its aircraft production and branched out into acquiring IT companies with customers in law enforcement and the government. Since that year, the company has acquired and integrated 40 companies. One of the largest acquisitions in company history came in May 1999, when GD announced the $5 billion purchase of Savannah-based luxury jet maker Gulfstream. In one fell swoop, GD had revitalized its aviation services and added a flagship line.

Two years later, GD proposed a $2.1 billion merger with Newport News Shipbuilding. However, federal regulators threatened to block the deal in October 2001, claiming a GD-NNS deal would create a monopoly on nuclear submarine construction; the two companies shortly called off merger talks. Northrop Grumman, GD's rival for Navy shipbuilding contracts, seized the opportunity and bought NNS, shortly overtaking GD in the shipbuilding field.

Long war, big profits

The September 11 attacks put defense spending back into a higher gear, significantly quickening the pace of new contracts. In October 2001, as federal authorities were blocking GD's bid to buy NNS, other federal authorities were beckoning for the company's help. GD's National Steel and Shipbuilding Company got a $700 million contract to build a new class of logistics ships; its electric boat and land systems subsidiaries received contracts totaling $90 million for a variety of projects; and a number of other subsidiaries landed a slew of contracts totaling $107 million.

Ongoing operations in Iraq and Afghanistan have provided steady sales of GD combat systems, such as the eight-wheeled Stryker armored vehicle. Stryker debuted at the beginning of the 2003 U.S. invasion of Iraq, and as of 2007, the company has made about 2,000 of them. GD hasn't just profited from manufacturing, either, as repair work on armored vehicles returning from the battlefront has provided excellent business. Since the 2001 attacks, GD's combat systems unit's revenue and profits have tripled, and net sales and earnings each jumped by 18 percent from 2005 to 2006.

Diversifying—that old word again

In response to increasing demand, the company has recently added small-caliber ammunition to its product offerings; the U.S. Army has so far signed contracts for it to supply up to 500 million rounds. GD has also generated revenue by selling armored trucks to Canada for use in Afghanistan, and from technology the company has developed to protect against roadside bombs.

Wireless post-war

GD is preparing for a slowdown in weapons spending, in form with the company's business model over the last 50 years. It first focused on IT as an avenue of growth in 1997, and information systems has indeed become the company's fastest growing unit—its revenue increased more than 34 percent from 2004 to 2006, to almost $9 billion.

Therefore, the April 2007 news that GD had edged out No. 1 defense contractor Lockheed for the next large governmental contract was especially welcome. GD won a contract (with an estimated value of $5 billion) to provide a wireless communication network (voice over Internet protocol, or VoIP), for several governmental agencies. The project will allow 80,000 agents of the Treasury, Justice and Homeland Security departments to communicate securely on a common network, and the system will be adaptable to advances in wireless technology. VoIP technology will also be installed in the Pentagon as part of the PENREN (Pentagon Renovation) project, for which GD is a main contractor.

GETTING HIRED

Have a Dynamic career

General Dynamic's employment site, at www.generaldynamics.com/employment, provides job seekers with information about hiring at the company. Each of GD's four main business units (aerospace, combat systems, information systems and technology, and marine systems) operates its own independent hiring and recruiting division.

However, the main web site does have a searchable list of jobs throughout the company. At press time, the company was seeking painters and airplane mechanics at Gulfstream, accountants in the IT division; administrative assistants, engineers and financial analysts were sought by other divisions. The web site also stresses the company's commitment to equal employment and diversity in the workplace.

OUR SURVEY SAYS

Join the General

One hire says that his interview was "very easy. One informal conversation with HR rep over the phone. One formal interview. A brief review of where I was stationed in the military and what jobs I held." "Due to required security clearances through the Canadian government, it is often difficult to hire people who have just immigrated to Canada," an insider points out.

Once in the office, contacts report that the company has "very much a military culture." "Many of the employees are ex-military (Canadian, British, etc.). Many layers of management as a result," adds a source. "Lots and lots of red tape in this company … it can take forever to get something small done," wails an insider. "Business casual dress code. Fridays, more relaxed. Cube environment," says a co-worker. Another points out the "flexible benefit package and the option to work a compressed workweek to get every other Friday off." "The company is very big on career development, training and taking external courses," adds a respondent.

Harman International Industries, Incorporated

1101 Pennsylvania Avenue, NW
Suite 1010
Washington, DC 20004
Phone: (202) 393-1101
Fax: (202) 393-3064
www.harman.com

LOCATIONS

Washington, DC (HQ)
Bedford, MA • Burbank, CA • El
Paso, TX • Elkhart, IN • Farmington
Hills, MI • Franklin, KY • Fremont,
CA • Middletown, CT • Martinsville,
IN • Northridge, CA • Phoenix, AZ •
Prairie Du Chien, WI • Rancho
Cucamonga, CA • Salt Lake City,
UT • Upper Saddle River, NJ •
Woodbury, NY

45 locations in 16 countries
worldwide.

DEPARTMENTS

Accounting/Finance • Customer
Service • Electronic System
Integration • Engineering • HR • IT •
Materials • Production • Quality
Assurance • Sales

THE STATS

Employer Type: Private Company
Chairman: Dr. Sidney Harman
President & CEO: Dinesh Paliwal
2006 Employees: 10,845
2006 Revenue ($mil.): $3,247

KEY COMPETITORS

Bose
Pioneer AKC
Sony

EMPLOYMENT CONTACT

jobs.harman.com

THE SCOOP

Turn it up!

Harman International manufactures stereo and entertainment equipment for the consumer, professional and automotive industries. Harman Industries brands include Harman-Kardon, JBL, Infinity, and Revel. Cars with Harman's systems include those manufactured by DaimlerChrysler, BMW, Land Rover, Porsche GM and Saab; Harman also makes aftermarket parts for car stereos so that consumers can add features like DVD players, navigation systems and louder speakers.

Consumer products include iPod-compatible speakers and speaker systems for travel, home and boating. Professional products include mixing equipment for recording studios, headphones and microphones, loudspeakers and audio systems for theaters. The company's professional goods are sold under the brand names JBL Professional, Crown, Soundcraft, Lexicon and Studer. The company's very existence owes much to its namesake, the 88-year-old Dr. Sidney Harman, who brought the company back from the brink of extinction in the late 1970s, and still serves as its chairman.

Semper fidelity

Sidney Harman and Bernard Kardon were two engineers then working primarily on public address systems at the Bogen Company in the early 1950s. They were very interested in high-fidelity home stereo equipment, and launched their own company in 1953, when Bogen demonstrated no support for the idea. Harman-Kardon, as the company was known, released an FM tuner as its first product; a high-fidelity receiver and a stereo radio followed in 1954 and 1958, respectively. These offerings set the company's style in the 1950s: they were the first home stereo products to actually look like pieces of furniture, and are widely credited with bringing hi-fi musical enjoyment to the masses.

Harman-Kardon made another leap in high-fidelity marketing in the 1970s by selling its products separately. An implicit invitation for purchasers to assemble the disparate pieces themselves, it actively encouraged customers to become aficionados of the latest sound equipment, igniting many an amateur debate over which equipment resulted in the best sound. The marketing gambit paid off and revolutionized the culture of home stereos even further than the company's work in the 1950s. And it was good for business, as Harmon-Kardon was ahead of its

Visit Vault at **www.vault.com** for insider company profiles, expert advice, career message boards, expert resume reviews, the Vault Job Board and more.

VAULT CAREER LIBRARY   **219**

competitors in offering separate and interchangeable stereo technology. By 1976, the company was taking in annual profits on $9.1 million.

The political animal

The next year, President Carter appointed Sidney Harman as his Undersecretary of Commerce, and Harman sold a controlling stake in his company to the large Chicago corporation Beatrice Foods. Harman watched from Washington as his company was nearly run into the ground. As soon as he was out of office in 1980, he gathered an investment of $55 million and repurchased Harman-Kardon.

Henceforth, the company would be known as Harman International, but it was much smaller than three years before—only about 60 percent of its former operations remained, and its continued survival was far from assured. Harman quickly realized it could not compete for mass sales as it had done in the 1950s and 1970s, and targeted the high-end market for sound technology. It started to acquire small, quality-oriented companies, largely leaving them alone to manufacture their luxury stereo equipment.

The strategy was successful in keeping the company alive and growing throughout the 1980s, but a recession resulted in losses of $19.8 million in 1991. Harman, then 70 years old, promptly fired the president of the company and took over himself, moving to California to focus on the company (taking him across the country from his wife, who was elected to Congress in 1992.) Harman merged its 21 small subsidiaries into five main divisions and turned away from the high-end market, selling its wares in mass retail outlets such as Circuit City. The company soon rebounded, attaining 1995 revenue of $1.17 billion and profits of $41 million.

Perfect Harman-y?

Harman's products are still attractive to the public, as consumer division sales have risen 20 percent from 2005 to 2006, prompting a 7 percent rise in revenue to $3.2 billion. The company is in fine operating shape as well, as profits have risen 10 percent to $255 million.

The company's boardroom, however, isn't functioning nearly as smoothly; CEO Douglas Pertz abruptly left in August 2006, only four months after his hiring. (In what must be some sort of record for return on a company's investment, Harman cut him a $3.8 million severance check!) Bernard Girod, his immediate predecessor as CEO, agreed to step in until another CEO was found, and Harman named Dinesh

Paliwal, former CEO of ABB North America, as its new CEO and president in May 2007. Sidney Harman himself has remained the chairman of the company throughout this upheaval, so analysts are not worried about the turnover harming the company's bottom line. Harman's search for a successor to its founder, however, is a genuine and pressing concern.

Please, infotain me!

As consumers increasingly demand more high-tech entertainment options wherever they go, Harman is venturing beyond the subwoofer to focus on what it calls "infotainment"—devices that can provide customers with options for e-mail, phone, movies, music and web browsing—in the car, at home and anywhere in between. Although the technology fits in better with Harman's line of high-end products, the company is confident it will soon become a mainstream attraction. While such a gamble would be potentially disastrous for another firm, Harman's track record in popularizing similar high-end technologies commands respect.

Sound investment

In 2007, Harman Industries agreed to go private in a deal valued at $8 billion. Two private equity firms, Kohlberg Kravis Roberts and Goldman Sachs Capital Partners, teamed up on the deal. Unlike other private equity buyouts, however, shareholders will have a stake in the company at the end of this transaction, creating an odd new kind of public-private company.

Just under a third of the shares of the company will be offered to current Harman stockholders. While the shares won't be listed on an exchange and it is highly unlikely shareholders will have any say in the running of the company, they will still have a stake in the company. Perhaps Sidney Harman, ever the innovator, has one more trick up his sleeve at 88 years old.

GETTING HIRED

Harman calling

Harman's careers site, located at jobs.harman.com, provides information for job seekers on open positions at the company. To apply, candidates must first create a profile.

Harris Corporation

1025 West NASA Boulevard
Melbourne, Florida 32919
Phone: (321) 727-9100
Fax: (321) 674-4740
www.harris.com

LOCATIONS

Melbourne, FL (HQ)
Albuquerque, NM • Alexandria, VA
• Annapolis Junction, MD •
Arlington, VA • Chantilly, VA •
Chesapeake, VA • Cincinnati, OH •
Colorado Springs, CO • Columbia,
MD • Denver, CO • Long Beach, CA
• Mason, OH • Quincy, IL •
Rochester, NY • Calgary, Canada •
Chatenay-Malabry, France •
Mississauga, Ontario • Montreal •
Shenzhen, China • Thames Ditton,
UK • Toronto • Waterloo • Wien,
Austria • Winnersh, UK

DEPARTMENTS

Administration • Corporate
Development • Engineering •
Finance • Government Relations •
HR • Information Systems • Legal •
Marketing • Operations • Program
Management • Sales • Technical

THE STATS

Employer Type: Public Company
Stock Symbol: HRS
Stock Exchange: NYSE
Chairman & CEO: Howard L. Lance
2007 Employees: 16,000
2007 Revenue ($mil.): $ 4,000

KEY COMPETITORS

Alcatel-Lucent
General Dynamics
Motorola, Inc

EMPLOYMENT CONTACT

www.harris.com/harris/careers

THE SCOOP

Can you hear me now?

Harris is a provider of broadcast and communications technologies for use in peace and war. Although the company has long had the philosophy of adapting its government contracts into technology useful for the mainstream public, it derived most of its 2006 revenue from government contracts in the areas of battlefield communications, surveillance and data transfer.

In the government arena, its customers include the Department of Defense and a wide range of government agencies—including the FAA, Census, NSA and NOAA. The company's commercial segment customers include domestic and international broadcasters and public and private telecommunications companies. In 2007, *BusinessWeek* ranked the company No. 84 on its InfoTech 100.

On the cutting edge of communication since 1895

Harris was founded in 1895 by two brothers in the jewelry business, Charles and Alfred Harris. When not providing baubles for the ladies of Niles, Ohio, their first successful invention was an automatic sheet feeder for printing presses in 1890. This device, which could work 10 times as fast as a human, became the foundational technology for the Harris Automatic Press Company, incorporated in 1895. Harris soon made other breakthroughs, including the first commercially viable offset lithographic press and the first two-color offset press. The brothers' company quickly became a worldwide leader in manufacturing printing equipment.

In 1957, Harris Automatic Press merged with Intertype Corporation, a provider of hot metal typesetting, to form Harris-Intertype. Soon, the company decided to expand its focus to types of communication beyond the printed word, to electronic means of communicating.

Bound for Florida groves

By 1967, Harris had found the right partner to modernize its technology. That year, Harris-Intertype acquired Florida-based Radiation Inc. in a deal worth $56 million, but it was more of a merger between the two companies, as Radiation's focus in electronics soon became the norm, and the corporate headquarters was moved to Florida. Radiation was a very successful government contractor that provided communications technology for the burgeoning American space program, but it had wanted to move into the commercial sector for some time. It was stocked with many accomplished engineers, and their examination of Harris' technology soon resulted in a breakthrough—however, one of their initial studies of the Harris printing technology led directly to electronic newsroom technology.

The new company created from the merger with Radiation resulted in a "technology transfer" philosophy within the company, whereby Harris-Intertype hoped to adapt the products that had been built by Radiation for government contracts into ones with commercial viability. Throughout the 1970s, a number of other acquisitions led Harris ever further away from its mechanical origins. It purchased General Electric's broadcasting division in 1972 and moved into data processing the same year with the purchase of University Computing Corporation for $20 million. Two years later, the company changed its name to Harris Corporation while divesting itself of the last traces of its corrugated paper machinery, and also purchased Datacraft Corporation, further broadening into data processing.

I'll process your data

The company was highly interested in data processing as an avenue of growth and grew throughout the 1980s and 1990s, with pursuits ranging from semiconductor operations to telephone equipment to television communications. In 1999, the company underwent major restructuring, and sold off its semiconductor business (then called Intersil), and its commercial fax and printer business (Lanier Worldwide).

A trio of acquisitions then strengthened Harris' remaining pursuits. It added Orkand Corp. to its Harris Technical Services line in July 2004 and acquired Encoda Systems one month later. Encoda would complement the subsequent $500 million purchase of Leitech Technology Corporation in October 2005, as both were added to the Harris Broadcast Communications business.

On television

Harris took in $3.5 billion in revenue in 2006, a 15 percent increase over the previous year, and $240 million in profit, which represented a 17 percent increase. In 2006, Harris expanded its television broadcast business. In April, it paid $37 million for Optimal Solutions, a company that specializes in software for selling broadcast airtime to advertisers. Later that year, Harris bought Aastra Digital Video, a provider of video technology.

Radar love

In early 2007, a division of Harris merged with Stratex Networks, a company that provided transmitters for wireless voice and data. The merger combined Stratex with Harris' microwave communications division; Harris owns the majority of the resulting company. The merger makes Harris a major player in the microwave communications arena—the combined companies will have customers in 150 countries.

Stand up and be counted

In 2010, Harris will provide the census with 500,000 wireless, portable computers. This contract, valued at $600 million, is intended to reduce the amount of paper the census must deal with, as well as reducing the cost of the undertaking by $1 billion. The computers, which will transmit data over Sprint's wireless network, are expected to be part of the most accurate census ever taken, and Harris reports that the project is being completed on time and under budget—about 1,500 of the devices were already being tested in the field in 2007. While census data has been analyzed with a computer since 1890, this constitutes the first time it will be collected with wireless devices.

GETTING HIRED

Communicate your interest to Harris

Harris' growth in recent years means that the company is hiring—it will be adding more than 300 new positions in 2007, primarily for work on contracts for various government agencies. Harris' careers site, at www.harris.com/harris/careers, provides jobseekers with information about career opportunities at the company, benefits and job openings. The company hires people for both administrative (contracts, procurement, logistics, marketing sales) and engineering (mechanical, electrical, manufacturing, software, testing) positions. Harris allows job seekers to search for jobs based on department, level of education, job type and location. To apply, job seekers must first create a profile.

Benefits include medical, dental and vision insurance, health and dependent spending accounts, 401(k) with company match and some locations arrange schedules to allow for three-day weekends. Harris offers its employees a tuition reimbursement plan that allows them to pursue advanced degrees and access to over 500 specialized courses in-house.

In addition to recruiting through its web site, Harris seeks experienced hires at a number of job fairs aimed at people with security clearances and an interest in engineering, aerospace and defense. College students shouldn't feel left out, though, as Harris makes stops at several universities across the U.S. including University of Florida, Georgia Tech, North Carolina State University, Auburn University, Carnegie Mellon, Cornell, Florida Institute of Technology and the Rochester Institute of Technology, seeking students with majors in engineering, business, finance, accounting and allied disciplines. Harris offers internships to undergrads in engineering and business related disciplines.

To make the shift from all-night study sessions and beer pong to neckties and morning commutes as easy as possible, Harris operates a program called GRAD (Graduate Acclimation and Development). The program gives recent graduates a chance to socialize and network while attending sport events and performing community service projects. The campus section of Harris' careers site also provides information on career paths in engineering and finance at the company.

Inquiries can be directed to staffing@harris.com; for questions regarding university relations., contact college@harris.com.

Hewlett-Packard Company

3000 Hanover Street
Palo Alto, CA 94304
Phone: (650) 857-1501
Fax: (650) 857-5518
www.hp.com

LOCATIONS

Palo Alto, CA (HQ)
Atlanta, GA • Austin, TX • Baltimore, MD • Boston, MA • Chapel Hill, NC • Chicago, IL • Cincinnati, OH • Cleveland, OH • Dallas, TX • Detroit, MI • Honolulu, HI • Houston, TX • Jacksonville, FL • Kansas City, MO • Las Vegas, NV • Los Angeles, LA • Miami, FL • Milwaukee, WI • Minneapolis, MN • New Orleans, LA • New York, NY • Oklahoma City, OK • Philadelphia, PA • Phoenix, AZ • Pittsburgh, PA • Portland, OR • Raleigh, NC • Salt Lake City, UT • San Antonio, TX • San Diego, CA • San Francisco, CA • San Jose, CA • Seattle, WA • St. Louis, MO • Tallahassee, FL • Tampa, FL • Tucson, AZ • Tulsa, OK • Washington, DC •

DEPARTMENTS

Administration • Business Planning • Customer Service/Support • Engineering • Engineering Services • Facilities • Finance • HR • IT • Legal • Marketing • Marketing Support • Operations • Outsourcing Management • Public Affairs & Communication • Quality • Sales • Sales Operations • Systems Integration • Technical • Training

THE STATS

Employer Type: Public Company
Stock Symbol: HPQ
Stock Exchange: NYSE
Chairman & CEO: Mark Hurd
2006 Employees: 156,000
2006 Revenue ($mil.): $91,658

KEY COMPETITORS

Canon
Dell
IBM

EMPLOYMENT CONTACT

www.hp.com/jobsathp

Visit Vault at **www.vault.com** for insider company profiles, expert advice, career message boards, expert resume reviews, the Vault Job Board and more.

VAULT CAREER LIBRARY   227

THE SCOOP

Colossus bestride the industry

Hewlett-Packard is the top player in the PC industry, having beaten out its rival Dell in 2006 in terms of market share. The company produces printers, digital cameras and servers, provides IT and consulting services, and makes networking products and software, to tie everything together. Its computers range from desktops and laptops to handheld computers and calculators. Hewlett-Packard is ranked No. 14 on the 2007 Fortune 500, and *BusinessWeek* recently ranked it No. 35 in the 2007 edition of InfoTech 100.

Granddaddies of Silicon Valley

Bill Hewlett and David Packard founded their company in a Palo Alto garage in 1939. Starting with just $538 in capital, the duo produced their first product, an audio oscillator, the same year. Bill and Dave decided to name the device the HP200A because they thought it would make the fledgling company appear more like an established player.

World War II started within the year, and HP was soon manufacturing military equipment for the U.S. government, including microwave signal generators and radar jammers. The defense contracts contributed to a period of explosive growth, and the company generated annual revenue of $2.2 million by 1947. The company went public 10 years later, and, in the first instance of what would become a tradition of egalitarian treatment of its employees, awarded all its workers with six months' service a grant of the new HP stock. Also that year, the company finally outgrew its various temporary homes and began moving into its corporate headquarters in Palo Alto.

Growing into techie godhood

Throughout the 1960s, HP acquired smaller, startup businesses and moved into new markets such as graphics recorders, medical electronics and analytical instruments. The company also courted international customers by beginning to market itself in Canada, Japan and Europe. It founded HP Laboratories in 1966 to conduct basic research in the fields of physics, electronics, and medical and chemical instruments.

The company also introduced its first computer that year, the 2116A. The company's first scientific calculator followed in 1968—it was about the size of a typewriter—

and, in a testament to the company's expertise in miniaturization, it came out with a handheld version just four years later. In 1977, HP came out with the HP-01, a wristwatch that included calendar and calculator features.

Entering the PC era

The company introduced a low-cost laser printer and a personal computer at the advent of the 1980s, and went on to introduce a low-cost color printer for the consumer market in 1991. In the 1990s, HP focused on bringing computers to the consumer market as well, drawing on the technology it had been supplying to business, government and research interests. The company's computer offerings were also boosted by a number of acquisitions, including Apollo Computer in 1989 and Convex Computer in 1995. In 2002, HP bought the big one, and the computer market hasn't been the same since.

The 2002 acquisition of longtime competitor Compaq vaulted Hewlett-Packard into the upper echelon of high-tech companies. HP announced the mammoth $19 billion deal—the largest merger in computer industry history—in 2001. The deal was finally completed on May 3, 2002, after a year of opposition by many HP stockholders, including members of the Hewlett and Packard families. It has established the company as a world-class producer of computer hardware such as PCs, servers, storage components and printers, and software and computer consulting services, making it comparable in size to industry-leader IBM.

Back to buying

In July 2006, HP bought Mercury Interactive, a software company whose products help companies obtain a better return on investment from IT expenditures, for $4.5 billion. Mercury's offerings complement HP's stable of business hardware and software products.

Loose lips sink ships (and computer chips)

In 2006, HP Chairwoman Patricia Dunn faced a nagging challenge: someone on the board of directors was leaking critical pieces of company strategy and confidential information about board proceedings to the press. Dunn was determined to find out who the leaker was, and so she did what any embattled chairperson would: she hired a private investigator to do some snooping.

Visit Vault at **www.vault.com** for insider company profiles, expert advice, career message boards, expert resume reviews, the Vault Job Board and more.

VAULT CAREER LIBRARY   229

The investigator discovered the source: longtime board member George A. Keyworth II. Dunn revealed his identity to the board, angering board member Thomas J. Perkins, a friend of Keyworth's. He resigned in protest over the investigation, and demanded that Hewlett-Packard reveal its unorthodox sleuthing methods.

Under Lockyer and key

Perkins had his wish, as HP made its spying methods public news that September. The private detectives had used a process known as pretexting, in which phone records are obtained by feeding (ostensibly private) pieces of information to phone companies, such as Social Security numbers and addresses, in return for activity records. The practice of pretexting is legally considered a form of identity theft, and soon Bill Lockyer, California's Attorney General, was looking into the matter. Other parties grew legally interested, as several journalists' phone records had been accessed by pretexting as well.

A few days after the dubiously acquired phone records became news, HPers were apologizing and resigning all around. Dunn announced that she would step down in January, to be succeeded by CEO Mark Hurd, while Keyworth, the source of the leak, said he was sorry and left the company. Apologies and resignations notwithstanding, the company soon faced legal activity on the matter. One month after HP had revealed the pretexting, Lockyer brought criminal charges against four people for using false pretenses to obtain information, unauthorized access of computer data, identity theft and conspiracy. The people charged were Dunn, the private detective she hired, an information broker who had obtained the phone records, and a former HP lawyer who advised Dunn.

The damage is Dunn

Prosecutors had trouble proving that there was felonious intent, however, since Dunn received incorrect legal advice that her actions were aboveboard, and since she did not stand to gain financially from her actions—her only purpose was to expose the boardroom leaks. All charges against Dunn were dropped in March 2007. Her co-defendants were also released, after agreeing not to contest the remaining misdemeanor charge of fraudulent wire communications, which would be dismissed after 96 hours of community service.

Settling up

In December 2006, HP announced it would pay $14.5 million—about what the company takes in every 80 minutes—to settle the civil lawsuit brought by the state of California. A portion of the money will be used to compensate the state for its legal fees, while another part will be used to pay fines. The remaining $13.5 million will be put into a fund to protect consumer privacy and media copyrights.

Of more pressing concern to the company, the settlement also affects HP's internal conduct until 2011, stipulating changes such as the hiring of an expert to review the company's investigations and board appointments. This settlement ties things up with the state authorities, but the feds—under the auspices of the SEC, FCC and the Justice Department—have yet to resolve their legal actions against the computer maker. Also, some of the journalists and media companies who were subject to the pretexting have stated that they plan on filing lawsuits against the company.

It was the best of times, it was the worst of times

Despite the scandal and intrigue, HP's revenue increased by 6 percent in 2006 to $91 billion, some $6 billion of which was profit, an increase of $3.8 billion over 2005. The financial figures also place HP as the world leader of the technology market, surpassing IBM. Every single business unit increased in profit margins within the fiscal year, and the software division, in particular, put in 23 percent growth year-over-year (the PC division grew by 9 percent).

But in the midst of such success, industry analysts are watching the board closely for continued fallout from the pretexting scandal. CEO Hurd has admitted that HP is still looking to shake up its directorial management, including the replacement of some directors and sending others to ethics training. Dunn and Perkins, both no longer with the company, have recently sniped at each other in press outlets such as The New Yorker, keeping the scandal in the public eye.

Not over by a long shot

Also, its shareholders are not content to rest after HP's 2006 performance, and have impelled the company to seek out ever larger growth opportunities as it keeps expanding. Therefore, HP is reducing its expenditures and moving into new business areas. The company began a restructuring program in late 2005 that called for the elimination of over 15,000 jobs; 1,000 positions remain to be cut in fiscal 2007. HP is also moving to reduce its number of data centers from 85 to six, and has been closing or consolidating a number of other offices to reduce its real estate expenses.

The Gutenberg method

A company cannot prosper, however, by cuts alone. In 2007, HP began seeking other sources of revenue. Oddly enough for a company that specializes in hardware and software, ink is still one of HP's most profitable products. In order to peddle more of its black (and cyan, magenta and yellow) gold, HP has moved into the realm of commercial printing, where its products range from the standard printer-copier-fax-scanner for office use, to specialized devices for printing posters and books.

An established player in the consumer printer market, HP's move to manufacturing printers and copiers was a natural step. The company operates its printer division on a blade-and-razor business model, whereby printers are sold near cost, and ink cartridges make up the bulk of profits. In 2007 it acquired Tabblo, a company with technology that makes printing web pages easier, for those who prefer to do their web browsing like it's 1499, which should help sell more canisters. The company isn't just pinning its hopes on rivers of ink, however. Also in 2007 it opened Neoview, its data warehousing service, which allows companies to keep and analyze vast quantities of data.

GETTING HIRED

Compute your options at HP

Hewlett-Packard's careers site, located at h10055.www1.hp.com/jobsathp, provides information for new and experienced job seekers on company organization and culture, benefits and job openings.

HP is organized into five different groups: customer solutions (sales and customer service), technology solutions (servers and other equipment for large businesses), personal systems (consumer goods), imaging and printing (printers and scanners) and HP Labs, the R&D arm of the company. The site also offers a fairly extensive list of common positions in the company, including sales, business development and engineering. Job openings are searchable by location, function and keyword; to apply, candidates must first create a profile. HP also offers opportunities for interns and recent graduates.

In the U.S., HP employees are given a budget with which to purchase benefits. Options include a 401(k) with company match, stock purchase plan and a retirement medical savings account for medical expenses during the golden years. Employees can also select between various health plan offerings, wellness programs, adoption and tuition assistance, and discounts on HP merchandise.

OUR SURVEY SAYS

Submit your resume on an HP machine

Ready for your interview? An insider reports that "Questions are behavioral—for example: … describe a time when you had to deal with an aggressive negotiator who would not share necessary information with you? With follow-up questions like— would you change your approach in the future or how could you have improved that interaction?" "All I did was have a phone interview with the HR rep, then an in-person interview with my direct manager and two peers," says a source.

Chummy culture, and a lot of it

Culture at HP gets good marks, but many workers feel overwhelmed by the size of the company. "[We have the] ability to work from home, flexible hours, benefits upon start date (which include matching pension, and flexible health plan, employee discounts, etc.) … and a very teamwork-oriented culture," sums up one respondent. "Culture is easygoing overall and the people are welcoming and friendly," says a co-worker. "[We have] strong senior management … a keen focus on the future and [know] exactly what the competitors are doing," boasts a sales contact. Others are less enthusiastic. One gripes about the "slow, bureaucratic processes. Of course, with a company the size of HP, bureaucracy is pervasive." "The culture of this large organization results in easy anonymity ... a sea of employees strolling long walkways in jeans and tennis shoes on their way to places hidden by mazes of walls," notes a former intern.

However, size has its advantages, too. "Diversity is really good," notes an insider. "Without question, one regularly sees employees from different cultures, ethnicities, and backgrounds," concurs a co-worker.

Hours are the source of many gripes. "60+ hour weeks are the order of the day," notes one overworked respondent. His co-worker agrees. "Long hours (50 to 55 hours), too many meetings if you are in the business units." Expect the daily grind to go on a while. "Opportunities for advancement are virtually nonexistent. With constant re-organizations, and downsizing, there is little room for career growth," a

Visit Vault at **www.vault.com** for insider company profiles, expert advice, career message boards, expert resume reviews, the Vault Job Board and more.

VAULT CAREER LIBRARY 233

contact points out. "No clear paths for advancement seem to exist, and so much depends on reorganization," says another. "Political alliances are often the key to opportunity," advises an insider.

There are some perks, though. "Dress code is officially business casual, although in many offices you will likely see shorts, sneakers and T-shirts. Some habits die hard," notes a member of the finance department. "The Houston HP campus has a great gym with lunchtime classes and extended daytime operations. Most people, if they wanted to, can find the time to workout in the gym and take part in the available nutritional services and online healthy-living plans. The campus also has a three-mile enclosed, air-conditioned walkway with glass floor to ceiling walls that look out at extensive landscaping and beautiful trees. Many employees walk "the loop" during lunch or as a break in the afternoon. Also, while local restaurants are available in all the nearby strip malls, the campus itself has at least two cafeterias with pretty good food, from hand-tossed, made-to-order salads to traditional Tex-Mex and burger entrees," adds a source.

Hitachi, Ltd.

6-6, Marunouchi 1-chome
Chiyoda-ku
Tokyo 100-8280
Japan
Phone: 81-3-3258-1111
Fax: 81-3-3258-2375
www.hitachi.com

LOCATIONS

Tokyo (HQ)
Anaheim, CA • Atlanta, GA •
Basking Ridge, NJ • Bentonville, AK
• Birmingham, AL • Brisbane, CA •
Charlotte, NC • Chula Vista, CA •
Detroit, MI • Eatontown, NJ •
Elkridge, MD • Fremont, CA •
Houston, TX • Huntsville, AL •
Itasca, IL • Mountain View, CA •
Norcross, GA • Norman, OK •
Philadelphia • Reston, VA •
Rochester, MN • San Francisco, CA
• San Jose, CA • Santa Clara, CA •
Somerset, NJ • Tarrytown, NY •
Torrance, CA • Twinsburg, OH

THE STATS

Employer Type: Public Company
Stock Symbol: 6501
Stock Exchange: Tokyo
President & CEO: Kazuo Furukawa
2007 Employees: 384,444
2007 Revenue ($mil.): $89,915

KEY COMPETITORS

Fujitsu
Matsushita
Siemens
Toshiba

EMPLOYMENT CONTACT

www.hitachi.us/Apps/hitachicom/con
tent.jsp?page=index.html&&path=js
p/hitachi/aboutus/Jobs

Visit Vault at **www.vault.com** for insider company profiles, expert advice,
career message boards, expert resume reviews, the Vault Job Board and more.

VAULT CAREER LIBRARY   235

THE SCOOP

Microchips and bullet trains

Hitachi—its name means "rising sun"—is a major conglomerate in Japan, where it is involved in a wide variety of industries. Its interests include everything from silicon to copper wiring, cast iron and cast steel; electronics equipment and microchips; electric plants that derive power from nuclear, thermal or hydroelectric sources; elevators and escalators; and a raft of consumer products from cell phones to LCD displays to white goods. The company also provides logistical and financial services. Sometimes called the General Electric of Japan, the company is considered one of Japan's five main electrical/electronics companies, the other four of which are Fujitsu Limited, Toshiba Corporation, Mitsubishi Electric Corporation and NEC Corporation.

As evidenced by its 2006 financial reports, Hitachi is a versatile company. In 2006, it derived 23 percent of its revenue from its information and telecommunications systems division, 11 percent from electronics, 25 percent from power stations and industry, 12 percent from digital media and consumer products and 15 percent from electrical components. Remaining revenue was generated from its logistics and financial services operations.

The dawn of Hitachi

Hitachi's origins date back to the dawn of the 20th century, when Japan was undergoing a period of rapid industrialization. It was founded in 1910 as a repair shop for electrical equipment, but soon moved into manufacturing small electric motors. A 1915 order for turbines provided a breakthrough for the company, and they moved from the fans they were then making to bigger pursuits. Hitachi was incorporated in 1920 under the name Kabushiki Kaisha Hitachi Seisakusho, and four years later it was producing electric trains.

What don't they do?

The company moved into the manufacture of elevators and refrigerators during the 1930s, while the following decade saw an emphasis on heavier machinery, including hydroelectric generators and construction equipment. In 1949, the company had its IPO. In the 1950s, Hitachi added scientific pursuits to its diverse resume, turning out an award-winning electron microscope in 1958 and a computer in 1959.

The company built a nuclear reactor in 1961 and cars for a bullet train in 1964. Ten years later, the company opened a microchip manufacturing facility, and performed experiments with nuclear fusion six years after that. It should have come as no surprise when, by the end of the 1980s, Hitachi had produced a robot. In 1993, Hitachi came out with a new bullet train, with speeds topping out at 170 mph.

A committed company

During the 1970s, Hitachi was one of the many Japanese electrical companies trailing America's IBM, the market leader. The Japanese Ministry of International Trade, desiring better results, sponsored a research and development effort to compete with America. Hitachi benefited greatly from the arrangement when they developed "plug compatible mainframes," a cheap model of computers compatible with IBM's technology; they became a big seller for the company. Later in the 1970s, however, the Japanese economy reeled during the 1974 OPEC oil crisis, and Hitachi executives drastically cut costs, including a 15 percent decrease of their own salaries.

Hitachi's zealous competition with IBM led to legal trouble in 1982, when the company, along with 11 employees, was indicted for commercial bribery and theft. Apparently, Hitachi was so obsessed with not losing ground to IBM that it had been stealing confidential trade secrets from the American company. The initial legal penalties were quite light—the company was fined $24,000, and only two employees saw any jail time. However, the public fallout in America was disastrous. Hitachi had just launched a huge marketing campaign overseas, and millions of orders were cancelled after the bad publicity from the trial. Also, IBM won $24 million in annual civil penalties for the next eight years against Hitachi and the right to inspect its software for the next five years.

Too much of a good thing?

After Hitachi resolved these legal matters, it woke up to stagnant sales and a worldwide lack of brand recognition. The company had historically concentrated in mature or cyclical industries, such as industrial equipment and semiconductor operations, which were excellent opportunities for research and development, but not profits. Also, Hitachi had never possessed a good marketing division and sold many of its products under different names abroad. Therefore, consumers didn't recognize the Hitachi name, except perhaps from negative coverage in newspapers of the day.

Under president Katsushige Mita, Hitachi began instituting cost-cutting measures such as increased automation and shifting production overseas to countries with

Visit Vault at **www.vault.com** for insider company profiles, expert advice, career message boards, expert resume reviews, the Vault Job Board and more.

VAULT CAREER LIBRARY   237

lower labor costs. Participation in the Very Large Scale Integration Project, another business initiative sponsored by the government, which fostered cooperation between large Japanese companies, helped Hitachi stay at the vanguard of semiconductor development. The company ranked as the top patent holder in Japan by the early 1990s and was even close to holding that title in America.

Lumbering giant

The 1980s and 1990s have been tough on Hitachi, as the company was forced to weather another period of stagnant and even declining sales after the Japanese economic bubble burst in the late 1980s. Its profits dropped 71 percent from 1991 to 1994. Simply put, the company was still focused on mature industries that put a premium on engineering and not on maximizing sales, and the Japanese economy was not helping.

The Asian economy went into crisis in 1997, and Hitachi was eventually hit very hard, posting a loss of $3 billion in 1999. The company spent several of the following years cutting jobs and restructuring operations to put itself on a firmer financial footing. Despite these efforts, the company posted a loss again in 2002. In recent years, Hitachi has been attempting with more urgency to streamline its operations and divest its less profitable sectors and is pursuing opportunities in more profitable areas such as data storage, hard discs and car parts.

Profit driven

In order to boost its position in the car electronics market, Hitachi acquired Clarion, a manufacturer of such devices, for $465 million in 2006. Clarion's offerings, which are primarily of the navigation system and audio variety, complement Hitachi's products in this area, which regulate functions like braking and engine emissions. The division is threatened by car manufacturers developing their own electronics systems; despite this, Hitachi hopes to increase the division's revenue by 50 percent by 2011.

Getting back in black

The year 2006 gave Hitachi reason to be financially optimistic. The company's revenue for the year was $80 billion and profits were $300. But the company's hard drive division is threatening the bottom line for 2007 and beyond—it lost over $300 million early in 2007, stemming from falling prices and competition from hardier, sleeker flash memory.

The company said that it was trying to bring the division back into the black. For the following fiscal year, it hopes to reduce costs by 22 percent by cutting more than 10 percent of its headcount and will increase production by 40 percent in order to boost profits. The division will also be moving the manufacture of its components to the Philippines from Mexico, in order to take advantage of lower labor costs.

Screen shot

About the time Hitachi announced these job cuts, it also announced the cancellation of an $800 million plasma display factory in Japan. While the plant was originally supposed to build more displays to leverage economies of scale, a glut of the devices in the market has depressed prices.

GETTING HIRED

Add some sun to your career

Hitachi provides a directory of its subsidiaries' careers sites at www.hitachi.us/Apps/hitachicom/content.jsp?page=index.html&&path=jsp/hitachi/aboutus/Jobs. Of particular interest will be Hitachi America, Ltd., (the headquarters in North America for parent Hitachi, Ltd.) Hitachi Computer Products America, which is involved in the production of printed circuit boards, Hitachi Software Engineering America, Renesas, a joint venture between Hitachi and Mitsubishi for the production of microchips, Hitachi Data Systems, Hitachi Electronic Devices and Hitachi Global Storage Technologies.

These divisions' careers sites provide information on job openings and benefits, which generally include health insurance, paid vacation and 401(k). Applicants can submit their resumes and cover letters to the indicated e-mail address on each site.

Hon Hai Manufacturing Co., Ltd.

Number 3-2 Chung-Shan Road
Tu-Cheng, Taipei
Taiwan
Phone: 886-2268-0970
Fax: 886-2268-7176
www.foxconn.com

LOCATIONS

Taipei, Taiwan (HQ)
Austin, TX
El Paso, TX
Fullerton, CA
Harrisburg, PA
Indianapolis, IN
Lexington, KY
Ontario, CA
Raleigh, NC
San Jose, CA
Santa Clara, CA

38 locations in about 15 countries
worldwide.

THE STATS

Employer Type: Public Company
Stock Symbol: 2317
Stock Exchange: Taiwan Exchange
Chairman & CEO: Terry Gou
2006 Employees: 200,000
2006 Revenue ($mil.): $10,381

KEY COMPETITORS

Flextronics
Sanmina-SCI
Solectron

EMPLOYMENT CONTACT

www.foxconn.com/location.html

THE SCOOP

The source for outsource

Taiwan-based Hon Hai Precision Industry is the major player in consumer electronics contract manufacturing sector; out-earning most of its closest competitors (and customers) by several billion dollars annually. The company makes parts for computers and communications devices (mainly cell phones) and other consumer electronics. Its areas of expertise include heat transfer and wireless technology. Currently, the company is exploring options in nanotechnology and environmentally-friendly manufacturing. Its customers include Apple, Motorola, Hewlett-Packard, Nokia and Sony.

Knobs and buttons

Hon Hai Plastics Corporation was born to make television parts in Taiwan in 1974, although it shortly changed its name to Hon Hai Precision Industry. In 1981, the company's founder Terry Gou (a/k/a Kuo Tai-ming) shifted the company's focus to computers and adopted the trade name Foxconn (although Hon Hai Precision Industry is the company's legal name; adding to the confusion, the company has two publicly traded subsidiaries, Foxconn International and Foxconn Technology Co.). The rechristened company's first order of business was to begin making plastic connectors for the rapidly-growing PC industry. In 1982, the company started fabricating wires to go between its connectors. Hon Hai went public in 1991; five years later, it moved into the manufacture of computer cases for such clients as HP and IBM. Although the company attracted criticism for its secrecy and lack of financial transparency, within another five years it was Taiwan's leading private sector computer company.

In 1998, Hon Hai Precision Industry appeared on *BusinessWeek*'s InfoTech 100 list for the first time, and Terry Gou was recognized as one of the world's 500-richest people. The firm made a number of acquisitions between 2003 and 2005 to boost its position as a major electronic parts manufacturer, while also acquiring many valuable personnel. It had 3,000 engineers and 100 PhD holders by the late 1990s, these staffers produced more than 5,000 patents by the turn of the 21st century.

Visit Vault at **www.vault.com** for insider company profiles, expert advice, career message boards, expert resume reviews, the Vault Job Board and more.

V∧ULT CAREER LIBRARY   **241**

Business is up

In 2006, Hon Hai Precision Industry had an excellent year, driven by strong demand for its products in the global market. It racked up $10 billion in revenue for the year, an increase of 63 percent over the year previous. Profits increased a healthy, if eye-popping 86 percent year over year. In 2006, the company merged with Premier Image, manufacturer of digital cameras, in a move gauged to improve the company's already vast offerings in that segment.

... and more is coming

In early 2007, Apple announced the impending arrival of the iPhone, its all-singing, all-dancing, wireless-Internet enabled cellphone/iPod combo. The device, which went on sale in June 2007, engendered demented levels of anticipation among Apple fans. Hon Hai Precision industry has the inside line on all this—literally; it manufactures the electronic guts of the iPhone, and is ramping up its production lines to make the 10 million devices that Apple expects to sell by the end of 2007.

Hon Hai Precision Industries came in at No. 4 on the 2007 InfoTech 100 list, ahead of customers Apple (No. 6) and Nokia (No. 17) and was ranked No. 145 on the Fortune Global 500.

GETTING HIRED

Out-Fox your career

Prepared to realize your precise ambitions? The company's careers site provides a list of job opportunities searchable by location. At press time, Hon Hai Precision Industries seemed to be searching for financial analysts, accountants and engineers who are fluent in Chinese. To apply, resumes can be sent to the e-mail address indicated in the job posting.

Ingram Micro Inc.

1600 East St. Andrew Place
Santa Ana, CA 92705
Phone: (714) 566-1000
Fax: (714) 566-7900
www.ingrammicro.com

LOCATIONS

Santa Ana, CA (HQ)
Buffalo, NY
Carlsbad, CA
Carol Stream, IL
Carrollton, TX
Jonestown, PA
Miami, FL
Millington, TN
Mira Loma, CA

108 distribution centers worldwide.

DEPARTMENTS

Accounting/Finance/Credit/Auditing
Administrative Support
Communications/Public Relations
Customer Service/Tech Support
Executive Management
Facilities/Maintenance
HR
IT/E-Commerce
International
Legal/Security
Marketing
Project Management/Program
Management
Purchasing/Product Management
Sales/Sales Operations
Warehouse/Operations/Logistics/
    Engineering

THE STATS

Employer Type: Public Company
Stock Symbol: IM
Stock Exchange: NYSE
Chairman: Dale R. Laurance
CEO: Gregory M. Spierkel
2006 Employees: 13,700
2006 Revenue ($mil.): $31,357

KEY COMPETITORS

Synnex
Tech Data

EMPLOYMENT CONTACT

www.ingrammicro.com/careers

Visit Vault at **www.vault.com** for insider company profiles, expert advice,
career message boards, expert resume reviews, the Vault Job Board and more.

VAULT CAREER LIBRARY   243

THE SCOOP

IT products for the whole family

Ingram Micro Inc. is the world's largest technology distributor and a leading technology sales, marketing and logistics company. As a vital link in the technology value chain, Ingram Micro creates sales and profitability opportunities for vendors and resellers through unique marketing programs, outsourced logistics services, technical support, financial services and product distribution.

It distributes more than 100,000 products ranging from PCs and networking equipment to consumer electronics or point-of-sale (POS) to about 160,000 resellers in 150 countries. The majority of Ingram Micro's sales comes from products made by Microsoft, IBM, Hewlett-Packard and Xerox. *Fortune* ranked the company as the top wholesaler on its 2007 list of Most Admired Companies, and No. 70 overall.

Ingram Micro is a member of the Global Technology Distribution Council. For more information about the GTDC, visit www.gtdc.org.

Distributor beginnings

Since its beginnings, Ingram Micro has connected technology solution providers with vendors worldwide, identifying markets and technologies that shape the IT industry. In 1979, Californian husband-and-wife team of Geza Czige and Lorraine Mecca founded Micro D. The company, located in Southern California, rapidly expanded nationwide and had a public offering in 1983. Ingram Industries became a majority stockholder of Micro D in February 1986 when it acquired all common stock held by the company's founders and acquired all remaining Micro D shares in March 1989. The combined company, which is based in Santa Ana, California, was renamed Ingram Micro at that point.

Avid AVAD fan

By 1993, Ingram Micro had expanded to Canada, Europe, Latin America and Asia through a series of acquisitions, and it became a publicly traded company three years later, listing its shares on the New York Stock Exchange.

In 2005, Ingram Micro made its foray into the world of consumer electronics by acquiring AVAD, a privately held alliance of 12 companies distributing high-tech home goods like LCD TVs and computers for media and gaming.

At your service

To insulate the company from future market slowdowns, Ingram Micro's management has been focusing on expanding the company's offerings in the realm of services, a strategy adopted by many tech firms in recent years. In 2006, Ingram Micro launched its North American services division in its Santa Ana location to provide Ingram Micro customers with warranty repairs, help desk services and IT staff.

An excellent year

Services aren't the only area into which Ingram Micro is expanding. In 2006, the company acquired SymTech Nordic, a provider of AIDC and RFID equipment, as part of its global strategy to expand into adjacent technologies. SymTech has a strong presence in Scandinavia and Northern Europe, and Ingram Micro hopes to realize gains from the acquisition, as it unites these customer relationships with its wide product line and distribution know-how.

In addition to its acquisitions that year, Ingram Micro posted record-setting revenue and profit. Sales jumped 9 percent over 2005, coming in at $31 billion, while profits increased to $265 million. If current market conditions persist, it is likely that good years are ahead for the company, and it hopes to see a sustained increase in revenue as tech consumption grows, and as people adopt Microsoft's more computer processor-intensive Vista operating system.

DBL your fun

In 2007, Ingram Micro expanded further in the consumer electronics sphere with its acquisition of DBL Distributing for $96 million. DBL Distributing is a $300 million-per-year concern specializing in the fast-growing market for consumer electronics. It primarily distributes about 17,000 products to independent retailers, and gives Ingram a leg up in the consumer market.

GETTING HIRED

Catalog your options at Ingram Micro

Aspiring employees can point their browsers to www.ingrammicro.com/careers, the online home of Ingram Micro careers. The site provides information about careers

Visit Vault at **www.vault.com** for insider company profiles, expert advice, career message boards, expert resume reviews, the Vault Job Board and more.

VAULT CAREER LIBRARY   245

in the 35 countries in which the company keeps offices. The company provides a searchable list of jobs at phx.corporate-ir.net/phoenix.zhtml?c=98566&p=irol-aboutIMCareers. Available positions can be sorted by function, location and keyword. To apply, candidates must first create a profile.

Intel Corporation

2200 Mission College Boulevard
Santa Clara, CA 95054
Phone: (408) 765-8080
Fax: (408) 765-3804
www.intel.com

LOCATIONS

Santa Clara, CA (HQ)
Austin, TX • Berkeley, CA •
Cambridge, MA • Chandler, AZ •
Chantilly, VA • Columbia, SC •
Dupont, WA • Folsom, CA • Fort
Collins, CO • Hillsboro, OR •
Hudson, MA • Pittsburgh, PA •
Raleigh, NC • Rio Rancho, NM •
Riverton, UT

Locations in 48 countries
worldwide.

DEPARTMENTS

Administration • Engineering •
Facilities & Site Services • Finance •
General • HR • IT • Legal •
Manufacturing • Marketing •
Materials/Planning/Purchasing •
Professional Services • Public
Affairs • Sales • Venture Capital

THE STATS

Employer Type: Public Company
Stock Symbol: INTC
Stock Exchange: Nasdaq
Chairman: Craig R. Barrett
President & CEO: Paul S. Otellini
2006 Employees: 94,100
2006 Revenue ($mil.): $35,382

KEY COMPETITORS

AMD
Samsung Electronics
Texas Instruments

EMPLOYMENT CONTACT

www.intel.com/jobs

Visit Vault at **www.vault.com** for insider company profiles, expert advice,
career message boards, expert resume reviews, the Vault Job Board and more.

VAULT CAREER LIBRARY   247

THE SCOOP

The best chips you'll never taste

Not only is Intel the world's leading manufacturer of semiconductors, it has also become one of the world's best recognized high-tech brands. For a company that sells the vast majority of its products to other businesses rather than to consumers, such widespread name recognition is unusual. But the company's marketing efforts have paid off, causing consumers to actively seek out products that advertise their Intel insides. Intel's relationship with Dell, its biggest customer and the world's second-largest PC retailer, has also cemented the company's hold on the market. Intel's products include processors and memory chips, for everything from cell phones to computers to routers and other networking hardware.

Birth of the microprocessor

Intel was founded in 1968 by three electrical engineers: Robert Noyce, Gordon Moore and Andy Grove, all former integral employees of Fairchild Semiconductor (Noyce and Moore were actually two of the eight Fairchild founders, in 1957). Noyce was already well known in the high-tech world as the co-inventor of the integrated circuit (IC)— just one of the reasons he'd earn the nickname "The Mayor of Silicon Valley." Moore would later become known for positing "Moore's Law"— the proposition that computing power would double roughly every two years. Grove, besides his brilliance as an engineer, is notable for his biography; he fled his native Hungary in 1956, in the midst of its thwarted revolution against Communist Russia. He would serve as Intel's CEO from 1987 to 1998, and as its president for even longer, from 1979 to 1997.

Intel, its name derived from the phrase "integrated electronics," set up shop in Mountain View, California, and began rolling out its first products: dynamic random access memory (DRAM) and erasable programmable read-only memory (EPROM) memory chips. The success of the fledgling company's memory products bankrolled research into new and more complicated semiconductors. Intel unveiled the world's first microprocessor, the 4004, in 1971 (the same year it had its first public offering). This new device was originally intended to power the Busicom scientific calculator. But Intel's processor chips had become powerful enough by the 1980s to spawn a new industry: the personal computer. Intel introduced the 8086 chip in 1978, and its cousin, the 8088, soon thereafter; IBM chose the 8088 to power its first PC in 1981.

86 times two, three, four

Intel introduced its next-gen 286 processor in 1982. It was the first processor developed by Intel that was fully compatible with software written for older chips. This backward-compatibility, along with Intel's decision to license the design to other semiconductor manufacturers, helped make the x86 architecture (which had first appeared in 1978 with the 8086 chip) an industry standard. By the time the company released its 386 processor in 1985, Intel had become the clear industry leader in tech innovation.

Intel and PC—a happy union

By 1989, Intel realized that its backward-compatible chips were too successful—customers could use its old chips on new technology without too much trouble. Of course, they wouldn't be as powerful or effective as the new chips Intel was always reeling out, so, with some advertising, the company wedded consumers to its own pursuit of progressive technology. It found the perfect growth partner in PCs; no other tech sector was growing so fast, nor better represented the future of technology.

The marketing was simple: computers work best when they work fast, they work fast with Intel chips. Intel's first ad campaign strongly urged customers to switch from the 286 to the 386, implying the older chip was obsolete. In 1991, the company began to stress the importance of its tiny products for the effectiveness of computer science, introducing the "Intel inside" logo. All the marketing was evidently a success—Intel's chips had overtaken 85 percent of the PC market by 1994.

PC or bust!

Inside the company, the focus on PCs was utterly dominant. A 2006 *BusinessWeek* profile declared that "engineers ruled the roost," and that "anyone not producing for the core PC business was considered a second-class citizen." The profile also illuminated some of the company's culture, stating that under CEO Grove, from 1987 to 1998, the company was a "rough-and-tumble place." Grove's motto was "Only the paranoid survive" and "managers frequently engaged in 'constructive confrontation,' which any outsider would call shouting." All was well while the PC rode high, but as early as the late 1990s, Grove appeared to recognize this could not last.

At this point, Intel started varying its offerings beyond PCs—in 1999 it unveiled prototype versions of its 64-bit Itanium processor, engineered to power enterprise-class high-end servers. However, the product was beset with manufacturing problems, and the full-scale Itanium rollout did not occur until 2001. Intel looked even further afield in 2000, when it began offering Intel-branded consumer electronics such as toys, digital cameras and Internet radios, but the line of products was discontinued in 2001. Company revenue continued to grow steadily through the 2000s—reaching a high of $38.8 billion in 2005—but profits stagnated, falling from their $10 billion peak in 2000 to $5 billion in 2006.

Andy Grove, in a 1999 interview with CNN, recognized the need to shift away from the company's PC-reliance. He called the current period a "strategic inflection point" for Intel, and admitted that although PC companies will remain Intel's primary customers, the PC industry "will be a subindustry." He then conceded that "if [Intel hangs] on too strongly to [its] old business model and [does] not make investments in the emerging part of the business, [it] will be on the periphery."

Time for a facelift

In the spring of 2006, Intel named Paul Otellini its new CEO—the first non-Ph.D. and non-engineer to run the company—and he embarked upon the revision of Intel's business model. His efforts have included an overhaul of the company's logo—which hadn't appreciably changed in three decades—and a new motto, "leap ahead," which replaced "Intel inside." His management style is vastly different from Grove's; his motto is "Praise in public, criticize in private," a far cry from anything confrontational.

But one key facet of Otellini's approach is rubbing some old Intelers the wrong way: marketing. As one of his first moves, Otellini hired Eric B. Kim away from Samsung, whose image he had done wonders to renovate from peddler of B-grade VCRs to maker of flashy cell phones. Kim has been charged with doing the same for Intel as its new chief marketing officer, and his solution has been to break away from the PC and break up engineer dominance from within company ranks.

Intel unrest

BusinessWeek says that "many high-level engineers working on PC products feel they've been stripped of their star status." The magazine quotes one employee: "The desktop group used to rule the company, and we liked it that way." Another engineer said, "There definitely are people who are highly skeptical (of the new approach), who think this is all fluff, all just gloss—that if you make good technology, you don't need the glitz."

Such engineers were surely even less happy with following developments—Intel's 2006 earnings reports revealed that the company's revenue fell by $3.5 billion year over year, and the company embarked on two rounds of job cuts. The first pink-slipped 1,000 managers in July 2006; the second cut its workforce by 10,500 jobs that September—10 percent of the workforce—and slashed costs by about $5 billion. The company assured investors that these spending reductions did not threaten major projects then underway to build factories for the new generation of 45-nanometer chips. Industry pundits agreed, saying the move was necessary for the relatively unwieldy Intel to remain competitive with its chief rival, the more lithesome Advanced Micro Devices.

Chips off the new block

In the new Intel, gone will be brand names such as the aged Pentium, in will be ones like … Viiv (rhymes with "alive"). The Viiv is still designed for PCs, but it hopes to bring them into the world of multimedia—basically like a TiVo system on a chip, it provides 7.1 surround sound, Intel Clear Video Technology and amped-up data storage (meaning that it can download movies and TV shows onto one's computer).

Another new chip is the Centrino, which Otellini conceived as a cooler and less power-consuming chip than the Pentium, as he wanted to decrease on the level of heat that laptops generate while running. It was ready for production in 2002 when Otellini recognized a further improvement—he wanted to wait until the company could add a wireless component to the chip as well. He told *Forbes* that many engineers were unused to factoring in such considerations while making microchips, and that, "Making that decision was tumultuous inside of Intel, to say the least. It was a cultural issue. We're a microprocessor company."

But Otellini and Kim want to move Intel beyond being just a microprocessor company and have identified a few tech allies to cultivate who have similarly thought outside the box. Apple is one of them, and Otellini has gone to lengths to cultivate Steve Jobs, much to the dismay of traditional Intel-customers Microsoft and Dell. Intel got its Apple wish in 2006, when Jobs announced at an annual shareholders meeting that a future line of the Macintosh computer would adopting the Centrino. He raved over the new chip, calling it "phenomenal—it blows away everything other suppliers have, including our former suppliers," and he added that Intel's future products are "the best I've ever seen in my life."

Try our chips, now better than ever!

Another tech friend solicited by Intel is Google, and Otellini designed another chip venture with it specifically in mind. In contacts with the company, he found out that Google's energy costs now exceed its equipment costs—the electricity charge for its servers (where it stores all of that search info) is greater than what it spends on new tech.

By March 2007, Otellini had produced a chip upgrade for the job at hand. The chip's new architecture will, for the first time, incorporate memory controller circuitry onto a regular microprocessor chip. This will enhance processing speed and enable it to conduct four times the number of chores that Intel's top chip is currently capable of handling. Furthermore, the reduced requirement for separate chips within a computer—which add to power requirements—will result in a smaller electricity bill.

The graphics-enabled microprocessors will also be smaller, since production will capitalize on etching technology that allows for circuits as narrow as 45 billionths of an inch. A forthcoming line of chips, called Penryn, will also feature the skinnier look. These chip upgrades also serve the purpose of ratcheting up pressure on Intel's main rival: AMD, which purchased graphics specialist ATI Technologies for the sum of $5.4 billion in 2006, and was able to amass a 25 percent market share by introducing a chip with similar attributes several years ago.

Chips served in China

In March 2007, Intel revealed plans to build a computer chip manufacturing plant in China. The new factory in Dalian, which will cost $2.5 billion, will be the company's largest investment in China to date; it has 16 locations in the country so far. The factory will produce 300-millimeter silicon wafers (the largest wafers in commercial production) and chips using 90-nanometer technology (that is, circuits on the chips are 90 billionths of a meter in width). When the plant opens in 2010, it will serve customers throughout China, which is projected to become the world's largest information technology market.

Attitude adjustment

Right now, Intel is still in the midst of restructuring—Otellini's unorthodox (or un-Intel) ideas have born some fruit, but analysts suggest that more layoffs could come at any time. But more than financial success, the company is looking to foster a new identity and approach to its business, perhaps finally accepting that the days when it lorded over the PC industry are over. Henri Richard, chief of sales at main rival

AMD, told *Forbes* that "Intel talks about being customer-centric, but it's not in their DNA. They've been brought up to rule the world. Intel tells the customer: 'This is the way it's going to be.'"

But Otellini's new regime is already drawing some good reviews. Mike Abary, vice president of Sony, has said, "I have seen more flexibility (from Intel), more of an open mindset than in years past. They realize that times have changed, that they don't have the answers. So it has been much more collaborative working with them." Others are taking a wait-and-see approach, as Intel is still an enormous company, with many engineers left over from old PC-centric fiefdoms. Russ Meyer, chief strategy officer for branding consultancy Landor Associates, has described Intel's situation thus: "In many ways, it's like trying to change the engines on an airplane when you're flying it." If he's right, it's time for everyone at Intel to strap in and hold on for the ride.

GETTING HIRED

Find your career inside

Intel's careers site, located at www.intel.com/jobs, provides information about job openings, student opportunities and benefits. Open positions can be searched by function, location and keyword. To apply for a job, candidates must first create a profile. There are also explanations of an assortment of career paths at the company.

Positions are available in 16 departments: integrated circuit engineering, integrated circuit manufacturing, software engineering, facilities, hardware engineering, hardware manufacturing, finance, research and development, IT and services, HR, sales and marketing, legal, materials, supply network and Intel capital, the company's venture capital arm.

Opportunities at Intel for students include internships and entry-level rotations in engineering, finance and marketing. Intel recruits at a number of colleges throughout the U.S., as well as seeking hires at diversity fairs like the Society of Hispanic Engineers (SHPE), the National Society of Black Engineers (NSBE), the National Black MBA Association (NBMBAA) and Women in Technology International (WITI). Intel also holds virtual recruiting events on its web site; to participate, candidates must first submit their resumes to the company. Intel generally screens applicants by phone before they are invited to an office location for a face-to-face interview.

Intel's U.S. benefits include stock for new hires and recently promoted workers, an employee stock purchase plan, bonuses, choice of health and dental plans, 401(k) with company contribution, on-site exercise facilities and flexible schedules. The company also offers tuition reimbursement for career-related courses and internal training. After seven years of employment, employees are eligible for an eight-week paid sabbatical.

Intermec Inc.

6001 36th Avenue West
Everett, WA 98203
Phone: (425) 348-2600
Fax: (425) 355-9551
www.intermec.com

LOCATIONS

Everett, WA (HQ)
Alexandria, VA • Alpharetta, GA •
Arlington, WA • Bridgeville, PA •
Cedar Rapids, IA • Charlotte, NC •
Cincinnati, OH • Cleveland, OH •
Collierville, TN • Dayton, OH •
Denver, CO • Fairfield, NJ • Fort
Washington, PA • Indianapolis, IN •
Kansas City, MO • Kennesaw, GA •
Las Vegas, NV • Minneapolis, MN •
Raleigh, NC • Richardson, TX •
Rolling Meadows, IL • Southfield,
MI • St. Louis, MO • Sunrise, FL •
Tempe, AZ • Vancouver, WA

DEPARTMENTS

Administrative
Customer Service
Finance & Accounting
HR
IT
Operations
Research & Development
Sales & Marketing

THE STATS

Employer Type: Public Company
Stock Symbol: IN
Stock Exchange: NYSE
Chairman: Allen J. Laurer
President & CEO: Patrick J. Byrne
2006 Employees: 2,407
2006 Revenue ($mil.): $850

KEY COMPETITORS

Casio Computer
Fujitsu
Motorola

EMPLOYMENT CONTACT

www.intermec.com/eprise/main/Interm
ec/Content/About/WorkAtIntermec

Visit Vault at **www.vault.com** for insider company profiles, expert advice,
career message boards, expert resume reviews, the Vault Job Board and more.

VAULT CAREER LIBRARY   255

THE SCOOP

A vital link in the supply chain

Intermec helps companies get the right stuff to the right people at the right time. The company's current name is a shortening of its original moniker Interface Mechanisms, but it has always specialized in exactly what its first name specified—computerized mechanisms that are accessible to each other through interfaces (i.e., specific entry points).

Its line of portable computers, automatic identification and data collection (AIDC) labels and readers (e.g., bar codes and bar code scanners) and radio frequency identification (RFID) products let companies know where their products are in the shipping process. Intermec's technology is used to organize everything from doughnuts to retirement plans, as its list of customers includes Krispy Kreme, NASA, the New York City Department of Sanitation and the Social Security Administration.

Name game

Intermec was founded way back in 1966 as Interface Mechanisms, and soon began contributing to bar code technology and other interface products. By 1972, the company had invented the first computerized cash register (the now-ubiquitous barcode scanner used in such registers is a perfect example of Intermec's products). Throughout the 1970s the company created Codes 39 and 11, still widely used in bar codes and telecommunications today. It shortened its name to the current Intermec in 1982, and continued working in the same field, making advances in "Smart Battery" technology and WLAN (wireless local area network) systems, among others.

In 1991, the Navy contractor Litton Industries, a large company with roots in the Cold War era, acquired the company. Three years later Litton split itself into military and commercial interests, and spun off the latter as Western Atlas; Intermec, along with some other companies related to factory and oilfield services, was included in this restructuring. Western Atlas made three acquisitions of other companies involved in AIDC and bar code technology in 1997, boosting Intermec's position over rival Symbol, whom it was already leading in the industry. But Western Atlas had more than recouped its investment by this time, nearly doubling its revenue. The same year, it spun off Intermec, which had been renamed Unova, as a separate firm. Intermec missed its old name, though, and officially reverted in 2006.

Tweaking the enterprise

The company is not immune to occasionally decreasing profits and restructuring, though. In 2006, Intermec reported revenue of $850 million, down 2 percent compared to 2005, and profits of $32 million, down nearly $30 million from the year before. It subsequently cut 65 jobs in March 2007, completely closing its design offices in Sweden. A few months later, the company cut 200 more jobs, or about 9 percent of its headcount, to save about $23 million per year. In 2007, Intermec's CEO, Larry Brady announced he would be retiring once the company chose his successor, citing no other factors besides reaching a customary retiring age.

RFID ... in spaaace

In 2007, NASA and Intermec said they were working to develop RFID tags that could withstand the rigors of space travel. NASA hopes to use RFID technology to keep track of parts and experiments in space. As such, Intermec sent several different tags, designed to handle harsh industrial environments, to be exposed to conditions on the outside of the International Space Station. The tags were exposed to ambient conditions outside the station, like extreme temperature fluctuations, lashings of solar radiation and charged solar wind particles, and bombardment with meteorites and other space debris. Once the tags return to Earth, Intermec will test them to ensure that they have survived their ordeal.

Change at the top

The Intermec board of directors elected Patrick Byrne as its new president and CEO in July 2007, only four months after former CEO Larry Brady announced his retirement. Byrne has a background in engineering, working for 24 years with Hewlett-Packard and Agilent Technologies, and he sits on the board of Auburn University's Samuel Ginn College of Engineering. The board also used the occasion to name a new chairman, selecting Allen Laurer, a director with the company since 2004.

GETTING HIRED

Integrate your career with Intermec

Intermec's careers site, at www.intermec.com/about_us/careers/index.aspx, provides information on company ethics and job openings. Jobs are searchable by location and keyword. To apply, interested parties must fill out a web form and upload their resume. At press time, several open positions were available in engineering, business analysis and administration.

International Business Machines Corporation

New Orchard Road
Armonk, NY 10504
Phone: (914) 499-1900
Fax: (914) 765-7382
www.ibm.com

LOCATIONS

Armonk, NY (HQ)
Atlanta, GA • Austin, TX •
Beaverton, OR • Bethesda, MD •
Boulder, CO • Burlington, VT •
Cambridge, MA • Chicago, IL •
Dallas, TX • East Fishkill, NY •
Endicott, NY • Kirkland, WA •
Lexington, MA • North Reading, MA
• Pittsburgh, PA • Poughkeepsie,
NY • Raleigh, NC • Rochester, MN •
San Francisco, CA • San Jose, CA •
Somers, NY • Tucson, AZ •
Waltham, MA • Wetford, MA

More than 1,000 locations
throughout the US. Offices in more
than 170 countries throughout
Europe, Africa, the Middle East,
Asia and Latin America.

DEPARTMENTS

Consulting & Services • Corporate
Operations • Engineering • IT •
Research & Development • Sales &
Marketing • Software • Systems &
Technology

THE STATS

Employer Type: Public Company
Stock Symbol: IBM
Stock Exchange: NYSE
Chairman & CEO: Samuel J.
  Palmisano
2006 Employees: 355,766
2006 Revenue ($mil.): $91,424

KEY COMPETITORS

Accenture
Hewlett-Packard
Microsoft

EMPLOYMENT CONTACT

www.ibm.com/employment

THE SCOOP

Big Blue

IBM is a dominant player in the tech industry. For many years, it was a provider of computer hardware, but as that market has become increasingly low margin, IBM has moved into the more lucrative areas of providing IT services, business consulting and business software under the Lotus, Tivoli, DB2 Rational and WebSphere names. It is also expanding into video game chip design and nanotechnology. It also supplies servers, supercomputers and outsourcing. A most prolific patent filer, its efforts were rewarded with 3,621 in 2006 alone, the most awarded to any company, a distinction it has maintained continuously since 1992.

A titan of American industry for much of the 20th century, the company has also been rewarded with a great amount of devotion over the years; at one point, company employees devoted themselves to writing songs about the company's noble pursuit of technological breakthroughs—"Ever Onward" was the most famous refrain. (If you want to check out these songs, they're posted online at www.digibarn.com/collections/songs/ibm-songs/index.html.)

Most recently, IBM was named the top tech company on *BusinessWeek*'s 2007 Best Companies to Launch a Career survey—ahead of Google, Microsoft and others. The magazine cited IBM for its employee-friendly programs, such as flexible work schedules and bonuses to work out at the gym, and for the way IBMers can collaborate through massive online brainstorming sessions called Jams (the last one involved 150,000 people). IBM is also using hot technologies, such as virtual worlds, to solve pressing issues that affect the planet, like finding ways to make drinking water more available in developing countries.

"Think"

IBM began as the Computing-Tabulating-Recording Company, a soon-floundering office machine firm. Thomas Watson, an executive from rival NCR, turned the company around by boosting sales and securing government contracts during World War I. He took over the company in 1914, and annual revenue had tripled by 1920. One of Watson's favorite mantras was "think," and the company adopted the moniker as its official slogan four years later; around the same time, it quickly established dominance in the office machine market, selling its tabulators, clocks and electric typewriters domestically and abroad.

The government and IBM worked well together, as IBM machines were chosen in 1935 to handle records for the newly instituted Social Security program. Two years later, the company invented a device that would terrorize students for decades to come—the test scoring machine, a device that could grade fill-in-the-bubble answers on standardized exams. IBM later proved invaluable to the allied forces during World War II; its factories made bomb sights and other weapons parts, and the company developed the Mark I, one of the first computers used to calculate the trajectories of explosives. In the meantime, the government also employed IBM's computational horsepower to crunch the numbers for the Manhattan Project.

The computer pioneer

IBM's inventions in the following decade eventually proved critical to the development of computers. In 1952, the company introduced the 701, one of the first computers with vacuum tubes, and by 1959 had an improved model that used transistors instead of the bulky, fragile tubes. (This latter invention was such a breakthrough it shocked other countries' computer industries into motion, notably that of Japan.) That decade, IBM also began to sell RAMAC, one of the first computers to use magnetic discs to store data, as well as FORTRAN, a programming language.

In addition to IBM's emphasis on customer service, these developments ensured that the company maintained control of about 80 percent of the mainframe market throughout the 1960s and 1970s. Other notable advancements in the 1970s included the floppy disc (1971) and an early version of the grocery-store laser scanner (1973). The mainframe also gave IBM its "Big Blue" nickname, as one of its famous models was large and ceruluean-shaded.

The famous missed opportunity

In the 1980s, however, the personal computer revolution blindsided IBM. At the time, IBMers were known for their adherence to classic engineer stereotypes, marked by conformity (down to a restrictive dress code), stagnation (employees were effectively guaranteed lifetime employment) and studying issues into the ground. In hindsight, industry experts wonder what could have been if IBM's computer culture had embraced the new PC technology more wholeheartedly, suggesting it could still control 80 percent of the market. But the culture was engineer-driven, and not as consumer-friendly as Apple and Microsoft would quickly prove to be. Mainframe

Visit Vault at **www.vault.com** for insider company profiles, expert advice, career message boards, expert resume reviews, the Vault Job Board and more.

VAULT CAREER LIBRARY   261

computer science was the more complex, scientific work at the time, and IBM stuck with what it knew. Simply put, it bet on the wrong horse.

IBM was easily able to produce its own PCs when it tried. After a slew of PCs launched by Digital Equipment Corp. and other competitors, it launched the (creatively-named) PC in 1981. Despite its late start in this market, the company had a banner year in 1985, with profits of $6.6 billion, making it the single most profitable company in the world. But in the fledgling PC market, Apple and a small Washington state-based outfit called Microsoft soon surpassed IBM, the latter of which IBM had hired to program its 1981 PC. IBM missed a crucial detail in its negotiations with Microsoft, who retained the licensing rights to MS-DOS, the operating system behind PC's success. Mainframe sales vastly outweighed those of PCs at the time, but PCs took off in the late 1980s, and the Microsoft-owned OS was the key technology.

Mainframe by the wayside

The 1990s were unkind to Big Blue, as mainframe sales would eventually drop by 90 percent, taking IBM's core business focus with them. In 1993, IBM announced, no doubt with reluctance, that it had suffered a $4.97 billion dollar loss for fiscal 1992, the largest corporate loss ever at that time.

The company began massive layoffs. From a peak of 405,000 employees in 1985, IBM dwindled to about 220,000 employees in 1994. The company also controversially restructured its pension plan, one of the first big companies to do so. Enraged employees sued the company and eventually won a $320 million settlement, although IBM successfully appealed to the Supreme Court in January 2007.

Making the elephant dance

In order to revive the flagging behemoth, nontechnical executive Lou Gerstner was lured from RJR Nabisco to lead the flailing IBM, joining the company as chairman and CEO in 1993 on April Fools' Day. Gerstner was a company outsider, but he proved successful in changing the corporate culture largely responsible for its fall from dominance. When it was taken over, IBM was planning to break up into "Little Blue" independent units, but Gerstner stopped the idea in its tracks, trusting in the still-strong IBM brand and the importance of a synergistic technology-driven organization that could deliver complete tech services to clients. The move was exemplary of his greater management approach, which emphasized breaking down intra-company competition between different departments, and focusing on the completion of tasks. Out was the old IBM habit of endless research and studying and in was the simplifying and finishing of projects. The changes showed results by 1998, when IBM's revenue again started to grow.

Gerstner also moved IBM toward a consulting model in 2002, purchasing the consulting arm of PricewaterhouseCoopers for $3.5 billion, outsourcing HR and selling the company's hard drive disk business to Hitachi. Current President and CEO Sam Palmisano succeeded Gerstner in January 2003 (he had previously been IBM president and COO). Then, in a final break with the past, IBM sold off its PC division to Lenovo in 2005—ending its long and tortured relationship with the PC market. Many of the changes Gerstner instituted at IBM brought the company back to profitability, but were ones an in-house executive could probably not have stomached; the more profitable services-based model Gerstner embraced is a far cry from the company's engineering legacy. (For more on Gerstner and IBM's turnaround, you can check out his 2002 book on the subject, *Who Says Elephants Can't Dance?*

Margin call

In 2006, IBM took in revenue of $91 billion, with a profit of $9 billion. Revenue increased $290 million year-over-year, and profits increased by $1.5 billion, marking IBM's eighth year of sustained revenue growth. The company is ever more aggressively moving into services and consulting to plump its margins. It's also been hiring in India, building up 53,000 workers as of 2007 in an attempt to beat competitors like Infosys and Wipro on their home turf. Ultimately, however, IBM's goal is to automate activities like data center management, to drive labor costs even lower and enable tech workers to spend less time on tedious tasks and more time working with clients and on R&D projects

You're not forgotten, engineering!

IBM is not forsaking its history in software, though: it spent nearly $4 billion acquiring small, high-margin technology companies in 2006 (the company has spent $11.8 billion on acquisitions overall since 2003). Notable purchases included CIMS, which provides technology resource allocation metrics, in January; Internet Security Systems, which provides just that, for $1.3 billion in August; and MRO, a software company specializing in consulting and tracking company property, in October. Then, in December, IBM bought Netherlands-based Consul, a compliance enforcement software company. IBM continued its purchasing of specialized tech

Visit Vault at **www.vault.com** for insider company profiles, expert advice, career message boards, expert resume reviews, the Vault Job Board and more.

V/\ULT CAREER LIBRARY   **263**

firms in 2007 with Softek in January, whose products help companies move data among servers, and Telelogic in June, another maker of security software.

The plan appears to be working: IBM announced in July that its software division's revenue had grown 13 percent from the previous quarter, to $4.8 billion. Together with the now impressive returns on its services offerings, the company is strong enough to rank at No. 21 on *BusinessWeek*'s 2007 edition of the InfoTech 100.

IBM also continues to make its computer engineering forefathers proud, as it recently announced a dramatic new advance in manufacturing microchips. In February 2007 the company touched off alarm bells at Intel when it announced the development of the fastest computer memory in history. At 10 times the performance level of current PCs, the breakthrough is loaded with potential for the computer industry, and raises the ante for Intel to develop equally fast technology. Apparently, the old master still has chops.

Cost-cutting everywhere!

It isn't all growth for IBM, though. The company announced the sale of its printer division, whose revenue has been steadily dropping since 2003, to Ricoh in 2007. Ricoh will pay $725 million for a controlling interest in the division, and acquire the outstanding portion by 2009. In March of the same year, IBM announced 1,570 job cuts to the technology unit, mostly in North America, after the division's profits fell 19 percent within the first quarter.

Two months later, the company announced a more unorthodox cost-cutting move: sponsoring a research project to make computer data centers more energy efficient. (The project won't cut costs in the short term, as it will require $1 billion in annual funding).

On to India

In 2006, IBM committed itself to an almost $6 billion investment in its Indian operations over the next three years; it plans to employ some 60,000 people in India by the end of in 2007. The company even went so far as to hold its annual investor conference in Bangalore in June 2006. It sees India as not just a source of inexpensive labor but part of what CEO Palmisano calls a "globally integrated enterprise" where sophisticated functions, such as product design, will be dispersed worldwide (i.e., in India, China, Brazil, Russia and other growth markets).

IBM currently has more of its resources invested in India than any other country besides America, and continues to expand its research labs and computing hubs in dozens of countries around the world. Aside from its massive Indian investments, IBM is also building its first research lab in Russia, which will employ 200 staffers by the end of 2008.

Patently disputed

In October 2006, IBM, holder of the largest patent storehouse in corporate America, filed two lawsuits against Amazon.com, alleging that the online retailer had used patented IBM technology in its rise to the top of the e-tail world. The suits, filed in Texas for unspecified damages, could amount to hundreds of millions of dollars, according to an IBM spokesperson. "These patents are core to modern electronic commerce," said John E. Kelly III, IBM's senior VP for technology and intellectual property. "Most, if not all of Amazon's business, is built on top of this technology."

IBM first sought compensation from Amazon in 2002, but the two firms could not come to terms. The five patents in question involve software-based methods for storing, displaying and receiving information on e-commerce web sites used to make buying recommendations and manage online transactions.

Look out, MySpace

IBM made a splash in the social networking sector with the January 2007 announcement of its Lotus Connections software. The program, intended for the business world, is designed to provide corporate clients with the kind of virtual world that other services (like MySpace and Facebook) have heretofore provided.

Lotus Connections is an entire social networking system set up to encourage intra-office relationships (former CEO Louis Gerstner, who stressed this sort of collaboration, would be proud). Its components, in addition to a blog, include activities, notes pages (called "dogears"), individual profiles and community pages. Moreover, the connections software, like Lotus Notes, will compete with Microsoft, which is releasing its own social networking system. Ray Ozzie, Microsoft's current chief software architect (and one of the heirs apparent to Bill Gates), should be familiar with Lotus Notes—he did create the technology, after all. It should be interesting to see the architect take on his old creation.

Visit Vault at **www.vault.com** for insider company profiles, expert advice, career message boards, expert resume reviews, the Vault Job Board and more.

VAULT CAREER LIBRARY   265

GETTING HIRED

Dive into the Blue

During a typical year, IBM hires more than 15,000 new employees in everything from accounting and data warehousing to Internet applications and software development. Hopeful IBM-ers should consult IBM's careers page (at www.ibm.com/employment), which provides information on job openings, career development and benefits.

Not surprisingly, IBM scans in the resumes it receives, and submissions should be in plain text format. Job openings are searchable by location, necessary degree, level and type. To apply, candidates must fill out a web form and upload a resume.

IBM's benefits include a bonus program, choice of fully-paid health and dental plans, flexible schedules, generous vacation, 401(k) and discount stock purchase plan, educational leave of absence and tuition reimbursement, and highly-regarded management training programs.

IBM maintains a database of employee functions, locations and costs that allows it to assemble teams to serve clients quickly. For example, the database was used to locate employees in the Southern U.S. who could be swiftly sent to help rebuild software systems after Hurricane Katrina. All IBM employees should be identified by their skills in the database by the end of 2007.

For college students interested in joining IBM, the company provides opportunities for co-ops and interns pursing bachelor's, master's and PhD degrees. IBM seeks students studying business, accounting, computer science, engineering and allied disciplines. MBAs and students of software development are welcome to apply to the Extreme Blue program, a hotbed of IBM innovation.

IBM has entry-level positions in accounting, consulting, finance, marketing, IT, HR, electrical engineering and logistics departments (though other opportunities open up as well). IBM recruits at the College of William and Mary, Cornell, NYU and the University of Pittsburgh, among other fine institutions. IBM also sponsors Project View and Project Able, programs that recruit diverse and disabled hires, respectively.

Intuit Inc.

2700 Coast Avenue
Mountain View, CA 94043
Phone: (650) 944-6000
Fax: (650) 944-3699
www.intuit.com

LOCATIONS

Mountain View, CA (HQ)
Calabasas, CA • Cleveland, OH •
Fort Worth, TX • Fredericksburg,
VA • Omaha, NE • Orem, UT •
Plano, TX • Reno, NV • San Diego,
CA • Shelton, CT • Stamford, CT •
Tucson, AZ • Waltham, MA •
Washington, DC • Edmonton •
Calgary • Berkshire, UK • Bangalore

DEPARTMENTS

Administration • Business
Development • Consulting • Contact
Center • Customer Service •
Documentation • Engineering •
Executive • Facilities • Finance &
Accounting • Government Affairs •
HR • IT • Legal • Marketing/Web
Marketing • Operations • Process
Excellence • Product Management •
Program Management • Quality
Assurance • Sales • Training &
Development • User Experience
Design/Web Technology

THE STATS

Employer Type: Public Company
Stock Symbol: INTU
Stock Exchange: Nasdaq
Chairman: William V. Campbell
President & CEO: Stephen M.
  Bennett
2006 Employees: 7,500
2006 Revenue ($mil.): $2,342

KEY COMPETITORS

H&R Block
Microsoft Business Solutions
Sage Group

EMPLOYMENT CONTACT

www.intuit.com/about_intuit/careers

Visit Vault at **www.vault.com** for insider company profiles, expert advice,
career message boards, expert resume reviews, the Vault Job Board and more.

V/\ULT CAREER LIBRARY   267

THE SCOOP

'Cause I'm the taxman

Intuit is one of the largest makers of accounting, personal finance and tax software in the U.S. In addition to Quicken, its marquee personal finance product, the company's offerings also include accounting software package QuickBooks, and TurboTax, a tax preparation application. The company is now primarily known for its tax software programs, which may not be a bad thing—since taxes are an annual inevitability for millions of Americans, Quicken and TurboTax have provided stable earnings growth.

Check and balance

Two young entrepreneurs, Scott Cook and Tom Proulx, founded Intuit in 1983. Cook was 23 at the time, and Proulx was even younger, still a computer engineering undergraduate at Stanford University. The next year, the company released Quicken, its flagship accounting program. In 1991, Intuit launched QuickPay, a program for handling salaries in small businesses. It was an almost identical program to Microsoft's Money, launched the same year, and by the year after its release, Quicken customers made up 70 percent of the market for this kind of software, despite Money's lower price (Microsoft sometimes gave it away). The following year, Intuit went public, and also acquired a company with a method for filing taxes online—thus TurboTax was born.

Banking on it

In April 1995, in the spirit of "if you can't beat 'em, buy 'em," Microsoft opened its checkbook and offered to buy Intuit for $2 billion—although it was forced to drop its bid in the face of antitrust concerns. Intuit, however, was free to open its own checkbook, making a number of acquisitions during the 1990s that bolstered its offerings in payroll services, business accounting and even streaming stock quotes. Intuit's 1994 acquisition of the National Payment Clearinghouse made the company a presence in the electronic banking industry, and it spread its wings in the second-largest PC market in the world with the 1995 purchase of Milky Way, a Japanese software company.

In February 1998, Intuit agreed to supply America Online with content for its personal finance channel and also announced a deal with Apple to prepare a new Mac version of its Quicken software. Intuit's acquisition of Rock Financial in late 1999 added mortgage-writing capabilities to its QuickenMortgage site—but Intuit sold off its mortgage loans business in 2002, as it focused on its core accounting products; thereafter, the company primarily expanded into software for companies involved in construction, the public sector, real estate management and wholesale durable goods.

Giving small businesses a leg up

In fiscal 2006, Intuit took in revenue of $2.3 billion, with a profit of $416 million—all while purchasing two more companies. Intuit's small business software got a step up in September when it acquired StepUp for $60 million, a company that directs online shoppers to brick-and-mortar stores; its offerings will be incorporated into a future version of QuickBooks. Then, Intuit acquired Digital Insight for $1.4 billion later that year. Digital Insight manages banking transactions such as bill payments, transfers of money between accounts and similar online maneuvers. By early 2007, Intuit had rolled out Personal FinanceWorks, a program based on Digital Insight's offerings that allows users to pay all their bills within one program, as well as view pending checks and payments.

Also in 2006, Intuit and Google teamed up to make it easier for consumers to find (and buy things from) small businesses. Small business owners who use QuickBooks 2007 will be able to note their location on Google maps, as well as purchase advertising from Google (on a service called AdWords).

Intuit got a little scare in 2006, too, as the company received a subpoena from the Securities and Exchange Commission about that familiar bane of Silicon Valley executives—the illegal backdating of stock options grants. But, after an independent review, the company found no unusual activity in August, and the SEC dropped its complaint. If anyone should have their finances in order, it's this company!

A little too turbo, perhaps

The 2007 tax season proved to be an eventful one for Intuit. As the filing deadline approached, Intuit's servers were slammed with an unusually large number of returns—the company estimated that a million people used TurboTax to do their taxes that year—and its servers were unable to cope, resulting in delays and late filings. Intuit apologized and, in a rare stroke of munificence, the IRS extended its deadline for TurboTax filers.

Visit Vault at **www.vault.com** for insider company profiles, expert advice, career message boards, expert resume reviews, the Vault Job Board and more.

VAULT CAREER LIBRARY   269

GETTING HIRED

Get a less taxing job

Intuit's jobs page, found at www.intuit.com/about_intuit/careers, lists open positions across the company. Jobs are divided into customer service positions and everything else. Customer service hopefuls have a number of hoops to jump through, beginning with filling out a web form, a multiple-choice test, Windows navigation test, role-play interviews and finally an actual interview. Those who aspire to hold another position with the company must create a profile to apply.

Students interested in positions at Intuit can read up on programs for interns as well as entry-level programs for college graduates and MBAs. Intuit recruits at MIT, Stanford, the University of California and the University of Wisconsin. An updated schedule of recruiting events is posted on its careers site. Internships are offered in the engineering, marketing, user-centered design, process excellence and finance divisions. Entry-level programs for recent graduates include a selection of rotational programs focusing on small businesses, finance and engineering. Intuit hires MBAs in the product management, brand management, finance and operations departments.

As far as benefits are concerned, the company has a full range of stock option and stock purchasing plans, as well as a 401(k) program with company matching and a choice of health, dental and life insurance plans; it even provides domestic partner benefits. The stock purchase plan allows employees working 20 or more hours a week to buy Intuit common stock at 85 percent of market value. The company offers gym subsidies, commuting alternatives like bike parking and van pools (only at headquarters), and matching employee donations to charities.

In 2007, the company's perks and benefits won it a place on *Fortune*'s list of the Top 100 Best Places to Work. Special mention was given to the CEO's quarterly web casts, in which employees are given free reign to ask questions and make suggestions, some of which have later been implemented as company policy. Not surprisingly, the employees seem really "Intuit."

OUR SURVEY SAYS

Great reviews

Intuit promotes a culture of "hard work and hard play," insiders say. The company stresses teamwork, and sources remark that their ability to complement each other's strengths and weaknesses is one of Intuit's leading assets. Respondents describe the office atmosphere as "casual and sometimes quite playful," and "still a mom and pop company at the core"; however, this can lead to some "petty politics and territorial infighting."

There's also "no dress code—written or unwritten—to speak of." While staffers often work long hours when a project nears completion, they say that Intuit "recognizes that its employees have lives outside of the office." Says one insider, "some employees take classes during the day or have regular work-at-home days. The emphasis is on judging people by their work output, not their face time." Intuit helps its employees play, note insiders: "We have private offices, an on-site cafeteria and gym." One contact offers: "The day-to-day atmosphere is laid-back. People keep flexible hours, dress casually and take time out to play. Some teams/departments have problems with employees having no life outside the company, but there are plenty of groups and managers who encourage people to take time off and pursue other interests."

The office space at Intuit seems to be pretty par for the course, where "Only people who give performance reviews get offices. Everyone else lives in a cube farm. It's pretty easy to get a window cube in some buildings, especially once you get the least bit of seniority; but it's still a cube." According to one source, each floor also has a kitchen area with a real fridge, dishwasher and icemaker to compliment the standard microwave.

Many praise Intuit for its generosity. Says one developer, "The company has generous stock option and stock purchasing plans. I would say that as far as salary goes, it is in line with the industry. If you are involved in product development, there are also ship bonuses." Another, less satisfied source has something of a different take on the bonus system: "There is a bit of a caste system with high performers getting a much bigger share of the goodies than average performers." However there are still "Yearly raises, yearly bonuses, twice yearly profit sharing, plus, in a well-managed group there are other opportunities for bonuses and additional stock grants."

Visit Vault at **www.vault.com** for insider company profiles, expert advice, career message boards, expert resume reviews, the Vault Job Board and more.

VAULT CAREER LIBRARY   **271**

Intuit is kind in other ways, too. One insider reports, "I recently became a single parent and the company has bent over backwards to allow me to have as flexible a work schedule as possible. I work from home for several hours a day." There also seems to be quite a bit of employee loyalty at the company, "Because people feel they have input into products and decisions, there is a lot less of the 'us against them' mentality that seems normal in other companies." Concludes another insider, "I do not personally know anyone who is dissatisfied at Intuit."

"This is the best company I've ever worked for as a woman," gushes one contact. "I've rarely been the only woman on the team and most of my managers have been women. Of course, ethnic diversity is high, as is normal for Silicon Valley. We have a lot of employees from other countries." One downside is that like other high-tech companies, blacks and Latinos are still in short supply. Another contact states that "Most gays and lesbians feel comfortable being 'out' here and I know of at least one instance where there was a problem and the previous CEO himself started a task force to make sure it was an isolated incident because he didn't want to run a company where gays and lesbians didn't feel comfortable at work."

Jabil Circuit, Inc.

10560 Dr. Martin Luther King Jr.
Street North
St. Petersburg, FL 33716
Phone: (727) 577-9749
Fax: (727) 579-8529
www.jabil.com

LOCATIONS

St. Petersburg, FL (HQ)
Auburn Hills, MI
Billerica, MA
Boise, ID
Louisville, KY
McAllen, TX
Memphis, TN
Poway, CA
San Jose, CA
St. Joe, MI
Tempe, AZ

41 locations throughout Europe,
Asia and Latin America.

DEPARTMENTS

Business Development
Engineering
Finance
HR
Information Systems
Logistics
Operations Management
Quality Assurance
Repair & Warranty
Supply Chain Management

THE STATS

Employer Type: Public Company
Stock Symbol: JBL
Stock Exchange: NYSE
Chairman: William D. Morean
President & CEO: Timothy L. Main
2006 Employees: 65,000
2006 Revenue ($mil.): $10,265

KEY COMPETITORS

Flextronics
Sanmina-SCI
Solectron

EMPLOYMENT CONTACT

www.jabil.com/jabilcareers

Visit Vault at **www.vault.com** for insider company profiles, expert advice,
career message boards, expert resume reviews, the Vault Job Board and more.

VAULT CAREER LIBRARY   273

THE SCOOP

The source for outsourcing

Jabil Circuit (the name is a portmanteau of its founders' first names, James and Bill), is a contract manufacturer of electronics. Jabil's customers include companies in a variety of industries, from cell phone manufacturers to aerospace companies, including Nokia, Hewlett-Packard and IBM. For these companies, Jabil provides a roster of outsourcing services—it sources parts and designs circuits, manufactures electronic products and encloses them in plastic cases, and finally ships them to distributors. Jabil also offers repair services.

Bill and Jay's excellent adventure

James Golden and Bill Morean founded Jabil in 1966 when they were contracted to build circuit boards by hand for a computer mainframe manufacturer near Detroit. Given the company's proximity to the car industry, it would inevitably be working on some project related to automobiles. Indeed, General Motors called upon Jabil to source, design and build circuit boards in 1976.

Due to the growth of the consumer market for computers during the 1980s, Jabil got a contract working for the PC division of IBM in 1982, and developed in-house designs for motherboards seven years later. The company had its first IPO in 1993, and by 2001 had become part of the S&P 500. Jabil then expanded its business by purchasing two electronics manufacturing businesses; it snapped up Philips Contract Manufacturing Services in 2002, and bought Varian's electronics manufacturing arm in 2005.

Geography of cost

In 2006 Jabil acquired Celetronix, an Indian electronics manufacturer with locations on the subcontinent in Chennai, Pondicherry and Mumbai. Shortly after purchasing the company, Jabil said it would, in a $250 million restructuring effort, be cutting an as yet unspecified number of jobs as it moved its production facilities to countries with lower labor costs. To further this end, Jabil constructed a production facility in Zarkappatya, Ukraine, to provide electronic parts both to the local and European markets.

Reaching the Green Point

In fiscal 2006, Jabil reported revenue of over $10 billion, up 36 percent over the results for 2005, which came in at $7.5 billion. In line with these results, management is looking forward to 2007, when they predict revenue growth of 20 percent. Meanwhile, in early 2007, Jabil acquired Taiwan Green Point Enterprises, a manufacturer of molded plastic parts for the consumer electronics and automotive industries, to provide the all-important plastic casings for its products.

Good morning, Vietnam!

In 2007, Jabil announced that it was building a factory in Ho Chi Minh City, Vietnam. The new digs will be ready to start churning out technology for Western markets in 2007. Jabil chose the location due to its cost-effective workforce and proximity to its other Asian facilities. The acquisition should also significantly boost Jabil's size, bringing approximately 10,000 new employees into the fold.

GETTING HIRED

Electrify your career at Jabil

Jabil's jobs site, at www.jabil.com/JabilCareers, provides information and a number of video clips for job seekers. The company is flexible about letting its workers choose to travel or to remain in one location. The site does not specify at which schools Jabil recruits; however, it does mention that it is primarily interested in candidates with business and engineering degrees. To apply, candidates must fill out a web form and attach a cover letter and resume. Benefits vary by location, and internships are available.

KLA-Tencor Corporation

160 Rio Robles
San Jose, CA 95134
Phone: (408) 875-3000
Fax: (408) 875-4857
www.kla-tencor.com

LOCATIONS

San Jose, CA (HQ)
Albuquerque, NM • Austin, TX •
Beaverton, OR • Boise, ID •
Chandler, AZ • Colorado Springs,
CO • Edina, MN • Essex Junction,
VT • Fremont, CA • Hopewell
Junction, NY • Lindon, UT •
Livermore, CA • Manassas, VA •
Milpitas, CA • Portsmouth, NH •
Richardson, TX • San Diego, CA •
Tucson, AZ • Vancouver, WA •
Westborough, MA • Westwood, MA

22 locations throughout Europe and
Asia.

DEPARTMENTS

Administration • Applications
Development/Field Applications •
Customer/Technical Support •
Electrical Engineering • Engineering
Support Services • Facilities •
Finance • HR • IT & Services •
Legal • Manufacturing & Engineering
Manufacturing, Production &
Operations • Marketing & Sales •
Mechanical Engineering • Optical
Engineering • Product Development
• Program Management • Research
Scientist • Software Engineering &
Development • Systems Engineering
• Testing

THE STATS

Employer Type: Public Company
Stock Symbol: KLAC
Stock Exchange: Nasdaq
Chairman: Edward W. Barnholt
CEO: Richard P. Wallace
2007 Employees: 6,000
2007 Revenue ($mil.): $2,731

KEY COMPETITORS

Applied Materials
Hitachi High-Technologies
Veeco Instruments

EMPLOYMENT CONTACT

https://ktcareers.kla-tencor.com

THE SCOOP

Towards a more perfect microchip

KLA-Tencor is aiding chip manufacturers in their battle against all the flaws that chips are heir to. The company provides software and devices with which to inspect chips during the manufacturing process, so as to reduce the number of flawed chips and increase chip-makers' revenue. KLA's products inspect wafers for microscopic defects prior to lithography, inspect lithographic plates for errors or damage, ensure that all of the designs to be printed on a chip have been correctly aligned, and inspect finished chips for mistakes. The company also offers a line of testing products to manufacturers of hard drives. In 2007, *BusinessWeek* ranked the company No. 95 on its InfoTech 100.

Microchip mashup

KLA and Tencor were both founded in 1976 to provide testing equipment to the semiconductor industry. KLA's products were mainly focused on the inspection of the plates that projected circuit designs on chips, while Tencor's devices detected dust on the chip itself and measured films that had been deposited onto the chip. KLA had its first stock offering in 1980, a few years after introducing its RAPID inspection system for lithographic plates in the late 1970s; it introduced the WISARD wafer inspection system in 1984. Tencor, meanwhile, brought out the Alpha-Step, a device to measure the depth of films deposited on the wafer, in 1976, and a dust-detection device in 1984.

These two companies decided to combine their expertise in a 1997 merger valued at $1.3 billion. But it was bad timing, as the merger occurred on the eve of a slowdown in the microchip industry, and 1998 was marked by a layoff of 20 percent of the company's headcount. Things turned around in 2000, though, enough for KLA-Tencor to make a number of acquisitions. That year, KLA-Tencor bought Fab Solutions, a developer of software used in chip factories; the following year, the company bought Phase Metrics, a company that specialized in memory chip inspection, and revenue was high as a result, at $2.1 billion. While KLA-Tencor's take decreased to a low of $1.3 billion in 2003 due to a slowdown in the market for its products, its revenue climbed back to the $2 billion range in 2005.

Visit Vault at **www.vault.com** for insider company profiles, expert advice, career message boards, expert resume reviews, the Vault Job Board and more.

VAULT CAREER LIBRARY   277

Three-letter purchase

The acquisitions kept coming in 2006 when KLA-Tencor bought ADE, a provider of measurement and test machinery for the semiconductor industry, in a deal valued at just under half a billion dollars. ADE's products include devices that determine the topography and flatness of silicon wafers.

Fall into the GAAP

In 2006, KLA-Tencor took in nearly $2.07 billion in revenue and $380 million in profit. This was only a small decrease from the previous year, when the company drew $2.09 billion in revenue and made a profit of $467 million. The company came under a cloud that year, though, when an investigation uncovered suspicious stocks options dating practices that had occurred between 1997 and 2002. KLA-Tencor believes that rectifying the situation will cost it somewhere around $370 million.

In October, KLA's founder and current chairman, Kenneth Levy, resigned, along with Stuart Nichols, the general counsel. The company did not specifically name anyone in its inquiry, but re-priced Levy's stock options, as well as those of Nichols and John Kispert, its CFO; it also canceled all grants to Kenneth Schroeder, a former CEO. KLA-Tencor has retained the services of Kispert and Richard Wallace, its current CEO, specifically absolving the two of any wrongdoing.

The company settled with the Securities and Exchange Commission in July 2007; the SEC still has fraud charges pending against former CEO Schroeder. The settlement dovetailed nicely with the June release of the company's financial results for 2007, wherein it seemed to recover from its legal troubles; revenue and profits were both up significantly, at $2.7 billion and $528 million, respectively.

Catching a wave

In spring 2007, KLA-Tencor purchased Therma-Wave, a manufacturer of devices that measure the effectiveness of the various steps in the manufacture of semiconductors, for $73 million. Therma-Wave will boost KLA-Tencor's measurement division.

GETTING HIRED

Test the waters at KLA-Tencor

KLA-Tencor's careers site, at https://ktcareers.kla-tencor.com, provides information about job opportunities and benefits for students, recent graduates and experienced professionals. The company typically conducts at least two rounds of interviews with a candidate, and may take as many as three weeks to reply to a resume submission. Once selected for an interview, candidates gain access to an area of the site with various forms and directions to company office locations.

Benefits at KLA-Tencor include health, dental and vision insurance, 401(k) with company matching and a profit-sharing plan. Perks at HQ include on-site sand volleyball and gym, pet insurance, commuter assistance, car detailing and a dentist's office (so that all employees can be equipped with two shiny grills).

Interns are accepted whether they're pursuing a BS, MS or PhD. Applicants must have at least a 3.0 GPA, and receive medical insurance and housing. The college careers site contains a spiffy "crystal ball" feature, so that students with degrees in physics, mechanical, chemical or electrical engineering, computer science or math can see what the future holds for them at KLA-Tencor.

The Powers that Hire at KLA-Tencor can be reached via e-mail at staffing@kla-tencor.com (for experienced hires) or ur@kla-tencor.com (for university relations).

Lam Research Corporation

4650 Cushing Parkway
Fremont, CA 94538
Phone: (510) 572-0200
Fax: (510) 572-2935
www.lamrc.com

LOCATIONS

Fremont, CA (HQ)
Austin, TX
Boise, ID
Fishkill, NY
Lehi, UT
Manassas, VA
Richardson, TX
South Portland, ME
Tempe, AZ
Vancouver, WA

Locations in 14 countries
throughout Europe and Asia,
including the UK, China, Japan and
Israel.

DEPARTMENTS

Corporate Administration
Corporate Communications
Corporate Legal
Corporate Marketing
Finance
Global Field Operations
Global HR
Global Operations
Global Products
Global Sales & Marketing
IT
Investor Relations
North America Field Operations
Sales

THE STATS

Employer Type: Public Company
Stock Symbol: LRCX
Stock Exchange: Nasdaq
Chairman: James W. Bagley
President & CEO: Stephen G.
 Newberry
2006 Employees: 2,430
2006 Revenue ($mil.): $1,642

KEY COMPETITORS

Applied Materials
Novellus
Tokyo Electron

EMPLOYMENT CONTACT

www.lamrc.com/careers_1.cfm

THE SCOOP

Etch-A-Sketch

Lam Research provides semiconductor manufacturers with machines that use plasma (a state of matter in which electrons have become disassociated from their atomic nuclei) in order to etch the design of circuits onto silicon wafers during the process of microchip manufacture. In order to prevent the entire surface of the wafer from being evenly removed by the plasma during the etching process, it is coated with a resist, a material that protects the underlying wafer. Etching is performed several times on successive layers of the chip during the manufacturing process. Lam's other products include devices to remove the resist used during photolithography and mechanical chemical cleaning machines that clean and level the surface of wafers following etching.

Fattening the Lam

Lam Research was founded by tech entrepreneur David Lam in 1980 to market the plasma etching process to the semiconductor manufacturing industry. Within three years, the company was profitable; within four, it had gone public. In the 1980s, the company began to sell its products in Japan, which was rapidly moving in to fill demand for chips. And Lam devoted itself not just to increasing sales, but to extensive research and development. By 1992, the company released its first product with its new Transformer Coupled Plasma (TCP) technology, which allowed for more efficient, consistent and high-quality chip production. The new technology quickly produced for the company, as its new TCP products yielded $33 million in sales in 1993 alone. Within another year, revenue climbed 86 percent to $493 million and sales doubled; the company was growing, and it added 400 new workers and four new facilities that year, in addition to expanding on its current locations.

On the Lam

But any success in the semiconductor industry is usually fleeting, and Lam's proved to be no exception. PC sales slowed in 1996, and the semiconductor industry suffered, bringing Lam with it—the company laid off over 500 employees in August 1996. The following March, Lam struck out in a different direction, merging with OnTrak Systems, a firm specializing in a chemical microchip cleaning process. But at the end of 1997, the newly enlarged company reported a $33.7 million loss, and the company went on a slew of job cuts. Lam cut 14 percent of its workforce in

February 1998, 20 percent of the remainder in June, and 500 more employees in November; it also closed two of its plants and consolidated other offices. Losses were even greater that year, totaling $145 million. At that point, the company was still the fourth-largest maker of technology for chip-manufacturing worldwide, and remained optimistic that it had weathered the storm. Successive financial results have been mixed so far, but the company has not suffered through losses as great as 1998 in the years since.

If you've got an etch, scratch it

In 2006, Lam posted revenue of $1.6 billion (an increase of $100 million over the year before) and profits of over $300 million. The stretch from 2004 to 2006 was the first time since the mid-1990s that Lam posted profits for more than two years consecutively. At the end of 2006, management announced that Lam would be moving into areas related to the etching process in the future to diversify its revenue base.

To bring this strategy to fruition, in 2006 Lam acquired facilities for the growing and fabrication of silicon from Bullen Ultrasonics for $175 million. These facilities, located in China and Ohio, are a source of the pure, flat silicon wafers that are the raw material from which microchips are made. An increased demand for microchips and solar panels (which also require pure silicon) has heated up the market for these wafers, meaning that the division should provide good returns in the near future for the company.

A rising tide lifts all ships

Industry analysts have noticed Lam's recovery: in 2007, Lam moved up 14 places on *Business 2.0*'s list of the fastest-growing technology companies to No. 7. It posted three-year revenue growth of 36 percent, and ranked in the top 10 for earnings growth. The company might not be out of the woods yet, though—it announced an internal inquiry into suspicious activity related to employee stock options grants in July 2007.

GETTING HIRED

Leave your mark at Lam

Lam's career site, at www.lamresearch.com/careers_1.cfm, provides information about openings for experienced candidates, college graduates and undergraduates. Jobs are searchable by function, keyword and location. To apply, job seekers must first create a profile. Benefits at Lam include a choice of health care plans, dental insurance, flexible spending accounts for health and dependent care, life insurance, sabbaticals, 401(k) plans and a discounted employee stock purchase plan.

Undergraduates and recent graduates are invited to apply for internships and entry-level positions, respectively. Internships are available for students pursuing a bachelor's degree, master's degree or PhD. Interns may participate in poster and essay contests, and are given ample opportunity to network with executives and other interns. Recent graduates are also invited to apply to positions in engineering, HR, finance, marketing, and R&D and software development, among other departments.

Lenovo Group Limited

1009 Think Place
Morrisville, NC 27560
Phone: (866) 458-4465
Fax: (877) 411-1329
www.lenovo.com

DEPARTMENTS

Accounting/Finance • Administrative •
Communications • Customer Service
• Engineering • Fulfillment • HR •
Inside Sales • IT • Legal •
Manufacturing • Marketing •
Procurement • Project Management •
Public Relations • Research &
Development • Sales • Sales Support
• Strategy & Operations • Supply
Chain

THE STATS

Employer Type: Public Company
Stock Symbol: 0992
Stock Exchange: Hong Kong
Chairman: Yang Yuanqing
President & CEO: William J. Amelio
2006 Employees: 23,500
2007 Revenue ($mil.): $14,590

KEY COMPETITORS

Acer
Dell
Hewlett-Packard

EMPLOYMENT CONTACT

www.pc.ibm.com/us/lenovo/about/
jobs.html

LOCATIONS

Raleigh, NC & Bejing. China (HQ)
Austin, TX • Baltimore, MD •
Beaverton, OR • Chicago, IL •
Cincinnati, OH • Cleveland, OH •
Dallas, TX • Denver, CO • Des
Moines, IA • Fort Worth, TX •
Houston, TX • Indianapolis, IN •
Jacksonville, FL • Kansas City, MO •
Nashville, TN • New York, NY •
Orlando, FL • Philadelphia, PA •
Phoenix, AZ • Pittsburg, PA • Salt
Lake City, UT • San Antonio, TX • San
Francisco, CA • San Jose, CA •
Seattle, WA • Tampa, FL • Tucson,
AZ

Rotating global headquarters with
executive hubs in Raleigh, NC, and
Beijing. 78 US locations; 8 research
centers and 6 manufacturing centers
throughout Asia. Sales headquarters
in Paris, Beijing and Sydney. Global
marketing hub in Bangalore, India.

THE SCOOP

New thinking for a new world

Lenovo is a Chinese manufacturer of desktop and notebook computers, servers, printers, monitors and peripherals. Its products also include cell phones and MP3 players. Lenovo is the top selling PC manufacturer in China and is now the fourth largest PC manufacturer in the world, after acquiring IBM's personal computing division in 2005.

Lenovo maintains rotating global headquarters throughout North America, Europe and Asia-Pacific with two executive hubs located in Raleigh, North Carolina, and Beijing. Lenovo has company offices in more than 75 U.S. cities as well as throughout 69 countries worldwide.

A new(ish) player in the computer scene

Lenovo was founded in 1984 as the Beijing Legend Computer Group by 10 colleagues from the Chinese Academy of Sciences. The company's first product, launched in 1987, was a Chinese character card and its first personal computer debuted in 1990. Four years later, the company offered its shares on the Hong Kong exchange, and it soon added laser printers to its product line in 1997.

In 2000, sales of the company's products had taken off to the point where it was included in the Hang Seng index, a list of the highest-capitalized stocks traded on the Hong Kong exchange. In 2003, Legend changed its name to Lenovo, and the following year rolled out a line of computers for low-income customers, a significant area of growth. In 2005, Lenovo became the world's third-largest computer manufacturer when it bought IBM's PC division. (It since slipped to fourth place, when Acer bumped it from its spot behind Hewlett-Packard and Dell.)

What's in a name?

Following its purchase of IBM's personal computing division, Lenovo was suddenly catapulted onto the world stage. Despite being a well-known brand in China—it's up there with Coca-Cola—Lenovo was comparatively unknown to the world at large. As such, in order to build up its brand equity and attract customers, Lenovo is sponsoring a number of high-profile sporting events and teams, including the 2008 Olympics, the Williams Formula One team, and the NBA.

Lenovo a-go-go

In the fiscal year ending 2006, Lenovo claimed revenue of $13 billion, up over 300 percent on the IBM acquisition, with a profit of $84 million, up 7 percent over 2005. Asia accounted for nearly half of the company's sales in 2006, while the Americas accounted for 30 percent. Also that year, Lenovo announced it would cut 1,000 positions in an effort to boost its margins.

Teaming up with the big guy

Lenovo and Microsoft have a close partnership and in April 2007, the two teamed up to develop a research center and also introduce FlexGo pay-as-you-go computing. This was the first such venture ever undertaken by Microsoft, however, Lenovo has been an innovator in this space in the APAC region since its early days of operation. The research center is located in Lenovo's Beijing research institute and will focus on portable computing technologies. FlexGo aims to bring computers to low-income families and provide Internet access for people who could not otherwise afford a computer.

Continuing to Evolve

While Lenovo announced additional layoffs in 2007, the restructuring plans are expected to save $100 million the following fiscal year. Financial results for 2007 have reflected higher margins already: revenue and profits each increased by over 9 percent, revenue to $14.6 billion and profits to $2 billion, with sales actually increasing in greater China by 2 percent.

GETTING HIRED

Grow with Lenovo

Lenovo's career site, at www.pc.ibm.com/us/lenovo/about/jobs.html, provides information about job openings and benefits. Open positions are searchable by location, type and functional area. When Vault investigated the site, there were a handful of positions listed, primarily in sales and HR. Benefits include a selection of medical plans, savings schemes for retirement, tuition reimbursement and financial planning services.

Lexmark International, Inc.

740 West New Circle Road
Lexington, KY 40550
Phone: (859) 232-2000
Fax: (859) 232-2403
www.lexmark.com

LOCATIONS

Lexington, KY (HQ)
Geneva (Europe, Middle East, Africa HQ)
Singapore (Asia Pacific HQ)
Coral Gables, FL
Richmond Hill, Canada

Manufacturing facilities in Colorado, Mexico and the Philippines.

DEPARTMENTS

Customer Services • Finance •
Human Resources • Information
Technology & Web • Legal •
Manufacturing • Marketing •
Purchasing & Vendor Management •
Research & Development • Sales •
Sales Operations • Service Delivery
• Site Operations • Supply Chain

THE STATS

Employer Type: Public Company
Stock Symbol: LXK
Stock Exchange: NYSE
Chairman & CEO: Paul J. Curlander
2006 Employees: 14,900
2006 Revenue ($mil.): $5,108.1

KEY COMPETITORS

Canon
Epson
Hewlett-Packard

EMPLOYMENT CONTACT

www.lexmark.com/employ

Visit Vault at **www.vault.com** for insider company profiles, expert advice, career message boards, expert resume reviews, the Vault Job Board and more.

VAULT CAREER LIBRARY   287

THE SCOOP

The future of print

As a leading manufacturer of printers and accessories, Lexmark bridges the gap between the "old" world of paper products and the new one focused on all things electronic. It produces laser, inkjet and dot matrix printers, multifunction machines (which also copy, fax and scan documents) and supplies. The company owns much of the technology used in its products (through its investment in research) and also works in intangibles, offering technical support and custom business solutions for small and medium-sized businesses, government agencies and educational resources. Lest all this sound incredibly pedestrian, the Lexmark name does carry some cachet: 75 percent of the world's leading banks, retailers and pharmacies use Lexmark, and according to research from the Gartner Group, a liter of Lexmark printer ink is pricier than the same amount of first-rate whiskey or Chanel No. 5 perfume.

Lexmark is ranked in the second half of *Forbes'* Global 2000 listing (at No. 1342), but still pulled in $5.1 billion in revenue in 2006. The appearance on an international roll call is telling, since more than half (56 percent) of the firm's intake comes from sales in over 150 different nations; two-thirds of that amount (or 38 percent of the total) is generated in Europe. The range of operating locations exposes Lexmark to a variety of differently-structured markets, at varying degrees of maturity.

The company's name makes a little more sense with a bit of explanation. The "lex" stands for "lexicon," while "mark" signifies the more obvious result of using the corporation's products. And the name also cleverly references company headquarters, located in Lexington, Kentucky.

Sold out

Later referred to as the "Rodney Dangerfield of printer companies" by one analyst (since it seemed to get no respect), the firm grew up as a division of IBM. However, management at the now $91 billion company hardly gave the printing operation a second look, ranking it more than a few steps down from its most lucrative businesses.

In 1991, the company issued the familiar statement that the division didn't fit with its core businesses, and sold it off to a venture capital firm in a leveraged buyout. That fact in itself was difficult for the segment's employees, but more troublesome for Lexmark's management was its lack of supporting structure: when it was cut off, the company had departments for product development and manufacturing, but no sales or marketing functions (previously handled at the IBM corporate level).

After designing a framework for the new company based on four business units and small project teams, it let go about half its workforce of 4,000. At that point, Lexmark was still left with $1.5 billion in debt taken on as part of the deal. But thrust into a hot market, it was able to establish a place quickly, and attacked its competitors by targeting (and engineering products for) their bread-and-butter industries, like banking. Within six years, it made itself into the No. 2 presence in the worldwide laser printing business, and pared down its debt from $1 billion plus to $176 million.

Printing perspectives

Lexmark operations straddle two highly-competitive worlds, which sometime makes it difficult to address all its constituents. Its commercial customers primarily use high-volume, extremely fast laser printers (which create a document in a complicated process involving a laser image, a powder combination of carbon and polymers, and an electrostatic charge). The company owns a markedly smaller portion of this market than its main competitor Hewlett-Packard (HP), which owns about a 40 percent share of units sold. In this sphere, traditional "printing" companies compete with others mostly known for their copiers, so the likes of Xerox, Canon, Ricoh and Konica Minolta also offer similar services.

In the world of the home consumer, the inkjet all-in-one is king. (Inkjets are generally easy to use, low cost, and produce high quality output by shooting tiny ink drops onto the paper.) Lexmark's chief rivals—HP, Epson and Canon—hold fully 75 percent of the customer base. In a retail setting, though, all players are subject to the whims of the market and the vendor, and even small variations in the amount of shelf space, pricing, promotion, etc., can affect sales. Lexmark has often competed on a cost basis, and some of its printers can be snapped up for $40 to $80, which throws out decades of business common sense, as chairman and CEO Paul Curlander acknowledged in an interview with *The Australian* in April 2007, when he said that part of the company's strategy is to "reluctantly" lose money on some items.

He also admitted that Lexmark hasn't done quite enough to address the needs of the business space between the larger enterprise and regular consumer. He vowed to concentrate more on the medium-sized businesses that make strategic decisions in a big-business way, but may be subject to pricing concerns and other aspects of the Lexmark brand, like a home shopper. (The midlevel companies customarily use a mix of inkjet and laser products.) Curlander recognized that Lexmark needs to

Visit Vault at **www.vault.com** for insider company profiles, expert advice, career message boards, expert resume reviews, the Vault Job Board and more.

V/\ULT CAREER LIBRARY   **289**

develop a specific brand message for the market, and was considering more in the way of industry customization.

Dirty inky money

But Lexmark is still a profitable company (even if profits aren't as high as analysts and investors would like); how does it manage this? Counterintuitively, the printing market doesn't thrive on the sale of printers for home and commerce, but on the millions of modest-sized plastic cartridges used to dispense ink inside the devices, and Lexmark is no exception. The toner case business produces $30 billion in annual revenue worldwide, around 25 percent of which comes from reformed or refilled "empties," as they're known in the industry. Between 1996 and 2003, the number of empties that Lexmark collected increased by 800 percent.

Lexmark has invested a great deal of money in the specialized, prime-quality inks these cases contain, but even more on the technology used to construct the print-head, which controls the speed and precision with which the cartridge delivers the ink. According to the company, "Everytime you buy a new cartridge it's like buying a new printer."

Cartridge case

As a peripheral operator in the printer market, Lexmark relies on its cartridge business for a lot of its income, but its operating practices regarding empties have led to conflict and court cases in a number of instances. One, in particular, involved the company's packaging for its laser cartridges, which includes a sort of contract that states: "Opening this package or using the patented cartridge inside confirms your acceptance of the following license agreement." The sentence referred to Lexmark's "Prebate" program, a promotion that gives consumers a discount on the product if they promise to return the used casing to the company.

A trade group of remanufacturers, the Arizona Cartridge Remanufacturers Association (ACRA), took Lexmark to court, alleging unfair business practices. ACRA had two complaints. First it claimed that only retailers, not Lexmark, could control prices charged to users. Furthermore, it took exception to Lexmark's installation of a chip in the printer, which deactivated it if the "wrong" type of cartridge was used. The Ninth Circuit Court of Appeals found in favor of Lexmark in 2005, stating that the pricing "contract" was fair, and ACRA did not fully support its case regarding the cartridge chip.

Fewer cables, more cash?

Printer sales have slumped worldwide in the past couple of years, as consumers embraced the convenience of laptop computers, swinging away from bulky desktops and linked stationary printers. Battered by increased competition on price, the market shift and a "difficult" 2005, Lexmark announced some initiatives to tighten its belt in 2006; it consolidated capacity, slimmed general expenses and reduced some of its stock of inkjet printers. (For these products, it was not uncommon to lose money on each sale, making up for it with subsequent ink purchases.)

Just as significantly, it chopped 850 positions and outsourced 550 more to cheaper foreign locales to provide ongoing savings of $80 million a year. The move didn't produce enough of the right results, since the following quarters were peppered with warnings that the company's earnings and revenue would be weaker than expected. The company's stock is at a five-year low.

One thing on the horizon that may help the company gain some strength is wireless printing for the tech-savvy home consumer. Notebook and wireless router sales are growing at an estimated 20 percent per year, and research firm IDC projected that by 2010, approximately 93.6 million households will have access to a wireless network, up 70 percent from current totals.

Lexmark spent $370 million on research in 2006—50 percent more than five years earlier—in part to speed up the schedule for wireless, which was available on eight of 12 inkjet models released in 2007. (On six the convention is standard, and the remaining two have a wireless print option.) Moreover, the installation process has been super-simplified, so the average homeowner won't spend more than half an hour to set up the network.

In addition, Lexmark has said it would look to spur growth in its non-European international business in the coming years, using its other "traditional" products to "drive pages and supplies." That focus could lead to more problems, since according to one analyst, "Emerging markets often leapfrog technologies and they have problems paying for their supplies as well."

Visit Vault at **www.vault.com** for insider company profiles, expert advice, career message boards, expert resume reviews, the Vault Job Board and more.

VAULT CAREER LIBRARY 291

GETTING HIRED

Looking for Lex-elent employees

Openings at Lexmark's Kentucky headquarters and U.S. offices can be accessed through the company's web site. International opportunities are accessible through each divison's web page available through a drop down menu on the main Lexmark web site.

The company also offers opportunities for students interested in a Lexmark career during college. Interested parties can work as an AYPT (academic year part time), co-op (alternating periods of work and study), or an intern (summer only) student. Program participants must have a B average, be eligible to work in the U.S. on a permanent basis, and attend an accredited college or university on a full-time basis. All student positions are paid. Other benefits include free housing (fully furnished apartments) for individuals who live more than 50 miles away, paid holidays and personal days, company discounts, on-the-job training and access to a "wide range of social and sporting events."

If its finances haven't been the healthiest of late, at least Lexmark promotes the well-being of its employees. Besides life, health, dental, vision and disability coverage, Lexmark gives its staff the opportunity to purchase stock and contribute to a 401(k) plan with a generous company matching plan. It further encourages employees' character growth by granting three paid vacation days in exchange for volunteer work, and offers technical training and leadership development programs. Workers at the Lexington campus have access to convenient medical and banking services on site.

LSI Corporation

1621 Barber Lane
Milpitas, CA 95035
Phone: (408) 433-8000
Fax: (408) 954-3220
www.lsilogic.com

LOCATIONS

Milpitas, CA (HQ)
Allentown, PA
Austin, TX
Boulder, CO
Colorado Springs, CO
Fort Collins, CO
Fremont, CA
Gresham, OR
Irvine, CA
Mendota Heights, MN
Minneapolis, MN
Norcross, GA
San Jose, CA
Wichita, KS
Waltham, MA

Other locations in Canada, Japan,
Taiwan, South Korea, India,
Singapore, Thailand, China,
England, Ireland, Italy, Germany,
Israel and United Arab Emirates.

DEPARTMENTS

Administrative/Clerical • Customer
Service • Engineering • Finance •
General • HR • IT • Manufacturing •
Marketing • Operations • Program
Management • Purchasing • Sales •
Technician

THE STATS

Employer Type: Public Company
Stock Symbol: LSI
Stock Exchange: NYSE
Chairman: James H. Keyes
President & CEO: Abhi Talwalkar
2006 Employees: 4,010
2006 Revenue ($mil.): $1,982

KEY COMPETITORS

IBM Microelectronics
NXP
Texas Instruments

EMPLOYMENT CONTACT

www.lsi.com/about_lsi/careersl

THE SCOOP

Large-scale integration for small-scale devices

LSI Logic Corporation (the LSI stands for "large-scale integration") manufactures semiconductors for the digital audio and video, communications, networking and storage markets. LSI's storage chips are designed to allow computers to communicate with storage devices, such as tape drives and data storage servers. The company also offers custom-designed "systems-on-a-chip" for the consumer market. LSI does not have its own fabs, but rather designs chips and then contracts with foundries for their production.

ASICS—not just snazzy shoes

Wilfred Corrigan, a British chemical engineer whose father had worked on docks in Liverpool, founded LSI in 1981 after working for 20 years in the semiconductor industry, including a failed attempt as the CEO of Fairchild Semiconductor from 1974 to 1980 to diversify the offerings of the industry giant. His time at Fairchild also earned him a somewhat dictatorial reputation, including an anecdote about his penchant for screening the opening scene of the movie *Patton* during sales meetings. When Schlumberger Ltd. purchased Fairchild in 1979 (only two years after Corrigan had started chairing the company's board meetings, in addition to serving as president and chief executive), Corrigan was shortly out the door.

He picked himself up by the boot straps and founded LSI Logic the next year. But for his new company, he headed in the opposite direction from his old one, aiming small and specialized where Fairchild had been large and general. LSI's initial product would be ASICs (application-specific integrated circuit). These would be partly-built microchips, which could then be finished to a customer's specifications in short order; an idea that differed from the industry standard of the time, which called for the mass production of essentially homogeneous microchips. Corrigan pitched his idea to a group of bay area venture capitalists, and they heard that magic investing catchphrase: small industry segment with great growth potential. They fronted him $6 million in startup capital, and LSI was off to the races.

Livin' on the edge

As a company dependent on an undeveloped niche in a high-tech field, LSI needed heavy investment in R&D before it could turn a profit, but Corrigan was innovative in cultivating support for his fledgling company. In August 1981, LSI won an invaluable boost when it announced a joint venture with Toshiba, who liked the idea of working with it on semi-customized, advanced circuits.

LSI's original bay area investors were pleased with this development, and committed the better part of $16 million to the company in March 1982. But LSI struggled financially that year anyway, earning $5 million in sales and posting losses of $3.7 million. Its investors everywhere smiled the next year, though—sales shot up to $35 million and profits to $12.5 million. But even though the company's revenue kept rising throughout the 1980s, profits were inconsistent, often lower than expected. Corrigan would have to rely on investor faith for many years to keep LSI afloat.

Going abroad early

LSI went public in 1983, and Corrigan took the company international almost immediately, recognizing that strong support from overseas investors would be crucial to the company's survival. Japanese investors, perhaps remembering LSI's recent partnership with Tobisha, had acquired much of the company's stock in the IPO, and Corrigan quickly appealed for their support with the 1984 formation of Nihon LSI Logic, a private Japanese offshoot of his company. Nihon's private offering raised $20 million, and left 28 Japanese investors with a 33 percent stake in the new venture. Months later, Corrigan tried the same thing with LSI Logic Ltd. in the U.K., raising $20 million there as well.

LSI's slim profits

The company's revenue and market share grew robustly throughout the decade; in 1986, it controlled 45 percent of the U.S. market and 25 percent of the worldwide market in ASICs, and its revenue reached almost $200 million. But R&D spending still chewed up a good deal of the company's loose change, as net income was only $3.8 million that year. Profits fluctuated until 1989 and 1990, when the company lost money for two consecutive years, totaling close to $60 million. By 1991, the company was profitable again, earning $700 million in revenue and $8.3 million in profits. It was not a number to celebrate, though—the company had earned more in 1982, when its revenue was $35 million. That nasty word raised its head in corporate headquarters: restructuring.

The company cut 1,000 jobs in 1992, over 18 percent of its total headcount. By 1995, the company's profits for the first quarter alone came in at $45.3 million, more than double its previous best. That was a good year for the company in other respects: it

Visit Vault at **www.vault.com** for insider company profiles, expert advice, career message boards, expert resume reviews, the Vault Job Board and more.

VAULT CAREER LIBRARY   **295**

earned over $1 billion in revenue for the first time, and became the sole owner of its Japanese affiliate, changing the division's name to LSI Logic Japan Semiconductor.

Are we there yet?

But this was still the semiconductor industry, where nobody is ever quite out of the woods: by 1998, LSI was posting losses again, $139 million this time. But the company had first thrived on the faith of its investors, and it maintained its faith going forward; it was even diversifying beyond ASICs (the first time Corrigan had tried diversifying his company in nearly 20 years). Corrigan had long insisted that all of LSI's products work together, and the company started producing ASSPs (application-specific standard products), a broader line of standard chips that used LSI's various chip desgns with each other—although some competitors, such as Broadcom and Cisco, had already bitten into a large chunk of the market.

The company also moved into system-on-a-chip technology, which put multiple functions (such as processors, memory and logic) all on the same microchip, a critical development in the race for increasingly small and feature-packed gadgets. From its losses in 1998, LSI recovered nicely the next year, earning profits of $67 million and exceeding $2 billion in revenue.

Cheers, mate!

LSI entered into a period of intense soul-seeking, both inside the company and without, at the turn of the new millennium. Two top executives left for other firms in 2000, including John Daane, who had long been Wilfred Corrigan's likely successor as LSI's top dog. Corrigan was now in his early 60s, and seemed ready to ride off into the sunset; moreover, the bursting of the tech bubble didn't spare LSI, and its operations suddenly looked very traditional and expensive. It cut hundreds of jobs and sold off some manufacturing facilities in 2003, while industry analysts started talking about the executive suite's need for an infusion of new blood. In May 2005, Corrigan introduced his successor as president and CEO, Abhi Talwalkar, a former executive in Intel's enterprise and storage division. Corrigan has since stayed involved with the company as the non-executive chairman of its board.

The plaudits immediately started to pour in for the company's departing Liverpudlian founder, whose company had largely created the ASICs market; one LSI alumnus said, "There are good startup managers and executives who are good at growing medium-sized companies. And there are good large company executives. But you rarely see one manager who has taken the same company successfully through all these stages." The same employee also vouched for how far Corrigan had shed his dictatorial image from long ago at Fairchild Semiconductor: "(He) keeps it very clear who is in charge in the boardroom … but he didn't impose his view. I would say he is one of the best strategic thinkers I have ever worked with."

The future of memory

Talwalkar has received mixed reviews so far, as investors and analysts closely watch his ambitious plan to overhaul the now venerable LSI. He says he spent his first 90 days on the job just meeting with the company's customers and evaluating its best areas for growth. Then, in August 2005, he announced plans to turn the company fab-less, selling off its main American wafer manufacturing facility in Gresham, Ore., and returning to the company's roots as a producer of highly specialized ASICs for its customers (the fab sold in May 2006 to the newly independent ON Semiconductor for $105 million).

In December 2006, Talwalkar backed up his talk with a bold move—the $4 billion takeover of Agere, a semiconductor manufacturer that specializes in chips for storage, mobile devices and networking. The deal more than doubles the size of LSI (from 4,010 in 2006 to 9,100), and gives it a robust clutch of patents—10,000 of them, to be exact—which can be licensed to other manufacturers. Cost savings associated with the acquisition are estimated at $125 million for 2008. But they have come at a cost—LSI posted second quarter losses of $378 million in July 2007, $340 million of which were attributed to charges from the Agere deal. In August, the company announced worldwide layoffs of 2,100 people, which might be unrelated, as they came to the sort of manufacturing facilities Talwalkar planned on trimming before the merger. LSI continued divesting its chipmaking resources that same month, selling off its phone and satellite-radio chip business to Infineon for about $450 million.

GETTING HIRED

Add some Logic to your career

LSI's careers site, at www.lsi.com/about_lsi/careers, provides a wealth of information for job seekers about LSI company culture, opportunities for training and advancement and benefits. Jobs are listed in two categories—LSI positions and former Agere positions—though these two lists will soon be merged. Jobs are

Visit Vault at **www.vault.com** for insider company profiles, expert advice, career message boards, expert resume reviews, the Vault Job Board and more.

V**A**ULT CAREER LIBRARY   **297**

searchable by type, location and function. To apply, job seekers must first create an account.

LSI offers its employees ample opportunities for training, 360-degree reviews and even has a tuition assistance program. Benefits include 401(k) with company matching and a stock purchase plan, as well as a health plan and even insurance against identity theft.

Mantech International Corporation

12015 Lee Jackson Highway
Fairfax, VA 22033
Phone: (703) 218-6000
Fax: (703) 218-8296
www.mantech.com

LOCATIONS

Fairfax, VA (HQ)

Aiea, HI • Alexandria, VA • Bethesda, MD • Burlington, MA • Chantilly, VA • Colorado Springs, CA • El Segundo, CA • Ellicott City, MD • Fairmont, WV • Falls Church, VA • Glen Burnie, MD • Greenbelt, MD • Hinton, WV • Johnstown, PA • Lanham, MD • Lexington Park, MD • Miami, FL • New York, NY • Norfolk, VA • North Charleston, SC • San Diego, CA • Sarasota, FL • Sierra Vista, AZ • Springfield, VA • Tampa, FL • Vienna, VA • Wallops Island, VA

Additional locations in 42 countries worldwide.

THE STATS

Employer Type: Public Company
Stock Symbol: MANT
Stock Exchange: Nasdaq
Chairman & CEO: George J. Pedersen
2006 Employees: 6,000
2006 Revenue ($mil.): $1,137

DEPARTMENTS

Administrative Programs & Operations • Administrative Services • Consulting Services • Contracts, Purchasing, Legal & Proposal Services • Customer Service & Technical Training • Engineering & Support Services • Engineers/Scientists • Facilities • Finance/Accounting • HR • IT • Intelligence • Logistics Support Services • Marketing, Sales & Business Development • Multimedia Support • Networks & Telecommunication Services • Programs & Operations • Publications & Graphic Arts • Scientific/Analytical • Security • Systems Development • Technical Analysis • Technical Assistance & Support

KEY COMPETITORS

BAE Systems
Lockheed Martin
SAIC

EMPLOYMENT CONTACT

www.mantech.com/careers

Visit Vault at **www.vault.com** for insider company profiles, expert advice, career message boards, expert resume reviews, the Vault Job Board and more.

VAULT CAREER LIBRARY   299

THE SCOOP

Ladies and gentlemen, it's ManTech!

ManTech has a long history of providing its country with high-tech manpower (and, name notwithstanding, womanpower). The company provides information technology (IT) services to the U.S. government, including the DoD, DoJ, the Navy and NASA. ManTech's national security offerings include intelligence analysis, information security and communications system support. On the IT side, the company offers network security, computational forensics and code analysis. ManTech also provides engineering test services to NASA and the Navy. The overwhelming majority of ManTech's revenue comes from the U.S. government; about 1 percent comes from NATO countries.

From the Cold War to the war on terror

Staten Islander George Pedersen and mathematician Franc Wertheimer founded ManTech in 1968 as a two-person operation; at the time they had a single naval contract to provide war-gaming technology for submarines. Pedersen has remained with the company to the present day, and has steered the company into persistent growth by cannily focusing on acquisitions to "leapfrog forward into new areas" (his phrase), while remaining mindful of overreliance on federal contracts, diversifying its platform just before the end of the Cold War.

Acquiring minds want to know ...

The company's first acquisition came in 1971, when it bought a company that held a Navy Air Systems command contract; it had 18 employees, but it portended greater things. By the time of its IPO in 2002, ManTech had acquired 36 other companies. After going public, the company bought two companies every year, most notably Gray Hawk Systems for $90 million in 2005, a company with 500 employees, many of whom possess the security clearances necessary for highly classified contract opportunities. From its humble beginnings with submarines, ManTech has long angled to break into the intelligence field; its first foray came in the early 1980s, when it bought a small, intelligence-based interest from Raytheon Corp. for $1.25 million. And the hard work has paid off: in 2006, ManTech reported classified contracts worth $200 million.

Company picnic

ManTech went public in 2002, a move which several analysts attributed to the "War on Terror" in the aftermath of the attacks on September 11th, but Pedersen insisted that he had taken 71 of his key executives to a company powwow the year before. They discussed the trajectory of the business there, deciding to focus away from commercial pursuits to government contracting once more. Pedersen subsequently hired Quarterdeck Equity Partners, a mergers and acquisitions consultancy, who suggested the company go public. "To be honest, I didn't really want to do that," Pedersen has said, "but the logic was overwhelming. And it really was the right decision."

It was the right place and the right time: company revenue skyrocketed from $431 million in 2002, to $1 billion in 2006. *BusinessWeek* named ManTech to its 2004 list of Best Small Companies, but the company has by now outgrown qualifying for the honor, indicative of its rapid growth since its IPO. Pedersen is thrilled, saying he expects the company to someday reach $5 billion in sales. He also isn't leaving the company he started more than 30 years ago, and worked nights and weekends to bring to this point. "I will remain chairman of the board until the day I die," he says, "I love this business. I love what we do every day."

Man-sized tech deals

In 2006, ManTech brought in revenue of $1.1 billion, and profits of $50 million. Revenue increased 12 percent over the previous year, while profits were up 13 percent. ManTech's revenue is forecasted to rise in the near term, since intelligence is critical to fighting the war on terror and to heading off any potential terrorist attacks. The long term looks good for ManTech, too: government IT spending, which has not seen a decrease since 1980, is predicted to rise by 5 percent per year until 2011. Based on contracts awarded at 2006's end, ManTech expects to see revenue increase by 10 percent in 2007. It isn't done buying things, either, as it also strengthened its position in intelligence contracts provision that year with its $18 million acquisition of GRS, a Virginia-based provider of intelligence services.

Trade an MSM for an SRS

They say the military is awash in strange acronyms, and this deal is proof: in 2007, ManTech sold a subsidiary, MSM, which provided background checks for security clearances, to a company owned by its CEO for $3 million. ManTech had originally purchased the company in 2003, but the division failed to turn a profit. Then, a few months later, ManTech turned around and purchased SRS, a company that specializes in space communications and weapons detection.

Finally, in 2007, the Navy awarded ManTech a five-year, $50 million contract to develop a sonar system that will interpret the noises made by ships and submarines moving through the water, allowing the Navy to ID them. Also that year, the Army awarded ManTech a $160 million contract to remove mines in Kuwait, Iraq and Afghanistan.

GETTING HIRED

Work for the Man

ManTech's careers page, at www.mantech.com/careers/careers.asp, allows candidates to search for jobs by location and function. The site also has information about career fairs at which the company recruits, including job fairs aimed at helping people in the military transition to civilian careers and job fairs for people with security clearances. Benefits at the company include health and dental insurance, tuition assistance, 401(k) and employee stock plan, pet insurance and a college savings plan. Any questions can be e-mailed to Jobs@ManTech.com.

Maxim Integrated Products, Inc.

120 San Gabriel Drive
Sunnyvale, CA 94086
Phone: (408) 737-7600
Fax: (408) 737-7194
www.maxim-ic.com

LOCATIONS

Sunnyvale, CA (HQ)
Beaverton, OR • Champaign, IL •
Colorado Springs, CO • Dallas, TX •
Fort Collins, CO • Grass Valley, CA
• Hillsboro, OR • Irvine, CA • Los
Angeles, CA • Melbourne, FL •
Phoenix, AZ • Sacramento, CA •
San Antonio, TX • San Diego, CA •
San Jose, CA • Santa Rosa, CA •
Tucson, AZ • Waimea, HI

34 locations throughout Europe and
Asia.

DEPARTMENTS

Administrative/Clerical • Advanced
Technology • Assembly/Packaging
Engineering • Device
Modeling/Technicians • Document
Control • Facilities •
Finance/Accounting • IT • Legal •
Manufacturing •
Marketing/Communications • Non-
Technical Fab Support •
Personnel/HR • Planning • Product
Engineering • Production Control •
Purchasing • Sales/Customer
Service • Security • Technician •
Verification • Wafer Fab/Fab
Research & Development

THE STATS

Employer Type: Public Company
Stock Symbol: MXIM
Stock Exchange: Nasdaq
Chairman: B. Kipling Hagopian
President & CEO: Tunc Doluca
2005 Employees: 7,980
2007 Revenue ($mil.): $2,007

KEY COMPETITORS

National Semiconductor
STMicroelectronics
Texas Instruments

EMPLOYMENT CONTACT

www.maxim-ic.com/company/
careers

Visit Vault at **www.vault.com** for insider company profiles, expert advice,
career message boards, expert resume reviews, the Vault Job Board and more.

VAULT CAREER LIBRARY 303

THE SCOOP

Analog has not yet left the building

Maxim Integrated Products is a manufacturer of analog and mixed-signal semiconductors. Analog chips, while not as fast at crunching numbers as the digital variety, have their place in measuring quantities that vary continuously, like temperature and pressure, converting audio to digital signals (and vice versa) and power management. Maxim's chips are used in the automotive, industrial, medical, communications and consumer products industries.

The chips are down

A number of former General Electric employees founded Maxim in 1983, betting that analog circuits, which had been largely discarded in favor of their faster, snazzier digital cousins, might still constitute a viable segment of the semiconductor market. Indeed, demand for analog chips grew during the decade, largely for use in new technologies such as cell phones and portable electronics. The company posted its first profit on a year's sales in 1988, and immediately had its initial IPO.

Leftovers are yummy!

Despite fluctuations in the health of the industry through the 1990s, Maxim played its analog chips to its advantage—they don't become obsolete as quickly as digital chips, and they can be made for much cheaper in older factories, thus insulating the company from the slings and arrows of the industry's outrageous fortunes. Also, Maxim was one of the few companies that hadn't switched to digital chips, and benefited by picking up the effective leftovers in the market. It was a strategy geared toward the long term, and the company invested in its employees and its R&D in the meantime, quickly earning recognition as an excellent place to work and a well-run small tech firm. In both 1991 and 1992, it was named to *BusinessWeek*'s Top 100 Small Companies and *Forbes*' Top 200 companies overall.

While many of its semiconductor peers suffered through cycles of boom and bust, Maxim hummed along, seemingly immune. Revenue increased every year from 1989 to 1991, while the rest of the industry suffered from a recession. In fact, the company has not suffered a loss in any of the years since, and profits increased from $74 million in 1991 to $334 million over the next 10 years. Maxim then hit a bit of a plateau, as profits didn't increase again until 2004. But, by 2006, *Fortune* noted that Maxim had the highest profits of any semiconductor company in the Fortune 1000. Suddenly, Maxim wasn't so small anymore, and it was soon encountering the sort of trouble befitting a large tech firm.

Considering the options

In 2006, the company took in revenue of $1.8 billion and made a profit of $400 million, down $78 million from the previous year. That year, Maxim was investigated by the SEC for questionable options practices. The audit of the options practices delayed the filing of several forms with the SEC, which caused the governing body of Nasdaq to threaten to delist the company's shares from the exchange. A Nasdaq board heard Maxim's case, and trading of the stock continued through the spring of 2007. To date, the company has still not filed its annual reports with the SEC for 2006 and 2007, citing a full audit of past records to ensure their accuracy in the face of accounting irregularities. The company has reported, however, $2 billion revenue for 2007—no word yet on its profits.

Gifford bows out

In 2006, Maxim's CEO and leading co-founder, Jack Gifford, elected to step down from his post for health reasons, in favor of a consulting role. (With the company facing investigation by the SEC on top of its usual operations, the stress exacerbated some of Gifford's health problems.) He was replaced by veteran Tunc Doluka, who has been with the company for 25 years.

Maxim goes to Irving

In 2007, Maxim bought an eight-inch wafer factory in Irving, Texas, for $38 million. The factory is capable of forming 20,000 silicon wafers, upon which microchips are created, per month. The factory, on 39 acres, has plenty of room to expand, and can max out at a capacity of 30,000 wafers per month.

GETTING HIRED

Take your career to the Max

Maxim's careers site, at www.maxim-ic.com/company/careers, provides information about job openings, culture, benefits, internships and entry-level programs. Jobs can be searched by keyword, function and location; to apply to a position, a candidate may fax, mail or send in contact information and a resume via a web form. The company's culture is distinguished by an emphasis on clear, efficient communication and continuous improvement of job functions. Benefits for Maxim employees include a choice of medical and dental plans, subsidized gym membership, 401(k) and a discount stock purchase plan. The company only acknowledges two holidays per year, but employees accumulate 12 floating holidays in addition to their annual allotment of vacation time.

Hopeful interns and aspiring college graduates are invited to submit their resumes to Maxim's database. College graduates may apply to positions in engineering, IT, finance, legal, accounting, customer service and purchasing.

McAfee, Inc.

3965 Freedom Circle
Santa Clara, CA 95054
Phone: (408) 988-3832
Fax: (408) 970-9727
www.mcafee.com

LOCATIONS

Santa Clara, CA (HQ)
Beaverton, OR • Mission Viejo, CA
• Oakbrook, IL • Plano, TX •
Sunnyvale, CA • Westborough, MA
• Amsterdam • Aylesbury, UK •
Slough, UK • Bangalore • Brussels •
Madrid • Milan • Paris • Tel Aviv •
Vienna • Waterloo, Canada • Zurich

DEPARTMENTS

Accounting/Auditing •
Administrative & Support Services •
Advertising/Marketing/Public
Relations • Building & Grounds
Maintenance • Computer Services •
Computers, Hardware • Computers,
Software • Customer Service & Call
Center • Executive Management •
Financial Services • HR/Recruiting •
IT • Legal • Manufacturing &
Production • Project/Program
Management • Sales • Sales—
Account Management

THE STATS

Employer Type: Public Company
Stock Symbol: MFE
Stock Exchange: NYSE
Chairman: Chuck Robel
President & CEO: Dave DeWalt
2006 Employees: 3,290
2006 Revenue ($mil.): $1,142

KEY COMPETITORS

CA
Microsoft
Symantec

EMPLOYMENT CONTACT

www.mcafeecareers.com

Visit Vault at **www.vault.com** for insider company profiles, expert advice,
career message boards, expert resume reviews, the Vault Job Board and more.

VAULT CAREER LIBRARY   307

THE SCOOP

Network defender

McAfee provides software that fends off all manner of digital threats, like spam, spyware, phishing, identity theft and unwanted guests snooping about the network. McAfee's products are sold to both the consumer and business markets for use on everything from servers to cell phones. In 2006, the company enjoyed total revenue of $1.1 billion.

Immune system for the computer

McAfee began in 1987, when a Lockheed engineer named John McAfee started distributing his home-brewed antivirus software for free via an online bulletin board. News of his product spread by word of mouth, and soon satisfied users were convincing their bosses to pay to license the product for company computers. McAfee founded McAfee Associates in 1989, and left Lockheed in January of the next year, quickly turning a profit. Two years later, the company had its first stock offering.

Computer—and spiritual—healing

In 1993, John McAfee got back to playing with software design, and demoted himself to chief technology officer. He left the company shortly thereafter, re-emerging with Tribal Voice in 1997—the startup is considered by many to be the first social networking site, and one of the pioneers of instant messaging, but it failed after going head-to-head with AOL. However, McAfee found a lucrative career afterwards as a yoga teacher; he's also written many books on the subject, including *The Secret of the Yamas* and *The Fabric of Self: Meditations on Vanity and Love.*

Enter the acquisitor

McAfee found a new CEO in Bill Larson, a computer veteran from Sun and Apple—and the acquisitions began! Larson immediately set the company on a course beyond viral software, and targeted network management software as its best opportunity for expansion. The firm purchased two companies specializing in the software in 1994: Brightwork Development and Automated Design Systems. The next year's acquisition of Saber Software for $41 million, together with Brightwork, Automated

Design and a host of others, gave McAfee a hold on 41 percent of the U.S. market for this management software.

By now, McAfee's trademark antiviral software only accounted for 60 percent of company revenue, but the company still dominated the market; this enabled it to undercut competitors by selling its own products at a bargain rate. In addition, harkening back to its origins in John McAfee's living room, it sold nearly all of its software online, saving both packaging costs and cutting out the extra step of selling to retail stores. Larson and McAfee were flying high—some analysts considered it the most profitable company in the world.

Corporate catfights

Larson, however, soon earned a reputation for playing hardball with his acquisitions—and acquisition targets. In late 1995, McAfee attempted a $1 billion takeover of Cheyenne Software, who felt strongly against it. Cheyenne sued McAfee to prevent the takeover, alleging securities fraud, and McAfee gave up the attempt. Larson then developed a rivalry with Symantec's then-CEO, Gordon Eubanks. The two companies sued each other, with McAfee seeking $1 billion in damages for defamation. Larson capped the activity with his issuance of a press release calling Eubanks "an accused felon for trade-secret violations."

But McAfee's thirst for acquisitions was unabated by Larson's ugly dustups with other tech firms—it merged with Network General in 1997 for $1.3 billion, its biggest deal yet. The new company was known as Network Associates until 2004— after the tech bust, the company unloaded a number of ancillary businesses, and thereafter changed its name back to McAfee.

Fending off the competition

McAfee's 2006 revenue rose 16 percent over 2005 numbers, bringing in more than $1.1 billion for the year. Profits, however, were down by 1 percent year-over-year (they came in at $137 million). The company expects demand for its products to lift revenue for 2007 to the neighborhood of $1.2 billion for the year.

Four more acquisitions in 2006 helped McAfee boost its offerings in online security—a necessary step for an industry where big competitors, like IBM and Cisco, are fast approaching. In April 2006, McAfee acquired SiteAdvisor, a company that created software that alerts web browsers to sites that may be sources of spyware, phishing scams and other forms of malware. Then, in June, the company bought

Preventsys, a company that specializes in network security. It followed that acquisition with a double-header in October, when it acquired Citadel, provider of security enhancements to businesses, and Onigma, a company that specialized in data security.

Mind the GAAP

While revenue at McAfee was rising, the company was weathering a number of financial scandals, dating back to CEO Larson's tenure. Larson, along with president Peter Watkins and CFO Prabhat Goyal, abruptly resigned in 2001, when fourth quarter revenue was $120 million below expectations. The next year, the SEC filed suit against Goyal and a number of other McAfee executives (but not Larson), alleging that the company inflated its sales figures by $622 million between 1998 and 2000.

George Samenuk, an old IBM hand, succeeded Larson as CEO and managed to settle the matter for the relatively small amount of $50 million in January 2006. Samenuk proudly heralded his "Ethics First" program at the company and said that it would be the "hallmark" of his tenure there.

But in June of the same year, the SEC started another investigation of McAfee, this time looking at its stock options policies. The commission eventually found that illegal securities enhancements had been going on since 1997—which may cost the company as much as $150 million. In the brouhaha, the general counsel was sacked, and CEO Samenuk expediently decided to retire.

What now?

Rumors abounded about the company's impending acquisition, but in 2007 McAfee hired Dave DeWalt as its new CEO to turn the company around. A former collegiate wrestler, he is a far cry from the company's yogi founder! Although he brings valuable tech experience as a former EMC executive, and early returns on his performance are good—quarterly earnings in 2007 have shown both profit and revenue growth. The financials are his most pressing concern, though; the company expects to take a hit between $100 and $150 million for the restatement of past earnings and it isn't out of the woods yet—it has failed to file its last two annual reports with the SEC, and is still sorting through its old records.

GETTING HIRED

Secure a job at McAfee

McAfee's jobs site, at www.mcafeecareers.com, provides information about job opportunities, training and benefits at the company. Job listings are searchable by location, category and keyword. To apply, candidates must upload a resume and provide contact information via a web form. The HR department can be contacted with any questions or comments at recruiting@mcafee.com.

McAfee's benefits vary by office location and job title, but the site assures candidates that flexible working conditions and emphasis on work/life balance are priorities for the company. McAfee also offers tuition assistance for job-related classes, as well as training in sales, technical skills, and management and leadership.

Visit Vault at **www.vault.com** for insider company profiles, expert advice,
career message boards, expert resume reviews, the Vault Job Board and more.

VAULT CAREER LIBRARY   **311**

MEMC Electronic Materials, Inc.

501 Pearl Drive
P.O. Box 8
St. Peters, MO 63376-0008
Phone: (636) 474-5000
Fax: (636) 474-5158
www.memc.com

LOCATIONS

St. Peters, MO (HQ)
Pasadena, TX
Sherman, TX

Chonan City, Korea
Hsinchu, Taiwan
Kuala Lumpur
Merano, Italy
Novaro, Italy
Tokyo

DEPARTMENTS

Accounting/Finance • Corporate
Development • Customer Service •
Human Resources • IT • Treasury •
Legal • Logistics • Manufacturing •
Marketing • New Products • Quality
• R&D • Sales & Marketing •
Subcontractor Operations

THE STATS

Employer Type: Public Company
Stock Symbol: WFR
Stock Exchange: NYSE
Chairman: John Marren
President & CEO: Nabeel Gareeb
2006 Employees: 5,500
2006 Revenue ($mil.): $1,540

KEY COMPETITORS

Shin-Etsu Handotai
Siltronic
SUMCO

EMPLOYMENT CONTACT

www.memc.com/jobs-splash.asp

THE SCOOP

Silicon jockeys

MEMC is a global leader in the manufacture and sale of wafers and related intermediate products to the semiconductor and solar industries. MEMC has been a pioneer in the design and development of wafer technologies over the past four decades. With R&D and manufacturing facilities in the U.S., Europe and Asia, MEMC enables the next generation of high performance semiconductor devices and solar cells.

MEMC is a major provider of silicon wafers, which are used by both the microchip industry and in solar panels. Wafers are slices of pure silicon crystals that have been grown from melted sand, and are the raw material from which microchips are manufactured. MEMC's customers include foundries where microchips are made.

Mr. Sandman, grow me a dream

The "M" in MEMC stands for Monsanto, which is also the name of a Midwestern agricultural company, founded in 1901. By the 1950s, the company was experimenting with chemicals, which dovetailed with Cold War era advancements in electronics. In 1959 it established the Monsanto Electronic Materials Company (MEMC) for the manufacturing and developing of silicon wafer technology. During the following decade, MEMC developed a number of wafer processing techniques, including mechanical chemical polishing in 1962 and epitaxial deposition, where a very pure silicon crystal is grown on the surface of a wafer, in 1966.

As wafer technology advanced and demand for microchips increased during the 1970s, MEMC was the first to move into selling wafers with a diameter of 100 mm in 1975. Four years later, it was able to produce wafers with a diameter of 125 mm and the company had upped the ante to 150 mm diameters by 1981 and 300 mm a decade later. In 1989, Monsanto sold off the division to a German company, who subsequently spun the company off through a stock offering in 1995, changing its name to the redundant MEMC Electronic Materials in the process. That year, MEMC also begin refining its own silicon. In 2002, a private equity firm purchased a majority stake in the company, but sold off all but a small portion of it in 2005.

The sun never sets on MEMC

In 2006, MEMC hashed out an eight-year, $1.6 billion deal to provide silicon wafers for solar panels to Motech, one of the world's largest manufacturers of solar panels.

To sweeten the deal, Motech is throwing in partial ownership of itself and the option to purchase a solar panel manufacturing facility.

High gas prices and an increased awareness of global warming are leading to an increased interest in environmentally friendly ways of generating electricity, like these solar panels. However, demand for the pure form of silicon needed to make such panels is extremely high, as microchip foundries compete with solar panel manufacturers for the material. Since refining silicon requires specialized plants capable of melting quartz, removing the oxygen atoms and purifying the resultant molten substance until it is 99.99999 percent pure (take that, Ivory soap!), demand is expected to remain high for the next few years—boding well for MEMC.

Grandaddy of polysilicon

In 2006, MEMC posted revenue of $1.5 billion and profits of $300 million. Revenue for the company has increased by around 40 percent year-over-year, and demand for MEMC's product should remain strong in the long term as use of microchips increases. Current levels of demand for silicon wafers may taper off in the near term, however, as more refining facilities are built. The market for polysilicon may experience a correction, as it did in 1999 and 2000, if demand for consumer electronics should fall off sharply.

Growth proposition

Indeed, in spring 2007 MEMC's stock got a modest bump, due to higher-than-forecasted earnings from the green energy sector. In addition, in May 2007, Standard & Poor's moved MEMC shares from the S&P MidCap 400 to the S&P 500 index.

GETTING HIRED

Chip in at MEMC

MEMC's careers page, conveniently located at www.memc.com/jobs-splash.asp, provides information on job openings by location. Candidates can apply either by filling out an online form, or by e-mailing a resume and cover letter to the address listed on the job posting.

Métier, Ltd.

3222 N Street NW, 5th Floor
Washington, DC 20007
Phone: (877) 965-9501
Fax: (202) 965-7600
www.metier.com

LOCATION:

Washington, DC (HQ)

DEPARTMENTS:

Business Development
Client Development Services
Consulting Services
Corporate Development
Knowledge Management
Marketing
Portfolio Management & Analysis
   Services
Product Development

THE STATS

Employer Type: Private Company
Chairman: Gerald McNichols
CEO: Douglas D. Clark

KEY COMPETITORS

IBM
Oracle
SAP

EMPLOYMENT CONTACT

www.metier.com/content/default.asp
?iContentID = 324

Visit Vault at **www.vault.com** for insider company profiles, expert advice,
career message boards, expert resume reviews, the Vault Job Board and more.

VAULT CAREER LIBRARY   315

THE SCOOP

Taskmasters for hire

Métier Limited, a leader in predictive project management software and services, derives its name from the French word "métier" meaning "specialty." The firm covers four main industries—commercial enterprise, systems engineering/consulting, government and defense—and provides services to high-profile clients, including BMW, IBM Global Services and Lockheed Martin, in addition to the FBI, U.S. Census Bureau, Department of Agriculture and Department of Energy. *Inc.* magazine has recognized it for the last two years as one the fastest growing privately held companies on its Inc. 500 list, and Métier proudly boasts some staggering statistics to back up the claim: as of 2005 it sported revenue growth of 353 percent, and its company website mentions that its staff has nearly doubled over the last 12 months.

A family affair

Douglas Clark and Sandra Richardson, a married couple, founded Métier in 1998— on the way back from their honeymoon, no less (the sunny Caribbean climes of St. Lucia, if you care to know). The Washington, D.C.-based firm has since billed more than 100,000 hours of consulting and analysis service. If the firm could be said to have a key software, it is WorkLenz—a predictive project management software designed by Clark, which syncs with Microsoft's Outlook e-mail application and helps managers track ongoing (or future) projects. Métier delivers its technology to private and public sector clients through partnerships with systems integrators and consultants. In 2005, the company collaborated with Lockheed Martin to create a system for handling the data from the 2010 census, which will be the first that allows people to fill out census forms online.

Not only do they do project management, they also do décor

In 2006, faced with an expanding employee roster, Métier opened a new office in the nation's capital. Located on N Street, Clark and Richardson designed the space with the tender, loving care of a family business—it incorporates elements of feng shui, which the company hopes will "inspire" its separate divisions to "further the growth of the company" (quotes taken from Métier's web site). The offices feature an open floor-plan, with no cubicles, full-length windows and balconies where employees can hold meetings or eat lunch. The office also sports such cushy details as glass-accented desks and conference rooms, wood floors and leather furniture. Despite the high-design atmosphere, dogs are welcome in the office.

Educational experience

In 2007, Métier announced that it had scored a deal to manage 200 investments for the Department of Education using its WorkLenz software. The deal with the DOE is structured so that the department doesn't have to invest in new computers, IT guys, or bandwidth, paying Métier for the use of its IT resources instead.

GETTING HIRED

Find your Métier here

Métier lists current job openings, and descriptions of those openings, on its careers site, at www.metier.com/content/default.asp?iContentID=324. (Or just go to the "About us" link from metier.com and select "Careers.") To apply for a position, candidates can submit a resume and cover letter to jobs@metier.com. Benefits include health insurance, 401(k), technical training and equity in the company.

Métier provides information about the company in two PDFs on its careers page. *A Job Seeker's Guide to Métier* allows hopeful Métierites to brush up on what exactly project management is, while *Inspiring Offices for Inspiring Individuals* provides information on Métier's tasteful décor and culture. But the company is looking for the cream of the crop, as CEO Clark has stated that 0.4 percent of applicants to the company get hired. They are looking for "sophistication," says Clark, by which he means people who "when you ask them a question, they'll pause for a minute, they're thinking. They're playing chess a few moves in advance."

Micron Technology, Inc.

8000 South Federal Way
P.O. Box 6
Boise, ID 83716
Phone: (208) 368-4000
Fax: (208) 368-4435
www.micron.com

LOCATIONS

Boise, ID (HQ)
Aguadilla, PR • Allen, TX • Lehi, UT
• Manassas, VA • Minneapolis, MN
• Nampa, ID • Pasadena, CA • San
Jose, CA

Avezzano, Italy • Bangalore • Beijing
Berlin • East Kilbride, UK • Erd,
Hungary • Fujian, China •
Giesshuebl, Austria • Kanata,
Canada • Munich • Nishiwaki,
Japan • Osaka • Oslo • Paris •
Seoul • Shanghai • Shenzhen, China
• Singapore • Stockholm • Taipei •
Tokyo • Tsukuba, Japan • Vantaa,
Finland • Virginia, Ireland

DEPARTMENTS

Accounting/Finance • Administration
• Business Development •
Communications • Engineering •
Facilities/Site Services • HR •
Information Systems • Legal •
Materials/Purchasing/Logistics •
Production • Sales & Marketing •
Technicians

THE STATS

Employer Type: Public Company
Stock Symbol: MU
Stock Exchange: NYSE
Chairman & CEO: Steve Appleton
President: Mark Durcan
2006 Employees: 23,500
2006 Revenue ($mil.): $5,272

KEY COMPETITORS

Hynix
Qimonda
Samsung Electronics

EMPLOYMENT CONTACT

www.micron.com/jobs

THE SCOOP

Small chips, big business

Micron Technology is a major manufacturer of semiconductors for the consumer electronics, medical and automotive industries, including memory chips for computers, cell phones, digital assistants, digital cameras and MP3 players, as well as image-sensing chips for cameras and medical equipment. Its subsidiaries include Crucial Technologies, which sells memory upgrades to the end-user market, and Lexar, which sells USB Flash drives, memory cards and card readers. Micron was ranked No. 438 on the 2006 Fortune 500.

Chip set

In 1978, a group of engineers incorporated Micron Technology in Idaho. At first, the company operated out of the basement of a dentist's office, but by 1981 the new company had established its first fab for building semiconductors in Boise. Three years later, Micron had its first IPO. The company weathered a drop in the price of memory that drove several competitors out of the market in 1985, and continued to expand its operations through the late 1980s and early 1990s. Its business got a big boost in 1998, when it acquired Texas Instruments' memory arm, and in 2001, when it picked up a stake in a Japanese semiconductor manufacturer. Following the burst of the tech bubble, the company was in the red for 2002 and 2003, but returned to profitability in 2004.

Braced for the plunge

In 2006, Micron acquired Lexar, a manufacturer of USB drives, flash memory cards and memory card readers for photography, industrial and communications customers, in a stock-for-stock merger. Micron's revenue that year was $5.2 billion, with profits of $400 million. The good times may not last, though, as a number of flash memory companies have struggled with falling prices early in 2007. Micron has started to pool its resources with major competitors, who are also struggling with the market downturn: in February it announced a joint venture with Intel, whereby each company's Singapore-based flash production operations will work together.

China fab

Micron is attempting to keep even with the memory industry, growing its global operations in a bid to stay competitive. In 2007, it announced its first fab in China, scheduled to be open for business in 2008. Micron already has sales offices in Shanghai, though this will be its first manufacturing venture in the country. The factory, which will employ 2,000 people, will be responsible for testing and assembling flash memory and image-sensing chips. The factory will be located in Xi'an, in midwestern China.

Domestic squabbling

As Micron grows increasingly global as a company, interests back in America—including its own board—have grown restless, especially as the flash memory market appears to be crashing. Micron's recent quarterly results, posted in July 2007, were disastrous: the company lost $225 million, and immediately cut 873 jobs in its Boise headquarters. Micron is the largest nongovernmental employer in Idaho, and watched closely by the state—its Boise outpost is even referred to as "Treasure Valley."

A swarm of local newspapers and politicians immediately voiced concern, and they were joined by two decade-long Micron board member Gordon Smith, who suggested that Chairman and CEO Steve Appleton step down, as he was leading the company astray and simply "not a moneymaker." A week later, Smith resigned his own post on the board, followed closely by news that the company had laid off another 107 from its Boise plant, and could be letting 130 more go by the end of August. Appleton didn't heed the calls to step down as chairman and CEO, but did resign his post as president in the aftermath of the hubbub.

GETTING HIRED

Chip in at Micron

Micron Technology's employment information is listed at www.micron.com/jobs. There is information on open positions at the company—searchable by function, location and keyword—benefits, college recruiting and internships. To apply for a position, candidates must first create a profile with a resume and contact information.

Benefits include health, life, dental and vision insurance, 401(k) with company matching, bonus programs, stock purchase plan, generous vacation allowance, a free on-site fitness center in Idaho (with trainers and classes offered throughout the day) and an on-site health clinic at HQ, for physicals, ergonomic consultations and any other health issues that may crop up during the workday.

The college recruiting section allows recent graduates to submit a resume, in addition to checking out potential positions at the company in information systems, engineering, technical jobs, production and operations management. Majors sought by the company include chemical engineering, computer engineering, computer science, electrical engineering, industrial engineering, manufacturing engineering/technology, mechanical engineering, microelectronics, physics and production operations/management, and allied areas of study. There is a helpful chart at www.micron.com/jobs/northamerica/college/majors that correlates major areas of study with job opportunities at the company. There are also helpful pieces of advice for composing a resume and acing the interview. The company recruits at its offices in Allen, Texas; Boise, Idaho; Manassas, Virginia; Minneapolis, Minnesota; Pasadena California; and San Jose, California.

To be an intern at Micron, you must be enrolled in a course of study to attain a bachelor's, master's or PhD, and have a GPA of at least 3.0. Previous work and manufacturing experience are both pluses. Internships are offered at the offices in Allen, Texas; Boise, Idaho; Manassas, Virginia; Pasadena, California; and San Jose, California, in engineering, technical, ops management and information systems.

The defining characteristic of interviews at Micron seems to be efficiency. One engineer reports, "The hiring process was very transparent and open. There was one HR interview and five to six technical interviews on ONE day. The interviews were very friendly," he adds. "They let me stay over the weekend to check out the city and real estate/apartment, which was nice," says an engineer.

OUR SURVEY SAYS

Get a little work at Micron

Once hired, workers have mixed reviews. "Micron is a very fast-paced company, though it's often without a well-defined direction," sighs one worker. "Corporate culture is laid-back," observes another. Just how laid-back is a matter of dispute. "You are monitored daily on when you leave and come in, it doesn't matter if you

Visit Vault at **www.vault.com** for insider company profiles, expert advice, career message boards, expert resume reviews, the Vault Job Board and more.

VAULT CAREER LIBRARY   321

were up until 4 a.m. on a support call, you better be at your desk at 8 a.m.," explains one source. "Extremely Big Brother-like," agrees her colleague. An early bird disagrees. "Hours [are] very favorable other than strict arrival at or before 9 a.m."

As befits a company named for an itty-bitty unit of measure, moving up in the ranks can be incremental at times. "Advancement is limited when trying to advance past 'middle-management,' you are really waiting for someone to die," notes a morbid contact. However, the company makes the journey worthwhile with good benefits. "401(k) matching funds, medical, dental, and vision insurance, stock purchase plan. Vacation time is given very generously," rattles off one manager.

Dress, naturally, varies by function. "Dress code can be casual to strict depending on where you work in the company. IT professionals working in satellite locations are allowed to wear shorts and sandals using discretion, while all main sites and manufacturing areas require long pants and close toed shoes."

Microsoft Corporation

One Microsoft Way
Redmond, WA 98052-6399
Phone: (425) 882-8080
Fax: (425) 936-7329
www.microsoft.com

LOCATIONS

Redmond, WA (HQ)
Atlanta, GA • Austin, TX •
Baltimore, MD • Boston, MA •
Chicago, IL • Cincinnati, OH •
Dallas, TX • Denver, CO • Des
Moines, IA • Detroit, MI • Honolulu,
HI • Houston, TX • Indianapolis, IN
• Los Angeles, CA • Miami, FL •
Milwaukee, WI • Minneapolis, MN •
Nashville, TN • New Orleans, LA •
New York, NY • Oklahoma City, OK
• Omaha, NE • Orlando, FL • Palo
Alto, CA • Philadelphia, PA •
Phoenix, AZ • Pittsburgh, PA •
Portland, OR • Raleigh, NC • Salt
Lake City, UT • San Antonio, TX •
San Diego, CA • San Francisco, CA
• San Jose, CA • Seattle, WA •
Tampa, FL • Tucson, AZ • Tulsa,
OK • Washington, DC

Over 100 US locations.

DEPARTMENTS

Consulting • Corporate Operations •
Finance • HR • Legal & Corporate
Affairs • Marketing • Sales • Support
• Technical

THE STATS

Employer Type: Public Company
Stock Symbol: MSFT
Stock Exchange: Nasdaq
Chairman: Bill Gates
President & CEO: Steve Ballmer
2006 Employees: 71,000
2006 Revenue ($mil.): $44,282

KEY COMPETITORS

Apple
Google
IBM

EMPLOYMENT CONTACT

www.microsoft.com/careers

Visit Vault at **www.vault.com** for insider company profiles, expert advice,
career message boards, expert resume reviews, the Vault Job Board and more.

VAULT CAREER LIBRARY   323

THE SCOOP

Colossus of the computer world

Microsoft is the world's largest software company and an undeniable force in the computing world. Its products are seemingly more ubiquitous than oxygen, and it takes some effort to avoid using them. It is the parent company of the Windows operating system, Explorer browser, the Office suite of programs (which include Word, Excel, Outlook, PowerPoint and Access), the Xbox game system and the Zune music player. The company was ranked at No. 49 in the Fortune 500 in 2007.

First-generation computer geeks

Microsoft traces its origins back to an unusual starting point for a market-leading corporation with a global presence—a high school classroom. The seeds of the company that would come to rule the worldwide software market were planted at Seattle's Lakeside School in 1968, when a group of forward-thinking mommies bought a computer for the students.

Tenth-grader Paul Allen and eighth-grader Bill Gates became friends while using the intriguing machine, and the two were soon accumulating massive computer-service bills. Their passion for computing quickly outstripped their economical resources as high school students, and the two boys began venturing downtown to a commercial computer center where they could use the hardware for free as long as they discovered and reported bugs in the center's system. After a great deal of bright-eyed pestering, they managed to learn the inner workings of the computers from the center's professional programmers.

Traf-O-Data days

Allen and Gates set up their first company, Traf-O-Data, in 1971. The plan was to use Intel 8008 chips to build computers that would analyze traffic volume for municipal traffic departments. The company "wasn't a roaring success," as Allen would later admit, but it gave the friends invaluable experience with microprocessors and software development. Soon, the two would be working with microcomputers, a compact-sized computer that uses microprocessors for the storage of data and the ancestor of the modern PC.

Allen and Gates spent the summer of 1973 working for TRW in Vancouver, Wash., managing hydroelectric dams with microcomputers. Gates enrolled as an undergraduate at Harvard but dropped out after his junior year to go into business with Allen.

At the time, BASIC (Beginner's All-purpose Symbolic Instruction Code) was an obscure 1960s computer programming tool invented to free computer programming from its mathematics-based, jargon-heavy past.

Allen and Gates had written a version of BASIC for their Traf-O-Data machine, and it was written well enough to serve as the programming language for the Altair 8800, the first commercial microcomputer. MITS, the maker of the Altair 8800, liked what it saw, and accepted Micro-Soft BASIC (the hyphen and the capital "S" would later be dropped) to program the 8800. The company also hired Allen and Gates for further programming and gave them office space in an Albuquerque strip mall.

Microsoft takes aim

In an office situated between a vacuum cleaner store and a massage parlor, the partners worked long hours, sometimes sleeping under their desks, only occasionally taking a break to catch a movie. Juicy contracts, like a $99,000 project writing programming languages for Texas Instruments, eventually came their way. By the time Microsoft relocated to Seattle in 1979, the company had 12 employees.

One year later, it was employing 35, one of whom was a Harvard friend of Gates'— marketing whiz Steve Ballmer. Today, Ballmer is Microsoft's CEO and one of the driving forces behind the company's lasting success (he also owns 4.8 percent of the company's common stock). One of Ballmer's first moves was to seal the deal on Microsoft's purchase of Seattle Computer Products' disk operating system, Q-DOS (the "Quick and Dirty Operating System").

Easy DOS it

Allen, Gates and Ballmer wouldn't have to wait long to reap the rewards of this purchase. IBM's new PC needed an operating system, and in 1980, the computer giant approached Microsoft about providing it. Allen and Gates quickly retooled the original Q-DOS software to create the new, fully functioning MS-DOS (now renamed the "Microsoft Disk Operating System"), designing it as an integral component of the new PC.

Microsoft made a relatively small amount of money directly from IBM for its programming, but Gates insisted that Microsoft retain licensing rights for MS-DOS. The new PC quickly became the industry standard, and, under its licensing contract,

Visit Vault at **www.vault.com** for insider company profiles, expert advice, career message boards, expert resume reviews, the Vault Job Board and more.

VAULT CAREER LIBRARY   **325**

IBM forked over royalties for MS-DOS to Microsoft for every single PC that sold. Microsoft's growth was assured, and it would soon supplant IBM as the major force in the computer industry. (Paul Allen left Microsoft in 1983, when he came down with Hodgkin's disease, taking over a quarter of the company's stock with him. Allen didn't do so badly out of the deal, as he is now the 19th-richest man in the world, according to *Forbes*.)

GUI wars

Microsoft used the popularity of MS-DOS to introduce the computing public to a number of clever technologies in the 1980s, although not all of them were original Microsoft creations. In 1983, Microsoft introduced its computer mouse and the first version of Word. Two years later, Microsoft introduced Windows, its first operating system with a GUI (graphical user interface), that is, a display which uses icons to symbolize different kinds of information.

The graphical user interface had first been developed by the Xerox Corporation in the 1970s, but it never got past the R&D phase. Apple Computers' Lisa was the first to release a GUI commercially in 1983, staking a lead in the fledgling PC market. The next year, as Microsoft rushed to release a PC with the same capabilities, Apple introduced the Macintosh, arguably the most influential PC of the 1980s, which made using the mouse and GUI even easier for the consumer.

Shocking the industry in 1988, Apple filed one of the most famous and controversial copyright lawsuits in tech industry history against Microsoft. In its complaint, Apple revealed that the two companies had entered into a secret pact in 1985, so that both could use the GUI pioneered by the Macintosh. When Microsoft released Windows version 2.03, however, Apple considered it too "Mac-like" and filed suit.

By 1994, both lower and appellate courts had found in favor of Microsoft in decisions still debated to this day. The legal victory cleared the way for Microsoft to become the world's dominant software PC software company, but Apple has made a comeback in recent years, renewing its rivalry with Microsoft in earnest (more of which below).

Everybody v. Microsoft

Microsoft emerged triumphant from the lawsuit with Apple in 1994, only to put out another fire. Netscape Communications Corp. had released its "Navigator" the same year, a technology that would soon be known as a web browser. The product was revolutionary in bringing the Internet to the public (only a few years before it was invented by Al Gore).

In 1995, Microsoft released its own web browser, Internet Explorer, along with the web portal MSN, to establish its own Internet presence. The company also launched its own cable news channel the same year, a joint venture with NBC called MSNBC. Two years later, Microsoft finally resolved the litigation with Apple, agreeing to invest in $150 million worth of non-voting Apple stock, and license Internet Explorer and MS Office to Apple PCs.

Microsoft soon started packaging Internet Explorer into its windows operating system, a practice that attracted the attention of lawmakers. The Department of Justice and a coalition of 20 state attorneys general charged Microsoft in May 1998 with violating antitrust laws by engaging in monopoly business practices. The suit claimed that Microsoft's incorporation of Internet Explorer into Windows unfairly limited competition.

The main contention was that the practice bundled two disparate technologies into one package at a price far lower than other manufacturers could offer, which effectively eliminated them from competing. As a result, innovation on this essential type of software suffered from being financially impractical.

In 1999, a judge ruled against the company, which settled with the 20 states between 2001 and 2004. Another antitrust case against the company was filed in Europe in 2003, where courts also ruled against Microsoft and ordered the company to pay a fine of $643 million and release a version of its operating system without Windows Media Player.

Beware the Google and Apple

Somewhere in the first decade of the new millennium, Microsoft's hegemony over computing services, both online and off, ceased to be absolute. The main challenge Microsoft now faces can be summed up in two words: Google and Apple. The companies are very different—Google is an upstart online-based company, less than 10 years old, while Apple is Microsoft's old rival from the 1970s and 1980s, seemingly back from the dead. Microsoft had a history of coming late to emerging technologies, most notably with the mouse and the web browser, but Google and Apple both exploited the maturation of the Internet to surpass Microsoft like never before.

Google developed a highly effective (and mysterious) algorithm which provides accurate results for online information searches and paired it with specific online advertising so as to turn a profit. Apple has partnered a product with a revolutionary design (the iPod personal music player) with iTunes, a music store that can be accessed over the Internet.

After a few years of gestation, both companies have grown extraordinarily successful, in the process seriously exposing growth areas of the computer industry that Microsoft has been neglecting. Apple has convinced five major record labels to sell its music through iTunes, thereby reinventing the music industry and making the iPod an enormous seller, especially during the holiday season. And Google, as it has grown, has most directly impinged on Microsoft's business, offering services for free online that Microsoft used to sell through Windows, signaling a shift in the industry from systems to services.

Neither Google nor Apple threaten Microsoft so much for their profits or market share, but for what they represent. Their success implicitly suggests that that Microsoft suffers from a lack of imagination—a notion reinforced by Microsoft's history, its copyright lawsuit with Apple being the most famous example. In fact, Apple has reinforced the concept of Microsoft as unimaginative and past-the-times with its famous ad campaign, personifying its Macintosh as a hip young dude and the Microsoft PC as a square businessman who hates being out of the office.

If you can't beat 'em, join 'em

In March 2005, Microsoft acquired Groove Networks; its founder, Ray Ozzie, developed the pioneering Lotus Notes application in the 1980s. After the Groove acquisition, CEO Steve Ballmer instantly turned to Ozzie, who was named Microsoft's chief technical officer, to spruce up Microsoft's Internet-based services, reorganizing management structure in September 2006.

Microsoft also whittled its business units from seven to three—Microsoft Platform Products and Services, Microsoft Business and Microsoft Entertainment and Devices. These units will account for, respectively, the Windows' server and tools, the organizing of MS Office products as well as software for small- and midsized businesses, and games, mobile phones, handheld devices products and the Xbox.

Also, Microsoft has moved to address its tendency of falling behind the waves of changing technology, as it tapped Craig Mundie to be its chief research and strategy officer in June 2006. His specific assignment is to keep track of emerging technologies, and not let Microsoft fall behind them, as it did with the Macintosh, the Netscape web browser and more egregiously with Google and the iPod.

Mundie has been with the company since 1992 and worked mostly on non-PC products such as video games and cell phone software. As Microsoft's biggest losses recently have come to Google and Apple outside its traditional PC market, Mundie was an attractive candidate for the job. He is currently focused on advancements in semiconductors, specifically the potential to increase their power without overheating, a new kind of "multicore" microchip.

The end of an era

In June 2006, Gates announced plans to leave his day-to-day job at Microsoft in 2008, while remaining chairman of the company. Ray Ozzie, who was promoted the previous year to oversee Internet offerings, immediately succeeded Gates as chief software architect. Gates and CEO Steve Ballmer outlined a plan in a news conference for other Microsoft execs to assume Gates' duties, and they also indicated that recent internal restructuring would shift more responsibility to lower-level execs, helping the company make strategic adjustments more rapidly.

The world's richest man announced he was leaving to focus on The Bill & Melinda Gates Foundation, the largest philanthropic organization in the world. He also expressed the desire to give away the majority of his estimated $50 billion fortune through the foundation's work, devoted primarily to fighting HIV, malaria and tuberculosis worldwide.

Redmond calling

In July 2006 Microsoft and Nortel announced that they would join forces to expand technology in the communications arena. Microsoft software will be married to Nortel phone networks in order to produce something the firms call "unified communications," a combination of e-mail, voice communications, conferencing, instant messaging, video and data transfer.

The two firms want all this to work with the existing phone system infrastructure so that companies wishing to upgrade to progressively more of their unified communications offerings can do so a la carte. Nortel and Microsoft intend their joint venture to be on the bleeding edge of communications technology. They are throwing their weight behind VoIP and expect unified communications to work on desk phones as well as the next generation of wireless devices. The partnership expects to ship products in 2008 or 2009.

Zune-r rather than later

In September 2006, Microsoft finally lifted the veil on its long-rumored portable music player, seeking to challenge iPod's hegemony over the market. The Microsoft Zune is a decidedly more social instrument than the iPod, featuring wireless technology that enables Zune users to beam music and playlists to each other. Zunesters can send full-length songs to their friends, which the recipient can listen to three times within a three-day time frame.

Like the iPod, the Zune also plays video files and MP3 songs from a user's personal stash, as well as files in those formats downloaded from a new service called Zune Marketplace. The player will come preloaded with music by artists on Virgin Records and EMI Music's Astralwerks. Despite the heavy hype surrounding the device, it claimed only 12 percent of the 2006 holiday shopping season's market for hard-drive players.

Clarity for the digital world

After five years, another heavily hyped Windows product also recently debuted, albeit to less clear-cut disappointment. Windows Vista was plagued by problems during its development (most notably when the entire project was scrapped in 2004), but the consumer version of the software finally went on sale in January 2007. The reception from tech reviewers has been a resounding "eh," despite some spiffy new features. The nuts and bolts haven't appreciably changed in Office 2007; despite rumors of an overhaul in 2003, the OS still depends on the NTFS file system, which was introduced in 2000.

There are bells and whistles aplenty, however. The most noticeable addition is Aero, a sleek, graphics-card-intensive GUI that sports transparent windows and a nifty feature in which several windows can be turned sideways and manipulated like a stack of cards. The interface has been updated to make it simpler to find files, while other new additions include items called Gadgets—sticky notes, a customizable desktop information bar with weather, RSS feeds and stock quotes.

If some of this sounds like old hat to Mac users, well, that's because it is—Mac OS X has a feature very similar to Gadgets called Widgets, for one. Microsoft has historically repacked pre-existing technology into more attractive forms, but they might finally have gotten too obvious about it.

Show me what you got!

Useful Vista security features include various anti-phishing and identity theft-prevention measures, as well as spyware-blocking software. Other (Microsoft-beneficial) features include anti-file-piracy elements and a software piracy-defeating "kill switch," which disables the use of any programs after one hour of browsing the Internet for copies of the OS that don't have a valid registration key. (According to the Business Software Alliance, a third of all software is obtained illegally.)

Analysts predict that Microsoft will see a bump in its revenue as people upgrade to Vista—but due to the computational horsepower needed to support it, the company expects the change to be gradual. In fact, Microsoft has extended its update service for XP until 2014.

The rollout of Vista and Office 2007 will have a unique effect on the IT sector of the economy, indicative of how much clout Microsoft still has. An industry study estimated that Vista will create 157,000 technology jobs in the U.S., as software gets tweaked to work with the new operating system and IT departments everywhere are called upon to reteach executives how to open documents in Word. The study went on to estimate that for every dollar Microsoft makes on Vista, the Windows world collects $18—which should lead to $70 billion in 2007 revenue alone.

In order to nudge this technological juggernaut into motion, Microsoft has embarked on a unique marketing campaign for Vista. Seeking to promote its benefits to tech-happy twentysomethings, Microsoft created a site with the indie comedian Demetri Martin, at clearification.com. The idea is that Martin finds clarity in his life when committed to some sort of institution, while Vista (which is rarely mentioned in the campaign) presumably performs the same process on one's computer. In another marketing stunt, Gates is making his copy of one of Da Vinci's notebooks, the Codex Leicester, available to Vista users who visit the British Library web site, where Da Vinci's Codex Arundel is available online.

... And audibility too

Although Microsoft already possesses a strong division of speech-recognition software, it added the private firm Tellme Networks, a California-based company specialist in the field, for $800 million in March 2007. Tellme has some huge

Visit Vault at **www.vault.com** for insider company profiles, expert advice, career message boards, expert resume reviews, the Vault Job Board and more.

VAULT CAREER LIBRARY   331

customers including Merrill Lynch and American Airlines and its Internet, data and voice services are utilized in calls made by more than 40 million people a month. Tellme services should enhance and elevate interfaces in existing Microsoft products, including those that use LiveSearch software in mobile phones.

Two months later, Microsoft brought aboard the online advertiser aQuantive for a whopping $6 billion. In line with CEO Ballmer's desire to compete with leading online ad-firm Google, he promoted aQuantive's CEO Kevin McAndrews to the head of Microsoft's own online ad department. The company has poached other executives from rival firms as well, including COO Kevin Turner (who came over from Wal-Mart in 2005) and CFO Chris Liddell (from International Paper Co., also in 2005). In October 2007, Ballmer stated that 25 percent of Microsoft's business would be devoted to advertising within a few years.

The Chessboxin' Xbox

In the summer of 2007, Microsoft's entertainment division got a new boss from outside the company: Don Mattrick, who came over from Electronic Arts, famous for its EA Sports' *Madden NFL Football* game. Mattrick is looking to revitalize the Xbox, which killed the competition back in 2004 with its second installment of *Halo*, a critically acclaimed video game in which a supersoldier attempts to prevent the destruction of humankind. *Halo 2* recorded $125 million in sales in its first 24 hours on the market, easily exceeding Microsoft's predictions, and ranking as one of the top-five computer games in history for first-day sales.

Halo 3 arrived in September 2007, and Microsoft badly needed it to be a hit for the sake of the company's gaming future. The new game is exclusive to the Xbox 360, which will need it for battle against its rivals in the latest gaming console war: Sony's Playstation 3 and Nintendo's surprising Wii. As of September, Microsoft had sold nine million 360's since their release in 2005, followed closely by the Wii, which had sold a nearly identical amount only since late 2006. (The Playstation trails behind, with about four million sold.)

Early reviews are good, although some contend that it is merely *Halo 2* with more bells and whistles. *BusinessWeek* has rebutted this criticism by pointing out that the game's new features include a potentially revolutionary social networking component. Microsoft engineers have made the game compatible with one of the largest *Halo* fan web sites, Bungie, and now gamers can record their own footage within the world of the video game, sharing clues and tips with each other online.

Halo 3 also allows gamers to design their own landscapes and weapons arsenals—all the better for killing aliens.

BusinessWeek quoted Dan Hsu, editor-in-chief of *Electronic Gaming Monthly*: "(Microsoft) bring(s) such a fine level of polish to their games—and no one else is as ambitious about connecting every piece online." But perhaps the most important reviews for the game are its sales figures: on its first day of its release, *Halo 3* easily beat its predecessor's record, racking up about $170 million in sales.

Shedding some Silverlight on the issue

In 2007, Microsoft announced that it would introduce Silverlight, a browser plug-in for Explorer, Firefox and Safari, that allows web sites to add music and animation. This new product brings the company up against Adobe's Flash, up until recently the only software of web sites featuring gratuitous and distracting animation, irritating music and loading times that would try the patience of sedimentary rock.

Adobe is fighting back with the launch of a music and video player that can run in any operating system—clearly Windows Media Player's turf. Adobe is also taking Flash to the next level with a project codenamed Apollo. Letting Flash loose from the boundaries of the browser, it will provide users with an all-singing, all-dancing, constantly-updated experience—right to users' desktops.

Most recently, Ray Ozzie has dismissed as "wrong-minded" the idea that computer hardware and software can one day be exclusively reachable over the Internet, a clear rebuke to Google. In fact, it looks like Microsoft will stick to its guns, as it has begun charging for subscriptions of computer security services and will soon be introducing a set of customer services, such as photo sharing. The company is very proud of its most recent operating system, trumpeting the sale of 60 million copies of Windows Vista so far, which it says is the strongest debut performance of any of its operating systems yet. *BusinessWeek* agrees, citing the high-paced sales of Vista as reason to place Microsoft at ninth on the 2007 edition of its InfoTech 100 (Apple is ranked higher, at sixth place).

GETTING HIRED

The basics

Microsoft's careers page, at www.microsoft.com/careers/ provides information about job openings in the U.S. and abroad, career trajectories and benefits. Jobs are searchable by location, title, category and product; in order to apply, applicants must create a profile. Career paths include operations, human resources, marketing, support, financial, legal and sales.

For recent graduates who might wish to work for Gates, the company provides information on internships and entry-level opportunities for graduates with bachelor's and MBA degrees. Internships are offered in core technical, software design, finance, HR, IT, marketing, user assistance and publishing. Interns get subsidized housing and assistance in purchasing a car or bicycle, subsidized health club membership, discounts on software, parties and bus passes.

Microsoft's college recruiting bandwagon makes stops at Brigham Young, Brown, Harvard, MIT, Princeton and the University of Florida, among other colleges. Hopeful Microsofties are also welcome to check out the Jobs Blog, at blogs.msdn.com/jobsblog.

Heaps of praise

The company landed in the No. 50 spot on the 2007 *Fortune* list of the Top 100 Best Places to Work. (Internationally, it's also been recognized as one of the best places to work in Canada, Greece, Sweden, Spain, Ireland, and the entire continent of Europe overall.) The company has some heavily pampered employees—perks offered by the company include generous health benefits—and at its Redmond campus, valet parking and on-site dry-cleaning.

Recently, new HR chief (and former product manager) Lisa Brummel, dubbed the "consigliere of happiness," has implemented a number of changes to the company's benefits, human resources and internal communications policies to improve employee morale and retention rate (turnover hovered at 10 percent in 2005; it has since dropped to a more comfortable 8.3 percent in 2007). The grocery delivery service (which very few employees utilized) is gone, and in its place Brummel's newest program, Mobil Medicine, dispatches doctors to pay house calls to sick employees. Brummel also spearheaded an overhaul of the employee performance review process, revamped the recruiting center's décor (to be more hip and inviting, of course!) and is looking into opening satellite offices and implementing a telecommuting policy.

On campus

The Microsoft campus, in Redmond, Wash., is like a small city unto itself. From lawns and a quad to "a really good cafeteria that has everything from junk food to gourmet" (not to mention a 20-foot-tall chunk of the Berlin Wall and a manmade stream climaxing in a waterfall), Gates spared no expense to make his 30-building headquarters breathtaking. It features 4,000 pieces of artwork, 25 cafeterias, a company store and facilities for running and hiking, basketball, soccer, baseball and volleyball.

Aside from the great corporate campus, the company offers its U.S.-based employees a choice of medical plans, dental and vision coverage, 401(k) with company match, discount stock purchase program, discounts to health clubs in some areas, discounts on merchandise and event tickets and free beverages, including Starbucks coffee.

To compete with Google's benefits, Microsoft recently started offering employees discounts towards the purchase of hybrid vehicles. The cafeteria, often noted as sadly institutional and subpar to Google's, is now colonized by popular area restaurants dishing up their specialties. Microsoft also has plans to offer a shuttle bus like Google's, complete with Wi-Fi.

One unique culture-with-a-capital-C feature of life at Microsoft: the company is home to a 60-piece employee orchestra that holds six free concerts a year (on the Microsoft campus and at other venues around Seattle) and rehearses on Monday nights. It does not, however, count its trombone-playing chairman among its ranks.

Students, take note

Undergraduates interested in technical positions should be pursuing degrees in computer science or a related discipline, as well as be highly proficient in C/C++. Any internships or work experience in programming is useful, though not necessary. "For a college grad," says one insider, "if you haven't done any intern work at Microsoft, the only thing they'll have to go by is your GPA and the college you attended. Unless you can whiz-bang the interview, most likely you'll start in quality assurance (QA) as a developer." Microsoft's policy of hiring the brightest people it can possibly find means certain sacrifices; the company's bias is "toward intelligence or smartness over anything else, even, in many cases, experience."

Visit Vault at **www.vault.com** for insider company profiles, expert advice, career message boards, expert resume reviews, the Vault Job Board and more.

V/\ULT $_{LIBRARY}^{CAREER}$ **335**

For MBAs

Degrees from highly competitive schools are a must for the MBA who wants to work at Microsoft, though real experience, such as "work on web sites and internships at software companies" will get you far. Even for those applying to business-oriented departments, "technical skills are very well received," since "there's a glut of people applying who look very similar on paper." We hear that the company avoids hiring employees who cannot follow the technical side of the business. This does not mean that every marketing whiz is expected to know how to write sharp code, though; Ballmer knew little or nothing about computers when he joined up. "Passion for technology" is extremely important, but applicants need "not necessarily [have] a background" in it. "Problem-solving skills and intellectual horsepower" are more prized assets.

The war for talent

Insiders say Microsoft likes to confirm what you claim on your resume. Therefore, be ready to demonstrate any skills you have included in it. "If you said that you can write C/C++ or Basic code, your interviewer may sit you down at a terminal and ask you to do it," say insiders. We also hear that interviewees up for software engineering positions are also sometimes called to write code on whiteboards. If you listed relevant coursework, the interviewer may ask you problems related to those classes, so brush up on all your old subject matter.

Still, Microsoft is perpetually locked in battle with other leading firms, in technology and finance, for access to top candidates. This is particularly true when it comes to minorities, as Microsoft strives to maintain a diverse group of employees. By most accounts the firm is fairly successful, although with some exceptions. One recruiter admits, "In recruiting outstanding MBAs of color, we have the same problem as other companies—the pool is not that big."

The battle with rival firms has grown more heated in recent years as new CEO Steve Ballmer is responding to the Google threat by bringing in more executives from outside the company, a departure from the company's tradition of grooming homegrown talent. Brian McAndrews, who ran Microsoft's online advertising department in 2007, insisted on some engineering oversight in addition to his other responsibilities

Not for the faint of heart

The Microsoft interview is difficult to prepare for; its interviews are known for being as quirky as Google's, with the company asking abstract questions to see how interviewees think. The process even spawned a book, *How Would You Move Mount Fuji?* about the questions interviewers ask. Wild stories abound in which unfortunate applicants are confounded with questions that are confusing, or even unsolvable. The best way to prepare for these out-of-left-field queries is to expect them, and not to be rattled when one pops up. The interviewer wants to learn how you think.

According to one employee who made it through the process, "They are pretty harsh at interviews. However, they aren't necessarily looking for the correct answers to their questions, but to see how you handle the question, and your thought process to come up with an answer." Microsoft wants employees who can brainstorm numerous possible ways to solve the knottiest problems, and then work through the possibilities rationally to find the best one.

For the more traditional interview questions, learn as much as you can about the company. Familiarize yourself with all of its products, so that if the interviewers ask you to analyze a Microsoft business issue in a "case interview" (as they often do), you won't be stuck asking them to explain what the product you're supposed to improve or sell is.

One insider reports that Microsoft interviewers like to throw curveballs at job seekers by asking them to come up with repeated examples to questions like, "Tell me about a time you came up with a creative solution." "They'll ask you that, and then go, 'Tell me about another time ... and another ... and another.' They want to make sure that you're not just spewing out canned answers."

One easy part is the dress code, however: The company advises candidates to dress casually for interviews, rather than break out their Sunday best.

The MBA Interview

For newly-minted MBA grads, who rate the company as one of the top-10 post-graduation destinations according to a 2007 survey conduced by *Fortune*, Microsoft "targets the top-15 or so [schools]," and conducts "early recruiting in November." Applicants go through "an on-campus interview first with two people. If you pass that, they fly you to Redmond." For the second round, candidates "interview with four to six people in one day. Three peers, one to two people in the lead position and one group manager, who is always last." Each meeting lasts for about an hour, and applicants may interview with members of two different groups. The process is "very intense," but somewhat disorganized, we hear. The interviewers do not agree ahead of time on which areas each will probe. As a result, the session entails "a lot of overlapping questions." Most of these are "questions about software. They are not too lofty. They are [just] looking for enthusiasm for software." The occasional real-world case question may sneak its way in somewhere during the day.

The last hurdle

Insiders warn that "the interviewers e-mail each other" to find out candidates' strengths and weaknesses during the session. While the meetings are "very much improvised," those conducting the later interviews may try to tailor them according to the information they glean from their colleagues. The last interview is a good indication of where you stand: "If you don't see the group manager, [they have decided] you're not worth it." Either way, candidates do not have to wait very long to find out their fate. "The decision is made at the end of the day, based on the feedback of the recruiter and all the people who met with [the candidate]. There's no mulling it over, though it may take a couple of days" before the lucky phone call comes, adds one source.

One last thing

When entering into pay negotiations, several sources warn that the company lowballs acceptable candidates with below-average salaries that are supposedly boosted by the company's stock packages and benefits. One source points out, however, that "Microsoft no longer gives options, it's all stock grants that vest over five years."

OUR SURVEY SAYS

Results-oriented

"The culture here is focused around doing great work, but it is also a very fun place to be," said one insider. The atmosphere is far from stuffy, leaning if anything toward sophomoric: "We have a basketball hoop in our hall, which is a great way to blow off steam in the evenings or the middle of the day."

While average citizens probably would like to write off software programmers as lonely and uncommunicative, Microsoft sources usually disagree. As one designer says, "The atmosphere is very social—a good-sized chunk of my day is spent in other people's offices or with other people in my office discussing how to accomplish some task or the latest industry news." Unlike at Hewlett-Packard or Intel, where cubicles dominate, Microsoft's full-time employees work in real offices, "which is so much nicer than rows of cubes and zero privacy."

And while the company has swiftly grown into a multibillion-dollar monster, insiders say they "don't feel lost." One respondent says that "we work together in small feature teams, so even though this is a large company, you always feel like an important part of a team."

Lovin' it

Job satisfaction runs extremely high at Microsoft, and employees generally agree that "it is a pretty great place to work." While one might expect a somewhat mercenary attitude, focusing on dreams of stock-fueled fortunes, employees more often cite the importance of their work. "It's been very exciting to be a part of shaping the future of computing," says one respondent. Despite an uptick in turnover in recent years, and complaints about job advancement ("vertical and horizontal movement is unusual at and most won't see more than a couple of promotions in a five year period," claims one source), Microsoft has managed to hold onto many employees for the long run. "I plan to stay here as long as I can because I enjoy the work and the people I work with," says one 10-year veteran.

A lot like being in college ...

A bunch of people in jeans and T-shirts banging away on computers late into the night? Shaky hands from heavy-duty caffeine buzzes, with empty soda cans littering the desks? Throw in a few togas and you know exactly what we're talking about. Microsoft is late adolescence extended, and proud of it. "From music groups to jugglers to various team sports, there's always something going on in the buildings or around the campus," says one Microsoft contact. "In many ways, it's so much like being in college." Although there are "lots of group activities," the number of people who participate is "actually pretty small," which is also not dissimilar to college life. And the attitude? "How can people be so intense yet so laid back at the same time?" asks one insider, perhaps a bit cheekily.

Home? This is my home.

Microsoft employees had better appreciate their lovely Redmond campus; they certainly spend enough time there. "Programmers work about 80 hours a week," we hear. But

Visit Vault at **www.vault.com** for insider company profiles, expert advice, career message boards, expert resume reviews, the Vault Job Board and more.

V∧ULT CAREER LIBRARY   **339**

this doesn't mean that there is a consistent grind, week in, week out—like in so many other ways, schedules at Microsoft often resemble college. "Work hours vary based on where we are in a project cycle—it is a lot like being in school—you slack off after finals and at the beginning of the semester and work extra hard right before finals." Insiders cite the flexible hours as a plus, but with a wink and a laugh: "There's a joke around here that goes, 'There are no set hours at Microsoft, you can work any 12 hours of the day you want,'" one insider tells us. Such long hours are "not mandatory," we hear, "but generally this is what you'll have to do, not only to get your job done but also time you'll need to obtain additional skills to advance your career." While most employees work at Microsoft because they genuinely enjoy doing their jobs, the long hours can affect them. "You have to actively manage your time to make sure you keep a good balance in your life," remarks one contact. Another says, "Make sure you get your downtime."

Diversity issues

In October 2000, a handful of employees filed a $5 billion discrimination suit against Microsoft, alleging inferior treatment of salaried black and female staffers. On the whole, though, Microsoft's track record with women and minorities is good, probably because "people are valued based upon their contribution to the company rather than some superficial quality. Sex and race are not a factor at all." Employee groups, including Blacks at Microsoft (BAM), are visible on campus, hosting career days for students of color. These factors weighed on the U.S. District Court's decision to prevent the plaintiffs from bringing a class-action suit. Nonetheless, minority representation is skewed toward "Indians and East Asians. Most of the blacks are not American but are from the Caribbean, South America or Africa." In what is a fairly progressive move, the company included sexual orientation in its nondiscrimination policy and has received recognition for its inclusion policies for transgender individuals.

While the company is responsive to women on the whole, there may be a glass ceiling in effect. Of the top-20 executives, two are female—but they represent HR and marketing. And only 17 of the 125 top executives are women (14 percent), but female sources "don't feel uncomfortable around the office." For a company founded in 1975 (with no holdover executives and partners from the days of fewer opportunities), this is still a surprisingly weak showing. One female respondent insists that it depends a lot on which department one is in: "I'm in marketing and I work with about 80 percent women."

National Semiconductor Corporation

2900 Semiconductor Drive
P.O. Box 58090
Santa Clara, CA 95052-8090
Phone: (408) 721-5000
Fax: (408) 739-9803
www.national.com

LOCATIONS

Santa Clara, CA (HQ)
Arlington, TX • Calabasas, CA •
Chandler, AZ • Federal Way, WA •
Fort Collins, CO • Grass Valley, CA
• Indianapolis, IN • Longmont, CO •
Norcross, GA • Phoenix, AZ •
Rochester, NY • Salem, NH • San
Diego, CA • South Portland, ME •
Tucson, AZ

17 locations throughout Asia and
Europe, including design centers in
Finland, India and China, with
regional headquarters in Japan,
Germany and Hong Kong.

DEPARTMENTS

Administrative Services • Business
Planning & Materials • Engineering •
Executive • Finance & Accounting •
General • HR • IT • Marketing &
Sales • Operator • Production &
Manufacturing Support • Standard •
Technician

THE STATS

Employer Type: Public Company
Stock Symbol: NSM
Stock Exchange: NYSE
Chairman & CEO: Brian L. Halla
2007 Employees: 7,600
2006 Revenue ($mil.): $1,929

KEY COMPETITORS

Analog Devices
Linear Technology
Texas Instruments

EMPLOYMENT CONTACT

www.national.com/careers

Visit Vault at **www.vault.com** for insider company profiles, expert advice,
career message boards, expert resume reviews, the Vault Job Board and more.

VAULT CAREER LIBRARY 341

THE SCOOP

Analog chips for the digital age

National Semiconductor Corporation manufactures analog and mixed-signal chips for the medical, wireless communications, LCD, automotive and industrial markets; analog chips are used in mobile phones, displays, communications infrastructure, and test and measurement devices, among other items.

National divides its business into several segments: chips for power management, that direct the right number of volts to the appropriate part of the device; chips for LCD displays; wireless Bluetooth chips; analog chips that turn data (pressure, temperature, sound) into digital data that can be handled by a computer, as well as data (pictures, sound) that can be handled by a human; and high-reliability chips for military and aerospace devices. Major clients include Nokia, Sony, Apple, Motorola and Toshiba.

Constitution state rivalry

In 1959, Dr. Bernard Rothlein was an employee of Sperry Semiconductor—the engineering offshoot of an old electronics company that traces its roots back to 1910. That year, he took seven other engineers from the company and founded National Semiconductor in Danbury, Conn., just 20 miles north of Sperry's headquarters in Norwalk.

The company initially produced integrated circuits, a new technology that had been invented the same year as the company's creation, and sales were slow to pick up— only $5.3 million by 1965. National Semiconductor was struggling, and its founder's old company, Sperry Semiconductor, also prevailed in a patent infringement in the 1960s, leaving National on the ropes.

Sprague and Sporck to the rescue

At this low ebb, the east coast financier Peter Sprague (whose family had been in the electrics business since the 19th century) swooped in and bought the company in 1966. He recruited Charles Sporck, head of Fairchild Semiconductor at the time, to be National's new CEO and president (taking a hefty pay cut with the move). Sprague and Sporck would preside over the company together until 1991.

Under its new leadership, the company moved its operations (and HQ) to the hub of the Silicon Valley in 1967: Santa Clara, California. As semiconductors became increasingly prevalent in consumer electronics during the 1970s, National started to focus on providing chips for consumer goods, and had grown by 1971 to become the fourth-largest semiconductor company in the U.S. (Sprague is waiting for a similar turnaround with another of his software properties, Wave Systems; now run by his son, Steven Sprague, the company has not turned a profit in two decades.)

During the mid-1970s, National began making inroads under Sporck into the cheap and productive Asian market. The company opened factories in Asia, one of the first semiconductor firms to do so. These factories helped Sporck enter into the emerging video game market, but they came to handle a bulk of National's production, and their low costs were key to the firm's survival and success during his tenure. Sales figures reflected the correctness of Sporck's management: revenue was $42 million in 1970 and $365 million in 1976.

Turnabout and fair play

Of course, National's pursuit of low margins left it vulnerable as the Asian semiconductor industry matured. Into the 1980s, the company attempted to produce consumer products, such as digital watches and video games, and even its own computers, but they could neither match the low costs of Asian competitors nor the high-end wares of its American ones.

Sporck, he of the penchant for cost control and low margins, was impelled by this spate of Asian competition to become an active agitator for federal trade protections. In fact, he was one of the main forces behind the 1986 "Semiconductor Trade Agreement" between Japan and America, which sought (in the name of "fair trading practices") to open the Asian market to American competition and restrict the dumping of extra-cheap Asian products on the American one. But National was still one of the giants in the field, and purchased Fairchild Semiconductor in 1987 (which must have made Sporck smile), further strengthening its position in the market, although the company was to be divested a decade later.

Networking, baby!

In the 1990s, when networking was all the rage, National started making network controllers. The company left the microprocessor market in 1999, where it could not compete with Intel and AMD, and focused on providing chips for communications and wireless applications. In 2001, National shifted its focus to creating products for expanding wireless Internet access and remote control of home heat and air conditioning. Two years later, it renewed its focus on analog and developed an energy-efficient chip that increased the running time for handheld electronics (like phones and music players) running off of battery charges, increasing chip activity times by as much as 400 percent.

The chips are (slightly) down

In 2007, National reported a revenue decrease from the previous year, from $2.16 billion to $1.93 billion as it further divested its digital businesses. Profits also decreased, from $449 million in 2006 to $375 million. Shortly before announcing its results for the year, National finalized the purchase of Germany-based Xignal Technologies, a manufacturer of high-speed digital to analog converters. The deal boosts National's product line of analogue-to-digital chips; Xignal's product line includes a power-sipping, high-speed, high-bandwidth analog-to-digital converter, among other converters.

An Apple a day

One of the benefits of working for a semiconductor manufacturer might be the ability to sample its products—once they've been turned into consumer electronics, that is. National is pioneering a novel way to keep its workers informed of company goings-on. In 2006, it handed out 30-gigabyte video iPods to all of its workers, to keep them abreast of the latest company news and training while driving or at the gym. The devices aren't freebies, however—the company asks that employees give them back if they leave, though they can be purchased for a reduced price.

GETTING HIRED

I have one word for you: semiconductors

National's careers page, at www.national.com/careers, provides a searchable list of jobs at the company. To submit resumes for consideration, applicants must first create a profile.

In addition to its openings for experienced hires, the company offers internships and co-ops, opportunities to gain academic credit while working, for college students—just search for positions with "co-op" or "intern." Opportunities for college graduates also exist in engineering, manufacturing, marketing, finance, HR and information services. Entry-level employees attend college club events, like baseball games, happy hours and movie nights, and can take classes at National Semiconductor University, which cover subjects from technical writing to speed-reading.

Benefits offered by the company include a generous stock purchase plan (85 percent off the lowest price of the quarter), 401(k) with lavish company matching, profit-sharing and bonus plans, health and dental insurance, on-site dry cleaning, and a gym and cafeteria at several sites.

Visit Vault at **www.vault.com** for insider company profiles, expert advice, career message boards, expert resume reviews, the Vault Job Board and more.

VAULT CAREER LIBRARY   345

NCR Corporation

1700 South Patterson Boulevard
Dayton, OH 45479
Phone: (937) 445-5000
Fax: (937) 445-1682
www.ncr.com

LOCATIONS

Dayton, OH (HQ)
Albany, NY • Atlanta, GA • Baltimore,
MD • Bentonville, AR • Boston, MA •
Charlotte, NC • Chicago, IL •
Cincinnati, OH • Columbia, SC •
Dallas, TX • Detroit, MI • Duluth, GA
• El Segundo, CA • Germantown, MD
• Houston, TX • Las Vegas, NV • Los
Angeles, CA • Morristown, TN • New
York, NY • Peachtree City, GA •
Raleigh, NC • Rancho Bernardo, CA •
Richmond, VA • San Diego, CA • San
Francisco, CA • Suwanee, GA •
Viroqua, WI

Locations in 61 countries throughout
Asia, Europe and Latin America.

DEPARTMENTS

Administrative/Office Professional •
Cross-Functional Management •
Customer Center • Customer
Operations • Customer Service
Delivery • Customer Services &
Support Services • Engineering •
Finance • HR • Legal • Management
Information Services • Manufacturing
• Marketing • Product Management •
Professional Services • Public
Relations & Communication • Quality
• Real Estate • Sales • Strategy •
Supply Chain Management

THE STATS

Employer Type: Public Company
Stock Symbol: NCR
Stock Exchange: NYSE
Chairman: James M. Ringler
President & CEO: William Nuti
2006 Employees: 28,900
2006 Revenue ($mil.): $6,142

KEY COMPETITORS

Diebold
IBM
Wincor Nixdorf

EMPLOYMENT CONTACT

www.ncr.com/about_ncr/careers

THE SCOOP

It's all about the cash ... registers

NCR Corporation is a major provider of cash registers, ATMs, point-of-sale scanners, bar code readers, and paper and other consumables. It also makes all manner of self-serve kiosks, from hospital and hotel check-ins and merchandise check-out to event tickets and ordering at fast-food restaurants. The company also provides data warehousing services under its Teradata division, which is set to be spun off as an independent firm in 2007.

The accounting machine

Former coal salesman and retail shop owner John H. Patterson bought a cash register manufacturer in 1882 and founded the National Cash Register Company two years later. The first office, with 13 employees, was located in Dayton, Ohio. In 1906, Patterson commissioned inventor Charles Kettering to design the first cash register powered by an electric motor. Within a few years, Kettering had developed NCR's Class 1000 register, which remained in production for four decades, as well as the O.K. Telephone Credit Authorization System, for verifying credit in department stores. Just five years later, the company had sold one million electric cash registers.

In 1921, NCR introduced the Class 2000 accounting machine, which proved to be an important money maker: models of it were marketed up to 1973, making the 2000 the longest-lived of any NCR product. Frederick Beck Patterson, the founding Patterson's son, became president of the company in 1922, one year before his father's death. At this point the company controlled more than 90 percent of the cash register market, and in 1926 the company went public. Although NCR was hit hard by the Depression (its stock dropped from $150 to $7), it had almost fully recovered by the mid-1930s.

Into the computer age

NCR didn't want to miss out on the wave of commercialized computer products and bought Computer Research Corporation in 1952. During the 1950s and 1960s, the company expanded its product line-up to include mainframe and disk-based computers, and it opened data processing facilities and research laboratories. Cash registers and accounting machines remained NCR's most important products, but the company's net income dropped as it failed to adapt quite fast enough to an increasingly automated industry—from almost $50 million in 1969 to barely staying in the black two years later. William Anderson became the company's president in 1972 and its CEO a year later—eventually saving the company from extinction by divesting more than three-fourths of its employees. By this time, the company had sold 10 million machines. In the late 1970s, NCR acquired a host of computer-related companies, allowing it to employ increased microfiche capabilities and access to high-quality video display computer terminals.

Ma Bell calling

Intent on becoming a leader in the PC market, AT&T took over NCR for almost $7.5 billion in September 1991. New products under the tech giant included a general-purpose computer, an imaging system based on microprocessor technology and open, scalable systems (e.g., a notebook computer and a pen-based notepad). But AT&T spun off NCR in 1996 after experiencing huge losses, and the once again independent company faced a challenge; at one point it was hemorrhaging as much as $2 million per day.

NCR slimmed down to fewer operating units and some 10,000 people were let go, shifting its platform exclusively to ATM, banking and retail automation products; it finalized the transfer and sale of its computer hardware manufacturing assets to Solectron by 1998. NCR wasn't just shedding weight, though: it acquired both customer relations management (CRM) provider Ceres Integrated Solutions and networking specialist 4Front Technologies in 2000 to bolster its services. Three years later, CEO Mark Hurd came on board at NCR. By the time of his departure five years later, Hurd had whipped NCR into financial shape with job and salary cuts as well as the outsourcing of manufacturing. His successor was Bill Nuti, formerly of Symbol.

Divide and conquer

In January 2007, NCR announced it would spin off its Teradata data warehousing division to its own shareholders later that year, so that each can focus more closely on their distinct growth strategies; NCR will target the self-service kiosk market and Teradata will go after larger data storage contracts. Teradata's current head, Mike Koehler, will remain in charge when it becomes a separate company.

While Teradata's market for data warehousing services undoubtedly has a bright future, NCR's prospects are nothing to sneeze at either. In 2006, NCR signed a deal with China's four largest banks to provide 5,600 ATMs. This will only add to NCR's

ATM hegemony in the country—it already controls about a third of that business. Furthermore, as the Chinese economy heats up and its middle class grows, demand for ATMs will only increase: this market is expected to grow by nearly 10 percent every year until 2011. NCR is looking to bring its services to a broader swatch of the population elsewhere in Asia, as well; in August 2007 it announced an alliance with Symstream Technology, using wireless technologies in India to reach as yet unbanked populations in the rural countryside.

GETTING HIRED

Ring up your career at NCR

NCR's careers page, at www.ncr.com/about_ncr/careers, provides information for both experienced and newly-graduated candidates. Openings are searchable by region, function and keyword; to apply, job seekers must first create a profile. NCR has diversity groups for women, Hispanic, African-American, Asian, disabled and GLBT employees. Benefits include stock purchase plans, health insurance, annual bonuses and retirement plans.

NCR also offers internships and entry-level positions for students and recent graduates in the U.S. and abroad. Interns must be pursuing a line of study related to one of the following business areas: IT, accounting and finance, procurement, engineering, professional services or marketing. Questions can be directed to University.Relations@ncr.com.

Visit Vault at **www.vault.com** for insider company profiles, expert advice,
career message boards, expert resume reviews, the Vault Job Board and more.

V/\ULT CAREER LIBRARY   **349**

Network Appliance, Inc.

495 East Java Drive
Sunnyvale, CA 94089
Phone: (408) 822-6000
Fax: (408) 822-4501
www.netapp.com

DEPARTMENTS

Administration • Engineering •
Finance • Global Service • HR • IT •
Legal • Manufacturing • Marketing •
NetApp University • Operations •
Sales • Training

THE STATS

Employer Type: Public Company
Stock Symbol: NTAP
Stock Exchange: Nasdaq
Chairman: Don Valentine
CEO: Dan Warmenhoven
2007 Employees: 6,814
2007 Revenue ($mil.): $2,804

KEY COMPETITORS

EMC
Hewlett-Packard
IBM

EMPLOYMENT CONTACT

www.netapp.com/jobs

LOCATIONS

Sunnyvale, CA (HQ)
Albuquerque, NM • Atlanta, GA •
Austin, TX • Baltimore, MD •
Beachwood, OH • Bellevue, WA •
Bentonville, AR • Bloomington, MN •
Boise, ID • Boulder, CO • Boston, MA
• Brookfield, WI • Charlotte, NC •
Chesterfield, MO • Chicago, IL •
Cincinnati, OH • Cleveland, OH •
Columbus, OH • Cranberry Township,
PA • Dallas, TX • Dayton, OH •
Denver, CO • Des Moines, IA • Detroit,
MI • Edison, NJ • El Segundo, CA •
Englewood, CO • Houston, TX • Irvine,
CA • Kansas City, MO • Kenosha, WI
• Lake Oswego, OR • Las Vegas, NV •
McLean, VA • Miami, FL • Milwaukee,
WI • Minneapolis, MN • Nashville, TN
• New York, NY • Omaha, NE •
Orlando, FL • Overland Park, KS •
Philadelphia, PA • Phoenix, AZ •
Pittsburgh, PA • Pompano Beach, FL •
Portland, OR • Raleigh, NC • Redwood
City, CA • Research Triangle Park, NC
• Richmond, VA • Rochester, NY •
Salt Lake City, UT • San Antonio, TX •
San Diego, CA • San Francisco, CA •
San Ramon, CA • Scottsdale, AZ •
Schaumburg, IL • Seattle, WA •
Sherman Oaks, CA • Short Hills, NJ •
Southfield, MI • St. Louis, MO •
Tampa Bay, FL • Trevose, PA • Tulsa,
OK • Tysons Corner, VA • Waltham,
MA • Washington, DC

Locations in more than 40 countries
abroad with 113 worldwide office
locations.

THE SCOOP

No, they don't sell USB-enabled fridges

Network Appliance (or NetApp) provides specialized data storage devices to companies involved in pharmaceutical and biotech research, computer animation, telecoms and manufacturing. Its customers include Yahoo!, Boston University, Chevron, BMW, Industrial Light & Magic, AGFA, the U.S. Army Command and the Swiss Federal Government. NetApp's products are sold in 120 countries, and its stock is listed on the S&P 500.

Data-wrangling for the Internet age

In 1992, three computer engineers named Michael Malcolm, David Hitz and James Lau were working for Auspex Corp., a file server manufacturer, although before his career in the computer industry Hitz worked as a cowboy (yes, cowboy). According to the NetApp web site, the experience gave him "valuable management experience by herding, branding and castrating cattle." The company's web site also laid out the trio's initial goal: "to simplify storage the way Cisco simplified networking." You can find Hitz's reflections on herding cows and other musings at blogs.netapp.com, if you're into that sort of thing. Hitz, NetApp's current VP, takes his simplified data storage seriously: his July 20, 2007 post was titled *Lies, Damned Lies and Benchmark Results (The Ferrari versus the School Bus.)*

Out of the fire and into the Warmenhoven

Network Appliance started in 1992, when Malcolm, Hitz and Lau jotted down their business plan onto restaurant napkins, but they didn't get a product ready for retail until the following year. The young company went through the usual startup struggles; they posted losses their first few years in business and were having trouble finding investors. By 1994, it became clear that investors were growing wary of Michael Malcolm, the co-founder who was then serving as CEO. Hitz and Lau brought in Daniel Warmenhoven, a 13-year veteran at IBM, as its new CEO in October of the same year, and he immediately set about convincing wary consumers of the benefits of NetApp's unconventional new product.

The fridge

This product, the brainchild of Hitz, Lau and the departed Malcolm, was a wholly new and different type of technology, which is probably why it struggled out of the

Visit Vault at **www.vault.com** for insider company profiles, expert advice, career message boards, expert resume reviews, the Vault Job Board and more.

V/\ULT CAREER LIBRARY   **351**

starting gates. Where data storage had historically been encumbered by specific network requirements and limitations, NetApp had pursued a quicker and easier approach. CEO Warmenhoven grew fond of using a "refrigerator" metaphor to describe it: "The issues around data management were complex," Warmenhoven has said, "but they needn't be. We built a data refrigerator. Open the door. Take something out. Close the door. It's a simple, dedicated, data appliance."

The server was also faster and less expensive than its competition, and as networking became more common in the business environment, demand for NetApp products grew. In 1995, the company had its IPO, and its first four profitable quarters the year following. As demand for its products increased, they found their way into the offices of Walt Disney, Boeing and Citicorp by 1998. The company hit a rough patch in the early 2000s, as the tech bubble's bursting caused tech spending to fall off, and it posted a loss in 2002. But NetApp picked itself up and restructured, returning to profitability the following year. It is currently in good financial health, as revenue and profits have increased each of the last three years, clocking in for 2007 at $2.8 billion in revenue and $297 million in profits.

Sticky tape

Contrary to what many people believe, magnetic tape data storage, which dates from the era of UNIVAC-type mainframe computers, is still a viable technology and, used by many companies to keep their records on hand. And it has proven less expensive and more spatially efficient than storing data on discs until only recently—it isn't used for much beyond infrequently accessed archives, however, since it can be very slow. Network Appliance, having conquered much of the frequently accessed archives, has turned its sights on these poor, slow systems; it wants to unstick these companies from tape. As such, in 2006, NetApp introduced its Virtual Tape Library (VTL), a device wherein backup software functions like a tape library, while actually storing the data on discs. The data on the discs can also be compressed to allow for more efficient use of the space contained on the device, and it should save any IT department a bundle in costs.

Happy Feet, happy investors

In 2006, NetApp products were used in the creation of the computer-animated movie Happy Feet, about a dancing penguin. The company also helped design the America's Cup challenger for the BMW Oracle Racing team. Yearly revenue was up nearly 30 percent over 2005, and profits came in at $266 million. NetApp also purchased Topio, an Israel-based data recovery company, for $100 million. As demand for data storage increases, the company hopes to see revenue increase by 30 percent in fiscal 2007.

Need some A-SIS-tance?

In 2007, NetApp introduced A-SIS, (Advanced Single Instance Storage) a deduplicating program that reduces duplicate data entries, reducing the volume of data in a database by a factor of 20. By trimming the number of gigabytes a company has to mind, this program reduces the number of servers it must invest in and buy power for. The program is intended for companies that maintain large quantities of data, like telephone directories and medical records, or companies that need backup or store old electronic documents for legal reasons.

GETTING HIRED

Store your career with Network Appliance

The NetApp career's site, at www.netapp.com/jobs, provides information about job openings and benefits for prospective hires. To apply for a position, applicants must create a job account with the company. NetApp is welcoming applications; although the data storage sector is going through hard times recently, NetApp actually increased its headcount by over 1,000 employees last year.

In 2007, Network Appliance was named No. 6 on *Forbes* magazine's 100 Best Companies to Work For list. It was praised for its accommodation of parents with special-needs children and widespread use of flex schedules—only 5 percent of the company does the standard 9 to 5. The company's stellar benefits include health, dental and vision insurance, a generous vacation plan, stock options for new hires as well as established employees, and a 401(k) with company matching.

Novell, Inc.

404 Wyman Street, Suite 500
Waltham, MA 02451
Phone: (781) 464-8000
Fax: (781) 464-8100
www.novell.com

LOCATIONS

Waltham, MA (HQ)
Atlanta, GA • Austin, TX •
Cambridge, MA • Chicago, IL •
Dallas, TX • Detroit, MI • Herndon,
VA • Irvine, CA • Lebanon, NH •
Melbourne, FL • Memphis, TN •
Minneapolis, MN • New York, NY •
Philadelphia, PA • Portland, OR •
Provo, UT • Sacramento, CA • Salt
Lake City, UT • San Jose, CA •
Shelton, CT • St. Louis, MO •
Vienna, VA • Washington, DC

Locations in 41 countries
worldwide.

DEPARTMENTS

Administrative Support • Business
Management • Consulting Services
• Customer Services • Education •
Finance • HR • Information Systems
• Marketing/Product Marketing •
Operations/Facilities • Product
Management • Sales •
Sales/Systems Engineering •
Software Development • Software
Development Management •
Software Test/Quality • Technical
Support

THE STATS

Employer Type: Public Company
Stock Symbol: NOVL
Stock Exchange: Nasdaq
Chairman: Thomas G. Plaskett
President & CEO: Ronald W.
  Hovsepian
2007 Employees: 4,549
2006 Revenue ($mil.): $967

KEY COMPETITORS

IBM
Microsoft
Red Hat

EMPLOYMENT CONTACT

www.novell.com/job_search

THE SCOOP

It's all about networking

Novell has long been best known for network technology, but it is also making moves on the desktop environment. Its flagship NetWare network operating system connects desktops to corporate networks, integrating directories, storage systems, printers, servers and databases. Still the world's largest networking software company, with more than 80 million users around the globe, its offerings include e-mail (GroupWise), secure identity management (Novell Nsure), web application development (Novell exteNd) and cross-platform networking services (Novell Nterprise), all of which are supported by consulting and professional services (Novell Ngage). The company also distributes and supports SUSE Linux.

The network's the thing

The big name in Novell's early history is Ray Noorda, who served as the company's CEO and guiding light for over a decade after its founding, but he was not present at its birth. That distinction belongs to Dennis Fairclough, who was a computer engineer in the late 1970s, working at the Eyring Research Institute. ERI was a laboratory in Utah which contracted in computer research with the government, often hand-in-hand with Brigham Young University's computer science department. Fairclough was also student at BYU while working at ERI, as were his fellow ERI engineers Drew Major, Kyle Powell and Dale Neibauer; they joined Fairclough at the inception of Novell Data Systems in 1979. Major, Powell and Neibauer left to found SuperSet Software in 1981, whose technology derived, naturally, from work they were doing at BYU and ERI.

However, while Fairclough and co. were enjoying fruitful research opportunities while simultaneously working at Novell, SuperSet, ERI and BYU, it wasn't the most stable business platform, and Novell was struggling by 1982. Safeguard Scientific, a holding company that specializes in technology pursuits, had provided the venture capital for Novell, and was getting impatient for results—it was soon interviewing for Fairclough's replacement. Jack Messman was a principal executive involved in the CEO search, and he soon located Raymond J. Noorda, a 58-year-old veteran of the computer tech industry. With Safeguard's blessing, Noorda secured a $1.4 million investment for the company, which was reincorporated as Novell Inc., and he became the new CEO. Noorda soon steered the company to success with a product called Netware, which had been developed, ironically and predictably enough, by

SuperSet Software. (Dennis Fairclough is currently a computer science professor at Utah Valley State College.)

Casting a wide NetWare

NetWare, released in 1983, enabled computers to share data and devices like printers, a novel idea at the time. As companies increased their dependence on networks throughout the 1980s, Novell's market share grew accordingly, particularly when it introduced a version of NetWare in 1989 that effectively broke down barriers between operating systems—it allowed computers running Microsoft, Apple and UNIX systems to share data. The SuperSet engineers were interviewed about their groundbreaking work in 2001, and said that "(Many) of the things we did seem so simple and obvious, but at the time they were great leaps." From 1987 to 1989, the company's revenue jumped from $221 million to $421.9 million, and it enjoyed a 70 percent share in the networking market by the early 1990s.

Killing me Micro-Softly

In 1992, Microsoft released Windows NT, its own type of networking software—a clear attempt to cut into Novell's share of the networking market, then the fastest growing sector of the computer industry. Novell CEO Noorda zealously guarded his turf, but not by refocusing on networking; instead, he made two enormous acquisitions in 1994 that looked like an attempt to turn Novell into another version of Microsoft. The company purchased WordPerfect (another ERI product, coincidentally) for $855 million, and then acquired Quattro Pro for $145 million. WordPerfect and Quattro were a word processor and a spreadsheet, respectively, and also the main competitors for Microsoft's Word and Excel products at the time.

In a stunning turn of events, Noorda abruptly retired the same year, and his successor, Robert Frankenberg, immediately began trying to sell off these purchases. In 1996, he managed to divest WordPerfect, but he was forced to accept $750 million less than what Noorda had paid.

Count the CEOs

Frankenberg was so concerned with returning Novell to its old, networking self that he missed the emergence of the Internet as a major force in the computing industry. Microsoft immediately beefed up its Windows NT with Internet compatibility, and Novell's networking market share fell from 70 to 57 percent between 1993 and 1997. Frankenberg had already resigned by late 1996, though, and his interim replacement

John Young began (belatedly) focusing on the explosive Internet market, a strategy that only intensified when Sun Microsystems veteran Eric Schmidt took the helm permanently in 1997.

That same year, the company laid off roughly 18 percent of its workforce (nearly 1,000 employees). Schmidt battled through the turmoil, and the following year introduced a version of NetWare designed exclusively for the Internet (redubbed IntranetWare); but his tenure as CEO was not to last, either. In 2001, Novell acquired Cambridge Technology Partners for $266 million (Schmidt landed on his feet, heading off to the top job at Google). The acquisition of CTP soon looked more like a merger, though, as its CEO took over at Novell. Moreover, he was a familiar face: Jack Messman, who had helped hire Ray Noorda in 1983 and sat on Novell's board ever since. The company, however, was deep in the red by 2001 and suffered substantial income losses for the next two years.

Linux to the rescue!

Messman immediately struck out in a new direction, integrating Linux and UNIX into its business model. While the source code for Linux is free for anyone to inspect (or tinker with) on the Internet, Novell packages the software and provides clients with things like printer drivers and tech support, which intrepid Linux coders, in their pursuit of a better tea timer, don't necessarily provide. The company purchased Linux-based companies Ximian and SUSE Linux in 2004 as well, leading to a nice 68 percent bump in fiscal 2004 revenue. Messman also seemed to recognize why the company had suffered since its early days, when Dennis Fairclough was unable to translate the technology that would become NetWare into good business; in a 2003 interview with *ComputerWorld*, he said, "(Novell) never really appreciated the value of marketing. We thought if you created a great product, the world would beat a path to your door."

SC-uh-O

But when Novell decided to push Linux as the future of computing, it didn't count on being dragged through the court system—but that's just what happened. In 2004, a company called SCO filed suit against Novell, claiming that it owned the rights to UNIX. The background is quite hard to follow, but here goes: AT&T developed UNIX in the 1960s, but was forbidden to use it itself for monopoly reasons, so it basically gave it away to universities for research. (Berkeley developed its own version of the software, called BSD, which became the Internet's first operating system in the 1970s.) In 1984, AT&T was broken up, and no longer a closely-watched monopoly—it tried to commercialize UNIX, but three companies had already released commercial versions of it, Microsoft, IBM and the Santa Cruz Operation (SCO). Novell then bought the rights to UNIX from AT&T in 1991 and sued for copyright infringement, which resulted in much of the original UNIX being released to the public domain.

Meanwhile, an MIT professor named Richard Stallman was offended by the commercialization of these operating systems and all of the resulting litigation—he formed a group of engineers to develop a free operating system, similar to UNIX, which would eventually result in Linux in the early 1990s. (Novell and CEO Messman saw Linux as a potential savior for precisely this reason: it is a freer sort of operating system which can help the firm transform into an operation that specializes in services.) By this point, Novell sold the rest of its UNIX business to SCO, who in turn sold the rights to a Linux startup named Caldera. Then, as if to confuse everyone to no end, the original SCO changed its name to Tarantella and Caldera changed its name to SCO.

A Novell argument

SCO (the former Caldera) now owned large portions of both UNIX and Linux and suspected that Linux contained code that was suspiciously similar to UNIX. It sued IBM for $5 billion in 2003, claiming Big Blue had sprinkled UNIX code into Linux. Enter Novell, waving contracts and claiming that in its original agreement with Caldera it had only sold certain pieces of the UNIX code, and still owned the rights in question, a legal point that would effectively clear IBM, if true. What Novell didn't count on was for its strategy to backfire, as SCO dragged out contracts from 1995 that showed that Novell had sold it "all rights and ownership of UNIX."

The two have been locked in an epic legal battle ever since, but Novell just may have won the day. In August 2007, the judge overseeing the case ruled that Novell is the rightful owner of UNIX, and that it could force SCO to abandon its lawsuit against IBM. Novell's counterclaim suit against SCO for the licensing fees it had been extracting from other companies for UNIX code is was scheduled for trial in September 2007, but the court postponed the trial indefinitely. It now seems unlikely to happen, since SCO filed for bankruptcy shortly before the trial date. In its SEC filings SCO admitted it had "substantial doubt" about its future. Not surprising—its latest quarterly filing revealed the company has a paltry $10 million in assets, while the smallest amount Novell can be awarded in its suit is $30 million.

which could finally push SCO into bankruptcy (the company lost $17 million in 2006, and has struggled to right itself during the litigation). Since the August ruling, Novell has reassured nervous open sourcers about its intentions, stating that it will not "pull an SCO" and sue anyone for copyright infringement.

Hovsepian steps in

But perhaps a drawn-out legal battle was not what Novell's board had in mind when it promoted Messman to CEO—it fired him in June 2006, replacing him with COO Ronald W. Hovsepian. Novell hopes that Hovsepian's marketing mojo will help the company take a bite out of Microsoft's desktop dominance. For the year, revenue came to $967 million, with a profit of $18 million, both down from the previous year (when the figures were $1 billion and $376 million, respectively). The desktop software division seems to have benefited from Hovsepian's influence, racking up 26 percent growth for the year.

Patent pending

In November 2006, Novell and Microsoft hatched a deal to make Windows and Linux more compatible. Businesses frequently run Linux on their servers (for its stability and security) and Windows on the machines in the office (for its familiarity to the office-frequenting population). The deal will allow an increased level of compatibility between documents created in Word and those in OpenOffice formats (a Linux word-processing program), allow Windows to run more effectively in a Linux environment, and vice versa. A clause in the agreement, however, specified that Microsoft would be unable to sue Novell Linux users for patent violations. While Novell does not believe that Linux violates any patents, the agreement has those involved in open source software worried that Microsoft will start going after users of other versions of Linux for features that supposedly violate its patents.

Plays well with others

In the spring of 2007, Microsoft and Novell announced the particulars of their deal. The companies had adopted a two-pronged strategy that focuses on making Linux and Windows work together, including beefing up the web-based programs that manage servers of both operating systems and improving the reading of files created in Word and Open Office. Of course, all software needs hardware to run on, and Dell stepped in to provide its wares in May—it will provide customers with SUSE Linux servers.

The partnership has not been without controversy, many of the open source devotees Novell had courted in its love affair with Linux circa 2003 have criticized Microsoft, a notorious anti-open source party, as an unlovely bedfellow. One of Novell's open source developers even resigned in the wake of the deal's announcement. First quarter sales in 2007 didn't help, either: the company lost $19.9 million on sales of $229.6 million.

GETTING HIRED

Networking will serve your career well

Novell's careers site, at www.novell.com/job_search, provides a list of open positions by location and function. In order to apply to a position, interested parties must fill out a web form. Internship positions are also available. Novell's benefits include health, dental and life insurance, 401(k) with company match and tuition reimbursement.

Novell may be headquartered in Massachusetts, with offices in over 30 countries, but almost half of its employees (approximately 3,000) work at the company's Provo, Utah, campus—the company has been operating out of Utah Valley for almost two decades.

OUR SURVEY SAYS

Try a Novell option

The corporate culture is open and honest," says one source. His colleague, on the other hand, disagrees. "The one constant at Novell is change," says one respondent, parroting the latest management best seller. "Cost cutting and expense minimization create an atmosphere of stress and fear, as well as degeneration of morale."

Novell is trying to reinvent itself, so there is a good deal of culture change right now," says a more diplomatic co-worker. The constant changes mean that "opportunities for advancement are slim." Despite this atmosphere of fear, uncertainty and doubt, life at the company goes on.

Dress code is casual, except for the executive level, which is business casual or dress," says one source. "Hours are fairly flexible." Diversity is as diverse as working in Provo, UT can be," says his associate.

Visit Vault at **www.vault.com** for insider company profiles, expert advice, career message boards, expert resume reviews, the Vault Job Board and more.

V/\ULT CAREER LIBRARY   **361**

NVIDIA Corporation

2701 San Tomas Expressway
Santa Clara, CA 95050
Phone: (408) 486-2000
Fax: (408) 486-2200
www.nvidia.com

LOCATIONS

Santa Clara, CA (HQ)
Austin, TX
Beaverton, OR
Bedford, MA
Bellevue, WA
Berkeley, CA
Charlotte, NC
Durham, NC
Fort Collins, CO
Greenville, SC
Houston, TX
Kirkland, WA
Madison, AL

Locations in 11 European and Asian countries, including China, Germany and India.

DEPARTMENTS

Architecture • Hardware • HR • IT • Software • Systems & Applications • Technical Marketing

THE STATS

Employer Type: Public Company
Stock Symbol: NVDA
Stock Exchange: Nasdaq
Chairman & CEO: Jen-Hsun Huang
2007 Employees: 4,083
2007 Revenue ($mil.): $3,068

KEY COMPETITORS

AMD
Creative Technology
Intel

EMPLOYMENT CONTACT

careers.nvidia.com

THE SCOOP

You can see everything from here

NVIDIA makes graphics processors, computer chips and circuit boards. The company's chips—which offer both two- and three-dimensional graphics—are used in personal computers and in motherboards built by other companies. NVIDIA chips are used in products made by some of the biggest names in the computer industry, including Dell, IBM and Sony. Its circuit boards find application in cell phones, games and industrial design applications.

Caution! May contain graphics content

If you've enjoyed the subtle temptations of the lastest flashy video games, there's a good chance you've come across NVIDIA. Three tech veterans, Jen-Hsun Huang (LSI Logic), Curtis Priem (IBM and Sun) and Chris Malachowsky (Sun) founded the company in 1993. Huang was a director in LSI's system-on-a-chip (SOC) division—which sought to unite various functions (i.e., a system) onto one microchip—and he led this new venture as CEO, designing similar devices, such as graphics processors and multimedia devices.

These devices quickly found usefulness in video games, and in May 1995, NVIDIA launched NV1. The company's first multimedia processor, it featured a joystick, game port and 3D graphics, and was a direct descendant of SOC research and development. Two months later, the company partnered with Sega, which saw possibilities in the NV1 for its Saturn gaming system. Almost immediately afterward, Microsoft designed a new graphics interface centered on quadratic graphics, rendering the NV1 (centered on polygons) obsolete. Sega kept its faith in NVIDIA's work and sunk millions of dollars into the also polygon-based NV2 graphics system, keeping NVIDIA afloat.

If you can't beat 'em ...

Many industry analysts have suggested that if it weren't for Sega's investment, NVIDIA would no longer exist, but it didn't survive on luck alone. CEO Huang recognized the need for drastic change and started designing towards Microsoft's quadratic graphics standard; he also steadily abandoned his SOC-based approach. In 1997, NVIDIA released Riva 128, its first quadratic-graphics interface. The

Visit Vault at **www.vault.com** for insider company profiles, expert advice, career message boards, expert resume reviews, the Vault Job Board and more.

VAULT CAREER LIBRARY   363

company's presence in this market grew steadily thereafter, and *PC Magazine* rated it as its most influential 3D processor company by September 1998.

The company went public at the start of 1999, and later that year it launched GeForce 256, the first graphics processing unit designed for use in desktop computers. NVIDIA eventually moved beyond its traditional focus on PC graphics chips, even though the company held a majority of that industry's market share by 2001. In March 2000, Microsoft tapped NVIDIA, whom it had nearly wiped out, to make the graphics chips for its Xbox console (an account it would eventually lose due to cost disagreements). Three years later, NVIDIA was tapped by Apple to provide the graphics goodies for its latest Powerbook laptop.

The future is visible

In 2006, NVIDIA acquired two companies to strategically enhance its position in the industry. In February, it acquired ULi, a developer of chip designs with a strong presence in Taiwan. In March, NVIDIA acquired Finland-based Hybrid Graphics, a developer of graphics software for cell phones and other mobile devices.

In fiscal 2007, NVIDIA took in $3.1 billion in revenue, some $450 million of which was profit. Gross margin, a reflection of the company's profit, increased four points over the previous year, to 42 percent in 2007. NVIDIA is looking forward to a profitable fiscal 2008, as users begin to upgrade to Windows Vista, whose new graphical interface requires some heavy horsepower on the graphics card. CEO Huang is proud of the company's remaining independence—many of its competitors have by now been gobbled up by bigger companies. (For instance, VIA Technologies bought S3 at the turn of the millennium, and AMD acquired ATI in 2006.) It's to the credit of both NVIDIA and the industry that the company leads its sector—it's one of the only large specialists in graphics left standing.

Visual thinking

In 2007, NVIDIA teamed up with Hewlett-Packard and Fujitsu-Siemens to add a high-powered graphics setup to their computers. The graphics will allow artists, engineers and scientists—not to mention anyone else who has to deal with a graphics-intensive job on the go—to access high-powered graphics without having to haul around a desktop. The Quadro FX line of graphics processors allocates computing power to the pixels that are actively displaying complex graphics, and so manage to keep temperatures and power consumption down.

GETTING HIRED

Visualize yourself at NVIDIA

NVIDIA's careers page, at careers.nvidia.com/pljb/nvidia/nvidiaemployment/ applicant, provides information on job openings, benefits and college recruiting. Jobs are searchable by title, keyword and location; in order to apply, interested parties must first create a profile. Job candidates must also undergo a background check before an offer of employment is made. Questions or comments can be directed to hr@nvidia.com.

Benefits include a choice of health care plans, dental and vision insurance, 401(k), stock purchase plan, college savings plans, flexible spending accounts for health care and dependent care expenses, and a generous vacation policy. Employees can also receive chair massages, on-site hair cuts, laundry and dry cleaning pickup and delivery services, and subsidized café and health club allowance.

NVIDIA recruits students for internships, co-ops and entry-level positions. It visits Stanford, UC Berkeley, the Rochester Institute of Technology, Cornell and MIT. Interns and co-ops can join the software, hardware, architecture, systems and applications, technical marketing or IT departments. Interns' tenures are heavily subsidized, from transportation and food at the café to assistance in shipping things hither and yon. To apply, submit a resume to myfuture@nvidia.com.

ON Semiconductor Corporation

5005 East McDowell Road
Phoenix, AZ 85008
Phone: (602) 244-6600
Fax: (602) 244-6071
www.onsemi.com

LOCATIONS

Phoenix, AZ (HQ)
Austin, TX
Chandler, AZ
East Greenwich, RI
Gresham, OR

12 locations throughout Europe and Asia.

DEPARTMENTS

Accounting
Engineering
Facilities
Financial
General Management
HR
IT
Legal
Manufacturing
Operations
Quality
Sales & Marketing
Security
Supply Chain/Logistics
Support Administration
Technical Writing
Technicians
Technology & Development

THE STATS

Employer Type: Public Company
Stock Symbol: ONNN
Stock Exchange: Nasdaq
Chairman: J. Daniel McCranie
President & CEO: Keith Jackson
2006 Employees: 11,691
2006 Revenue ($mil.): $1,531

KEY COMPETITORS

Fairchild Semiconductor
Infineon
International Rectifier
National Semiconductor
STMicroelectronics
Texas Instruments
Vishay

EMPLOYMENT CONTACT

www.onsemi.com/PowerSolutions/content.do?id=1035

THE SCOOP

They're always ON

ON Semiconductor is a leading provider of integrated circuits and standard components that control the flow of power and data through everything from cars to computers, televisions to game consoles and communications servers to MP3 players and cell phones. With a catalog of more than 28,000 parts, the company's devices are used in electronics made by Intel, Continental, Siemens, Sony, Samsung, Motorola and Philips, to name a few. The company derives its name from its integral nature—without its tiny chips, one simply cannot turn many machines "on."

Chip off the old block

Motorola sold off its semiconductor products division to a private equity firm in 1999, which then became ON Semiconductor. The following year, ON Semiconductor acquired Cherry Semiconductor, a provider of power control chips to the car industry, for a quarter of a billion dollars. Also in 2000, the company raised $500 million in its IPO, which it used to pay down debt.

Less is more

In 2006, ON Semiconductor introduced its GreenPoint power supply reference designs for desktop computers, LCD and CRT televisions and 60-, 90- and 200-watt power supplies for other devices, like printers. These operate at an efficiency of 83 percent—12 percent more efficiently than a non-power saving chip—and consume 300 fewer milliwatts when turned off. The 80 Plus and EnergyStar programs have recognized the chips for their efficiency.

The other kind of green

The GreenPoint chips seem to be bringing in green of a more pecuniary sort, too: in 2006, ON's yearly revenue was $1.5 billion, up 22 percent over the year before. Profits also came in at a record-setting $272 million. Also in 2006, the company purchased a factory in Gresham, Ore. from LSI Logic, a semiconductor vet seeking to slim down and go fab-less. This factory and its employees have significantly enhanced ON's internal manufacturing capabilities.

Visit Vault at **www.vault.com** for insider company profiles, expert advice, career message boards, expert resume reviews, the Vault Job Board and more.

VAULT CAREER LIBRARY   367

Turning students on to microelectronics

In April 2007, ON announced that it had contributed $230,000, as well as necessary equipment for a clean room facility, to help establish a lab for research on semiconductor design and materials at Masaryk University, in the Czech Republic. The company also will help design the engineering curriculum, and ON employees will conduct lectures at the school.

GETTING HIRED

Turn your career ON

ON's careers page, at www.onsemi.com/PowerSolutions/content.do?id=1035, provides information on careers in the U.S., Asia and Europe. Jobs are searchable by function and location; to apply, hopeful ON-ers must first create a profile.

Internships and opportunities for recent college graduates are also available at ON. The former are available to rising sophomores studying electrical, chemical or computer engineering, computer information systems, business, marketing, finance, supply chain management or allied fields. Most internships take place in Phoenix, Arizona, though some are in Rhode Island or another office. During their tenure, interns are eligible for paid holidays and additional compensation for time worked in excess of a 40-hour week.

Oracle Corporation

500 Oracle Parkway
Redwood City, CA 94065
Phone: (650) 506-7000
Fax: (650) 506-7200
www.oracle.com

DEPARTMENTS

Business Operations • Computing • Consulting • Facilities • Finance • General Administrative/Secretarial • HR • IT • Legal • Marketing • Manufacturing & Distribution • Pre-Sales • Product Development • Sales • Support • Training

THE STATS

Employer Type: Public Company
Stock Symbol: ORCL
Stock Exchange: Nasdaq
Chairman: Jeffrey O. Henley
CEO: Lawrence J. Ellison
2007 Employees: 76,674
2006 Revenue ($mil.): $17,996

KEY COMPETITORS

IBM
Microsoft
SAP

EMPLOYMENT CONTACT

www.oracle.com/corporate/employment

LOCATIONS

Redwood Shores, CA (HQ)
Albany, NY • Atlanta, GA • Austin, TX • Berwyn, PA • Birmingham, AL • Boca Raton, FL • Boston, MA • Boulder, CO • Buffalo, NY • Burlington, MA • Cambridge, MA • Charlotte, NC • Chesapeake, VA • Chicago, IL • Cincinnati, OH • Cleveland, OH • Colorado Springs, CO • Columbia, MD • Columbus, OH • Costa Mesa, CA • Dallas, TX • Dayton, OH • Denver, CO • Detroit, MI • Edmond, OK • El Segundo, CA • Encino, CA • Grand Rapids, MI • Honolulu, HI • Houston, TX • Indianapolis, IN • Iselin, NJ • Kansas City, KS • King of Prussia, PA • Knoxville, TN • Lexington, KY • Memphis, TN • Miami, FL • Milwaukee, WI • Minneapolis, MN • Moorestown, NJ • Nashua, NH • Nashville, TN • New York, NY • Northborough, MA • Onalaska, WI • Orlando, FL • Phoenix, AZ • Pittsburgh, PA • Pleasanton, CA • Portland, OR • Providence, RI • Raleigh, NC • Reston, VA • Richmond, VA • Rochester, NY • Rocklin, CA • San Antonio, TX • San Diego, CA • San Francisco, CA • San Rafael, CA • Sandy, UT • Santa Monica, CA • Seattle, WA • St. Louis, MO • Tallahassee, FL • Tampa, FL • Tarrytown, NY • Teaneck, NJ • Trenton, NJ • Tulsa, OK • Washington, DC • Westchester, IL

Sales offices in over 145 countries worldwide.

Visit Vault at **www.vault.com** for insider company profiles, expert advice, career message boards, expert resume reviews, the Vault Job Board and more.

VAULT CAREER LIBRARY 369

THE SCOOP

Making the business world go 'round

Oracle is the world's No. 2 business software company and a leading provider of databases and programs for web development, enterprise performance assessment, supply chain, customer relationship and HR management. In 2007, *BusinessWeek* ranked the company No. 22 on its InfoTech 100 (its main business software competitor, Microsoft, ranked ninth).

A vision of the future

In 1977, Larry Ellison, a computer programmer at Ampex Corp., entered into a partnership with his current boss Robert Miner. Ellison's intention was to form a company that would capitalize on technology first developed by IBM. He put up $2,000 of his own money and called the firm Software Development Laboratories. The company was renamed Relational Software Inc. in 1979 and finally became known as Oracle in 1983 after its flagship database product. The Oracle database was prophetically designed. As it was especially adapted to work with IBM's structured query language (SQL), a computer language that relayed information on how to store and retrieve information on databases, use of the Oracle became widespread when the SQL language became an industry standard.

Throughout the 1980s, Oracle produced several types of databases, each varied to suit different customers' needs. In 1986, the company had marketing subsidiaries in 17 countries outside the United States and clients in 39 total countries, and it also issued shares for the first time. By the next year, Oracle was the world's largest database management software company, with 4,500 clients in 55 countries. It was listed on the S&P 500 as a tech sector representative of the economy in 1989.

Held to account

In 1990, the company's own shareholders sued for accounting irregularities, and an internal audit by the company found that it had overstated its revenue. Oracle was soon on the verge of bankruptcy and looked unable to keep up with its rapid growth. The company reported its first annual profit loss in 1991 and subsequently cycled through several CFOs and accounting standards before establishing an even keel two years later.

Good service from an Oracle?

Oracle needed a shakeup, and it got one when it switched to consulting services, a business model now in vogue all over the tech industry. By 1994, Oracle's consulting services were pulling in 20 percent of annual sales, and the company spearheaded a shift across the business world from the old model of storing corporate data on a mainframe to the new model of businesses contracting their data storage needs to a server company such as Oracle (the client/server model). The company also released its Oracle 7.1 database that year, which large corporations soon favored for their "data warehousing" needs.

Bring it on!

Oracle thought outside the box with its database design, as well, most notably with the failed "Network Computer" campaign. It introduced WebSystem software in 1995, which allowed data to be transmitted over the Internet or an intranet, and soon followed with the "Network Computer" in 1996, a sort of computer without any applications that was designed for purely data storage purposes.

Oracle liked its invention and puffed itself up about inventing "a new era in computing," only to see Microsoft and Intel fiercely defend their turf. By 1998, CEO Ellison refocused the company on producing bigger and better databases. By 2000, sales reached an all-time high of $6.3 billion, and *Forbes* deemed its small office suite as Best of the Web the next year.

I see many purchases in your past

Sales briefly declined in 2002 and 2003 following the burst of the tech bubble, but Oracle immediately embarked on a hotly contested takeover bid in 2004 for PeopleSoft, a competing business software company. After a year and a half of rejected offers, offended shareholders and an intervention by the Justice Department, PeopleSoft came to an amicable agreement with Oracle, and the deal, valued at $10.3 billion, was completed in 2005.

Oracle is still following an aggressive strategy of expansion through a combination of innovation and industry-focused acquisitions. In 2005 and 2006, the company snapped up 26 companies with complementary product lines to its own. Among its 2006 acquisitions were a company that provided supply chain management for retail operations, another that specialized in open source database management software, three companies that provided software to the communications industry, a supplier of demand management and trade promotion software, a content management software company, and ... well, you get the idea. The company's biggest acquisition, however, was the purchase of Siebel—a supplier of customer relationship management (CRM) software, a growing software area that's expected to reach $10 billion in 2009—for $5.8 billion in September 2005. The deal made Oracle the No. 1 supplier of CRM software, and puts the pressure on its key competitor SAP.

Hyperproductivity

The parade of acquisitions didn't stop at the end of 2006, however. In March 2007, Oracle acquired two more companies, Hyperion Solutions and Tangosol. The Hyperion deal, which went down for $3.3 billion, gave Oracle all of Hyperion's performance management programs and kept the pressure on SAP, many of whose customers use Hyperion products. It also increased Oracle's overall market share in the business performance software arena. The nearly simultaneous purchase of Tangosol gave the company access to Tangosol's method of rapidly processing transactions by increasing the speed at which frequently-used data is available. Oracle is even more of a giant now, as revenue, profits and employee headcount are all at record levels; in 2007, they clocked in at $17.9 billion, $4.2 billion and 76,674, respectively. (If you're wondering what 2006 looked like, the company earned $14.3 billion in revenue, $3.3 billion in profits and had 56,133 employees.)

GETTING HIRED

See your future at Oracle

Oracle's careers page, at www.oracle.com/corporate/employment, provides information on job openings, college recruiting and benefits at the company. Jobs are searchable by function and location; to apply, interested parties must create a profile.

The company encourages its employees to transfer between jobs after a year or two and also provides ample training sessions on technical and nontechnical subjects, as well as tuition reimbursement and opportunities for employees to attend college classes via a video link with Stanford.

Oracle offers its employees a flexible benefits plan. Each employee is given a certain number of benefits credits; unused ones can be added to a flexible spending plan or 401(k) program or taken as income. Benefits for purchase include health, dental and vision insurance; life, disability and flexible spending accounts. Oracle also offers some cushy employee perks, including discounts on auto and home insurance, computers, cars, amusements park admissions and more. Other benefits include the aforementioned 401(k), discount stock purchase plan, generous vacation and flexible working hours.

College students interested in joining Oracle can check out the page at www.oracle.com/corporate/employment/college, which provides a wealth of information about the hiring process and life at the company. There is a page of advice for students attending campus interviews (don't dress up, ask questions) and a virtual tour of Oracle's campus in California. Resumes can be directed to Larry Lynn, Director of Recruiting, (lslynn_us@oracle.com); any questions can be sent to college_us@oracle.com. Students who are still a few years out from their graduation day can apply for internships by sending a resume to interns_us@oracle.com; internships are available for computer science majors in product development. Oracle provides housing and transportation for interns.

Visit Vault at **www.vault.com** for insider company profiles, expert advice, career message boards, expert resume reviews, the Vault Job Board and more.

V∆ULT CAREER LIBRARY   373

Palm, Inc.

950 West Maude Avenue
Sunnyvale, CA 94085
Phone: (408) 617-7000
Fax: (408) 617-0100
www.palm.com

LOCATIONS

Sunnyvale, CA (HQ)
Andover, MA
Chicago, IL
Denver, CO
Missoula, MT
New York, NY
Philadelphia, PA
San Diego, CA
Washington, DC

Locations in 16 countries
worldwide, including the UK, China,
Argentina and Australia.

DEPARTMENTS

Administration • Business
Development • Customer Services •
eCommerce Sales • Engineering •
Facilities • Finance • HR • IT • Legal
• Manufacturing • Marketing •
Operations • Product Marketing •
Professional Services • Sales

THE STATS

Employer Type: Public Company
Stock Symbol: PALM
Stock Exchange: Nasdaq
Chairman: Eric A. Benhamou
President & CEO: Edward T. Colligan
2007 Employees: 1,247
2006 Revenue ($mil.): $1,560

KEY COMPETITORS

Hewlett-Packard
Nokia
Research in Motion

EMPLOYMENT CONTACT

www.palm.com/us/company/jobs

THE SCOOP

The power is in your hand

Palm Incorporated, a manufacturer of popular personal digital assistants (PDAs), has distinguished itself as a tech powerhouse and as a successful startup in a field rich with failures. Over the course of its development, the Palm has evolved from a device that kept phone numbers and to-do lists into an all-purpose address book, phone and e-mail device. It isn't just for salesmen and executives anymore, either. Stanley Access Solutions is using the device to calibrate and repair automatic sliding doors, and doctors employ it as a way to carry instantly updated medical records for their patients.

Hawkins and Dubinsky have it in hand

Jeff Hawkins and Donna Dubinsky founded Palm in 1992. U.S. Robotics acquired the company in 1995, and the two kept working to develop products Palm products until 1998. The company's first handheld device, the Palm 1000, hit the market in 1996, and 3Com snapped up U.S. Robotics the next year, also buying Palm in the deal. Hawkins and Dubinsky left in 1998 to found Handspring, a rival manufacturer of PDAs. Palm remained part of 3Com until March 2000, when it spun off from its parent, regaining its standing as an independent entity.

Palm's IPO in 2000 was impressive—its offering stock price of $38 was more than double what 3Com had expected. Furthermore, Palm replaced its parent company on the Standard & Poor's 500 stock index in July, just five months after going public. The reason for soaring stock prices? Among other factors, Palm at that time had a dominating 80 percent share of the PDA market. Moreover, Palm has brought in additional revenue by licensing the operating system it designed for its own devices, Palm OS, to other PDA manufacturers, like Sony and Handspring. In 2001, the first phone-equipped Palm hit the market, followed by the first phone and e-mail Palm, the following year.

In 2003, Palm acquired rival Handspring after each had posted losses in the previous quarter, and Hawkins and Dubinsky rejoined the Palm fold, Hawkins as chief technology officer and Dubinsky on the board of directors, where they remain to this day. The company briefly changed its name to PalmOne after the merger, but dropped the "One" in 2005. (A more detailed account of the Hawkins and Dubinsky saga is available in a book written by former Palm employee Andrea Butter and *New*

York Times reporter David Pogue, called *Piloting Palm: the inside story of Palm, Handspring and the Birth of the Billion Dollar Handheld Industry*.)

Teaching an old dog new tricks

As any tech-obsessed teenager will tell you, most cell phones these days come ready-equipped with all manner of extras, like cameras, music players, e-mail, text messaging, address books and calendar applications—not to mention QWERTY keyboards (no more little squiggles to learn!). Of course, this proliferation of extras from other phone manufacturers has stolen a big part of Palm's thunder.

And the barrage of feature-packed phones show no sign of stopping: now that most of the business portions of a cell phone can fit on a single chip, there's plenty of space left in even the sleekest case for all manner of bells and whistles. One researcher at consulting firm Creative Strategies estimates that phones with the aforementioned bells and whistles will account for nearly 20 percent of all phones sold in 2010. Palm's revenue reflects a similar trend. In 2004, the company derived six times the revenue from handheld computers than from smart phones; by 2006, customers were spending twice as much on smart phones as they did on handheld computers.

Filling the playing field

Also looming over the landscape of the mobile device market, a field already saturated with devices that feature full keyboards, such as Research in Motion's BlackBerry and Nokia's 682, is the release of the iPhone, Apple's all-in-one music player/web browser/phone/e-mail device. The iPhone is forecasted to move 10 million units in 2008—most companies count a launch successful when they sell a fifth of that quantity.

At an investor conference, Palm's CEO, Ed Colligan, revealed that Palm was revamping its production process in 2007, reducing the cost of parts by a third by standardizing the base on which its devices are built. This streamlining will also enable Palm to sell smart phones inexpensively, expanding the market for these gadgets. In 2006, Palm took in $1.5 billion in revenue, of which $300 million was profit.

Reinventing the laptop

In June 2007, Palm agreed to sell about a quarter of its outstanding shares to a private equity company for $325 million, in order to inject some capital into the company.

Palm also got Jon Rubinstein, the man who had spearheaded Apple's design and production of the iPod; he will attempt to inject the same sort of consumer-pleasing brilliance into Palm's product line as he did into Apple's.

A few weeks later, Palm announced its newest product, the Foleo (pronounce it like one of Shakespeare's, or fole-ee-oh, for the uninitiated). Billed as a "mobile companion," this $500 device, approximately the size of a small laptop, boasts a 10-inch screen and reasonable keyboard real estate. Instead of being a real computer—you can't play World of Warcraft on this puppy while waiting for your flight to board, tragically—it's designed to act as a scaled-up version of the Palm's keyboard and screen. Functions include not-so-you'd-notice boot-up time, e-mail, ability to read and edit attachments, display presentations and pictures and browse the Web. Whether businessmen will cotton to this device in lieu of their travel-battered laptops remains to be seen.

GETTING HIRED

Give your career a hand

Palm's careers site, at www.palm.com/us/company/jobs, has a search form for those seeking jobs; it can be sorted by location or function. Sales, marketing and engineering had several open positions at press time, as well as some listings for internships. To apply, interested parties must upload their resumes and cover letters on a web form. Benefits at Palm include sabbaticals and 401(k) plans with matching contributions.

QUALCOMM Incorporated

5775 Morehouse Drive
San Diego, CA 9212
Phone: (858) 587-1121
Fax: (858) 658-2100
www.qualcomm.com

LOCATIONS

San Diego, CA (HQ)
Austin, TX
Bedminster, NJ
Boulder, CO
Cary, NC
Concord, MA
Portland, OR
San Francisco, CA

Locations in 24 countries worldwide, including China, South Africa, India, Brazil, Japan and the UK.

DEPARTMENTS

Administrative
Business Development
Engineering
Facilities
Finance
HR
IT
Legal
Library & Information Sciences
Manufacturing
Marketing
Procurement
Project/Product Management
Public Relations
Sales
Technical Support

THE STATS

Employer Type: Public Company
Stock Symbol: QCOM
Stock Exchange: Nasdaq
Chairman: Dr. Irwin Mark Jacobs
CEO: Dr. Paul E. Jacobs
2006 Employees: 11,200
2006 Revenue ($mil.): $7,526

KEY COMPETITORS

Nokia
Samsung
Texas Instruments

EMPLOYMENT CONTACT

www.qualcomm.com/careers

THE SCOOP

Today, CDMA, tomorrow, the world!

QUALCOMM is a manufacturer of software and microchips for cellular phones. A communications pioneer, it invented the CDMA (code-division multiple access) code, which is a method of transmitting data over cellular networks. It currently controls around 4,500 patents—four out of five are related to CDMA—and issues licenses to other companies that allows them to use its technology. CDMA enables the transmission of a much greater quantity of data over cellular networks than is possible with other technologies and currently stands as the world's second most widely-used wireless tech standard, used by about a fifth of all cell phone networks.

QUALCOMM supplies about 95 percent of the chips for CDMA phones worldwide, collecting a royalty on each one. The firm is part of the S&P 500, has the No. 381 spot on the 2006 Fortune 500 and was No. 89 on *BusinessWeek*'s 2007 edition of the InfoTech 100.

Pepperoni is in its DNA

College professors Irwin Jacobs and Andrew Viterbi founded QUALCOMM in an office above a pizza joint in San Diego, California, in 1985. Its first product, OmniTRACS, released in 1988, was designed to coordinate truck transport in the U.S. via satellite. The company began selling OmniTRACS in Japan and Brazil in 1991, the same year as its IPO.

OmniTRACS wasn't the only trick up QUALCOMM's sleeve, though: in 1989, the company introduced its CDMA method of handling cell phone calls. The cellular industry, at the time, had largely decided on TDMA as the standard code for carrying calls, but tests conducted in 1991 by cellular carriers proved that CDMA could be a viable alternative.

CDMA-zing

Two years after its 1991 IPO, the funds from which QUALCOMM used to build a national CDMA system, cellular carriers began ordering QUALCOMM equipment in order to set up their own CDMA networks. By 1995, 11 of the 14 top cell phone carriers nationwide supported CDMA.

Within two years, nearly 50 cell phone carriers worldwide supported the technology. QUALCOMM then tried to spread the CDMA gospel in Europe and China but was rebuffed in Europe by lawmakers who wanted to maintain the TDMA standard there, until a ruling allowed CDMA in 1999. China was also resistant to CDMA until a 2002 agreement with that country's state-owned telecommunications provider opened the doors to QUALCOMM's technology. By the end of 2003, the company had 7,400 employees and yearly sales of $3.9 billion (up from $3 billion in 2002).

Technical KO

Like many tech companies, QUALCOMM's remarkable rate of growth in the late 1990s and early 2000s temporarily halted as the tech bubble burst. Thousands of employees lost their jobs in 2001 through 2003, when the company slashed its employee head count from a high of about 10,500 in 2001 to 7,400 in 2003. The company's formerly skyrocketing stock steadily declined in 2000 and continued to do so until late 2003, when wireless sales began to explode around the world.

As part of the company's cost-cutting measures in 2000, it sold its antennae infrastructure business to Ericsson and its handset manufacturing unit to Kyocera, decisions which looked short-sighted after each division returned to profitability within three years. QUALCOMM erred again in November 2001, investing $266 million of equity financing in Vesper, a Brazilian CDMA wireless service provider, only to turn around in 2003 and sell the unit to Embratel Participacoes SA for an undisclosed amount after failing to turn the company around.

Despite these snags, QUALCOMM was not exempt from the 2003 cell phone boom, reporting in 2004 that it expected the Indian and Chinese telecom markets to continue powering company growth, as demand for wireless services expands in the world's two most populous nations.

Adding a piece of flair

In 2006, QUALCOMM perked up its offerings with the $800 million acquisition of Flarion. Flarion's most compelling feature is its intellectual property related to OFDM (orthogonal frequency division multiplexing), a method of sending large files—such as music and video—over wireless networks. While QUALCOMM's latest-generation CDMA also works for sending large files, OFDM can be used when the potential for radio interference is particularly high.

The acquisition gives QUALCOMM more leverage in intellectual property pertaining to the transfer of large files, as consumers increasingly demand the ability to send and receive picture, audio and video files from their phones. Flarion wasn't the only piece of flair that QUALCOMM pinned to its suspenders that year. The company took in $7.5 billion in revenue—a hefty $2 billion of which was profit.

A bitter breakup

QUALCOMM depends to a great extent on its intellectual property to make money, and, as inventor of a widely used cell phone code, must spend a great deal of money defending its patents from potential interlopers—in 2006 alone, it spent at least $2.6 billion of its revenue in 2006 on legal matters. One of its chief sparring partners is Nokia, a former sweetheart of a trading partner who has long since soured at what it considers to be QUALCOMM's unduly expensive licensing fees. The conflict between QUALCOMM and Nokia has reached high levels of animosity, and cell phone chip competitor Broadcom has also thrown its hat in the ring. In 2005, Nokia and Broadcom each filed lawsuits against QUALCOMM. Then, Nokia let its licensing contract with QUALCOMM lapse in April 2007—a huge blow to both, as Nokia needs QUALCOMM's patents for a huge number of its products to function, and QUALCOMM lost one of its largest clients.

In August, four months after the Nokia/QUALCOMM contract expired, things got thorny for QUALCOMM very fast, as it lost three successive court battles against its competitor Broadcom. First, a San Diego judge presiding over the patent case "QUALCOMM v. Broadcom" found for Broadcom. Even though Broadcom had violated QUALCOMM's patents, he reasoned, QUALCOMM had "deliberately concealed" the patents in question from its competitor—a shady trade practice, at best. The judge rebuked QUALCOMM for an "organized program of litigation misconduct," as Broadcom attorneys had discovered over 200,000 documents that its competitors had hidden during the trial (first hiding patents, then e-mails, what next?). The very same day, the White House got in on the QUALCOMM/Broadcom fun, as President Bush upheld an International Trade Commission ban on QUALCOMM products that were found to have infringed on Broadcom patents. Finally, a week later, a judge in a third QUALCOMM/Broadcom patent case ruled for Broadcom again, ordering the former to pay the latter $39.28 million in damages.

Where does Nokia stand in all this, you ask? Also in August 2007, as Broadcom was taking QUALCOMM's lunch money in court, Nokia announced that it had signed a new licensing contract for cell phone microchips. Who would be its new partner, replacing QUALCOMM? You guessed it—Broadcom was to be Nokia's new partner

Visit Vault at **www.vault.com** for insider company profiles, expert advice, career message boards, expert resume reviews, the Vault Job Board and more.

VAULT CAREER LIBRARY

381

in all things microchip. The industry scrambled to adapt to the sudden change, which forebodes a different landscape with different corporate alliances.

QUAL-ity control

Although QUALCOMM's legal defeats amount to little superficial and financial damage (what's a $40 million fee in the face of $2 billion profits?), they are a huge public relations disaster, and it must not be easy to explain the loss of such a trusted trading partner as Nokia to stockholders. Almost immediately, QUALCOMM announced the resignation of general counsel Louis Lupin, the man responsible for "litigation misconduct" in court (it seems the company wants very much to move beyond that ugly incident). It hasn't yet found his permanent replacement, but QUALCOMM is bringing in big guns in the interim—Carol Lam, the former San Diego U.S. attorney (who might best be remembered for her involvement in the U.S. attorney firing scandal in early 2007). Industry analysts are painting QUALCOMM as a major loser in this case, but CEO Paul Jacobs has remained optimistic, stating in August that, "(f)rom a performance standpoint and a technology standpoint, we're still in the lead, (and) I think we're going to stay in the lead."

GETTING HIRED

Opportunities for the chipper

QUALCOMM's careers site, located at www.qualcomm.com/careers, provides plenty of intel for the aspiring employee. The site is elaborately customized, with particular information for interns, graduates and experienced hires. To apply for a position, job seekers must first create a profile with a resume and contact information. In addition to accepting resumes through its web site, the company recruits at the NSBE (National Society of Black Engineers) Convention and at the NSHMBA (National Society of Hispanic MBAs) Conference. If you get an interview with the company, note that interviewers will probably ask behavioral questions. The whole ordeal lasts about 45 minutes, and interviewees will meet with about five employees from various levels.

In 2007, *Fortune* named QUALCOMM to its list of the 100 Best Places to Work. According to the magazine, QUALCOMM allows any employee to make suggestions for new products (by cell phone, naturally)! Other perks include generous health benefits, job sharing, an on-site gym and dinner for employees working late. Other benefits for U.S. employees include 401(k) with company matching, a discount stock purchase plan, medical and dental insurance, an honor system for sick leave, tuition assistance, scholarships for children of employees, flexible spending accounts for health care expenses and a charitable donation matching program. Benefits vary by office location, however.

OUR SURVEY SAYS

Don't phone it in at QUALCOMM

Want to get your foot in the door at QUALCOMM? One source reveals, "I interviewed with four different people within the finance function, three out of the four I now work with closely, and one from the HR/recruiting function. The finance questions focused on my background and my working style; the HR questions focused on benefits and salary expectations." Others found the process quite amenable. "Interviews were not stodgy or too formal … the process was not painful—in fact, it was more conversational," says one hire. "Interview was much more technical than any I've had before," warns an engineer. "Don't expect easy questions from QUALCOMM," advises his colleague. If you don't hear anything from the company for a while, don't despair. "QUALCOMM can and does take up to a month to get back to you if they are interested in you," says a source.

Hires at QUALCOMM give their working environment high marks. "Good working environment and active team players," says one entry-level staffer. "The work culture is excellent," his co-worker gushes. "Questions are encouraged and teamwork is a must," adds another newbie. "Working at QUALCOMM is like coming to a university to study. Every day you get to learn something new," adds an engineer. "They're big on meetings ... so big that they often entice your attendance with food and drinks!" says one enthusiastic hire. If there's a downside, the employees aren't letting on, except for this insider: "QUALCOMM is a[n] e-mail-based giant corporation," he sighs. "QUALCOMM is hiring clever and competent yet laid-back guys," says an engineer. But don't despair, ladies! There's "loads of diversity."

Hours at QUALCOMM are long. "I have to stay at my office [late] in the evening," says one employee. "The hours are … flexible," says his colleague. The hours are flexible, and so is the dress. "The dress code seems to vary by employee and by group. Within my group we are business casual Monday through Thursday with the option to wear jeans on Friday," says one hire.

Opportunities for advancement are plentiful as well. "Pretty much once you're in the only way left is up," observes an insider in San Diego. One programmer notes that he has access to the company's "library, online courses, [and] technical training." An entry-level engineer in San Diego notes that "the living expenses here [are] extremely high."

Red Hat, Inc.

1801 Varsity Drive
Raleigh, NC 27606
Phone: (919) 754-3700
Fax: (919) 754-3701
www.redhat.com

LOCATIONS

Raleigh, NC (HQ)
Boston, MA • Mountain View, CA •
Tysons Corner, VA • Buenos Aires •
Munich • Singapore

Additional facilities in over 25
countries worldwide.

DEPARTMENTS

Accounting • Administrative •
Business Development • Channel
Sales • Consulting •
Documentations • Engineering •
Engineering Services • Executive •
Facilities • Finance • Human
Resources • IT • IS • Inside Sales •
International • Investor Relations •
Legal • Marketing • Product
Management • Project Management
• Quality Assurance • RH University
• Sales • Sales Engineering •
Software Engineering • Support •
Training

THE STATS

Employer Type: Public Company
Stock Symbol: RHT
Stock Exchange: NYSE
Chairman, President & CEO:
  Matthew J. Szulik
2007 Employees: More than 1,800
2007 Revenue ($mil.): $400.6
2007 Net income ($mil.): $59.9

KEY COMPETITORS

Microsoft
Oracle
Sun

EMPLOYMENT CONTACT

www.redhat.com/about/careers

Visit Vault at **www.vault.com** for insider company profiles, expert advice,
career message boards, expert resume reviews, the Vault Job Board and more.

VAULT CAREER LIBRARY   385

THE SCOOP

Hats off to Linux

Red Hat makes a living by causing the big guys to squirm: it helped pioneer the use of Linux, the open source computer operating system that has become a chief rival to Microsoft's Windows and Sun's UNIX. The term "open source" refers to a set of practices and principles that encourages access to the design and production of goods and knowledge. Most typically, the expression is applied to the source code of software. Through lenient or nonexistent intellectual property restrictions, the general public can access the source code. As a result, users can create software content through collaboration.

The company's open-source software solutions include Red Hat Enterprise Linux and the JBoss Enterprise Middleware Suite. Red Hat also runs a global training program that operates in more than 60 locations.

Although the company is relatively small potatoes compared to Microsoft, Novell and the like, that doesn't stop CEO Matthew Szulik from his self-described "bold mission" of implementing an aggressive strategy based on the tenets of the open-source architecture movement and wooing corporate customers by blending Red Hat Enterprise Linux, Red Hat Network, Red Hat Applications and Services, "to become the defining technology company of the 21st century."

Szulik does seem to have momentum on his side. During the 2007 fiscal year, the company's revenue reached more than $400 million. This represented a nearly 44 percent increase from FY 2006, when Red Hat's revenue was about $278 million. The company has also hired lots of new employees: in February 2007 more than 1,800 people worked for Red Hat, about 700 more than in early 2006.

Finn-ishing school

Finnish grad student Linus Torvalds was the mastermind behind the Linux operating system. Torvalds created the programming code as a hobby in 1991, and then he released it free over the Internet for anyone to use. It soon picked up a devoted following of programmers, who were excited to find a no-cost alternative to the Microsoft Windows monopoly and to make their own revisions. Chief among them was an IBM programmer, Marc Ewing, who began selling a new-and-improved version he'd developed called Red Hat (named after a cap his grandfather had given him).

In 1994 Robert Young bought the rights to Ewing's creation, and the two men formed Red Hat Software, Inc. The company began distributing Linux by CD-ROM as well as over the Internet (charging $50). Its main sources of revenue, however, were the manuals and technical support it sold to new users and businesses challenged by the original Linux source code, which was constantly changing as improvements were made. Red Hat opened its headquarters in Durham, N.C., in 1996. In 1997, Red Hat and its Linux code still remained on the fringes of the high-tech world, but the operating system was mainly utilized by a select group of programmers who recognized and understood its possibilities.

The company's IPO was in August of 1999, and the shares were priced at $14 each. In February 2000, the firm had a secondary public offering, and his time, it sold four million shares at $96 per share. Red Hat brought in about $250 million through the secondary offering, and the company used the proceeds to pay down debt and expand its geographic reach. In late December 2006, the company left the tech-heavy Nasdaq exchange and joined the New York Stock Exchange; the move punctuated the firm's rapid growth phase, putting it in league with other software heavy-hitters listed on the Big Board such as SAP, McAfee and CA.

Acquired by Red

Red Hat has grown rapidly in recent years through some big acquisitions. In December 2004 it purchased the Netscape Security Solutions unit from Netscape Communications Corporation and America Online. Through the $21.1 million purchase, Red Hat gained key products including the Netscape Directory Server and the Netscape Certificate Management System. The company said that it planned to start offering the products as part of its Open Source Architecture during the coming year, and that these products would enhance security, productivity and manageability.

In 2006, Red Hat scooped up JBoss, Inc. for more than $328 million. JBoss pioneered the "Professional Open Source" model, which, the company says, "combines the best of the open source and proprietary software worlds." This blend of open source and proprietary models makes JBoss a safe choice for enterprises as well as software providers. Companies that use JBoss' software include Continental Airlines, La Quinta and Travelocity. Red Hat's acquisition of JBoss contributed more than $23 million to the firm's revenue for the 2007 fiscal year.

Turning hats into sombreros

In addition to strategically acquiring new products, the company is funneling money into research and development efforts. In 2002, Red Hat opened an engineering and R&D facility in the Boston area, and in early 2007, it expanded operations in the Czech Republic, where the company is investing about $1.7 million in a new open source R&D center.

Red Hat is also growing through increasing its international reach. In May 2006, Red Hat expanded its geographical scope to Latin America by acquiring the local operations of one of its distributors in Argentina and Brazil. As a result of the purchase, Red Hat now has a direct corporate presence in Latin America for regional support, training and sales activities. The company will also be able to localize its offerings.

Following suit

But as Red Hat starts to run with the big boys, it will have to get used to some things: like lawsuits. Just weeks after Red Hat purchased JBoss in 2006, Massachusetts-based Firestar Software sued Red Hat. FireStar alleged that JBoss' Hibernate 3.0 (part of the JBoss Enterprise Middleware System) infringed on one of FireStar's patents. Red Hat's stock price fell on the news of the lawsuit.

And Red Hat has its hands full with another lawsuit: during the summer of 2004, shareholders filed 14 class-action lawsuits against Red Hat as well as some of the company's present and former officers. The investors—who all purchased the firm's securities during various periods between June 19, 2001 and July 13, 2004—claim Red Hat's IPO was dishonest and that the offering was set up to the advantage of insiders. Litigation is pending.

And filing suit

Although Red Hat has been sued, the company has also taken legal action against a company called SCO Group. In 2003, SCO Group filed a lawsuit against IBM, maintaining that IBM had illegally contributed source code to Linux. In August 2003, SCO broadened the scope of its suit beyond IBM, saying it would sue Red Hat's customers unless they shelled out licensing fees to SCO. Red Hat responded by suing SCO for, among other things, deceptive and unfair business practices.

Then, in 2004, SCO filed a lawsuit against one of Red Hat's competitors, Novell. SCO Group claimed that it owned the copyright to UNIX operating system software, since it bought the rights to UNIX from Novell back in 1995. In August 2007, a federal judge, Dale Kimball, ruled on the Novell lawsuit, saying that SCO Group did not own the copyright to UNIX, and Novell had the right to dismiss the suit. As SCO is by now teetering on the brink of bankruptcy, Novell's countersuit could knock SCO down for the count. Novell itself has recently tried to position itself as a friend to all kinds of open-source software, and as the probable winner of these court battles, it should help the sector sidestep a major legal threat.

Making a difference one laptop at a time

In January 2005, Nicholas Negroponte, co-founder of the MIT Media Laboratory, announced a new initiative at the World Economic Forum. The mission of the program, called One Laptop per Child, is to give every kid in the world an Internet-ready laptop. Red Hat is heading up software integration for the initiative. CEO Szulik said, "In the face of the massive wealth creation that the technology industry has created for so many, we have found it unconscionable that so many could be without the tools and resources to join the digital ecosystems of the 21st century. The One Laptop per Child initiative is another step in Red Hat's work to do defining work while making life a little better for others."

Red Hat is red hot

It looks like Red Hat is going to continue to thrive in the foreseeable future. According to a recent article in *The Wall Street Journal*, Linux operating systems, which many corporations already use on server computers, are finally starting to make their way onto employees' personal computers. In fact, one of the challenges Red Hat faces is its own growth. Red Hat believes that its culture fosters creativity, teamwork and innovation, and the firm hopes it will be able to hold onto its philosophy and startup company feel even as it gets larger.

GETTING HIRED

Top Hat

In 2007, *CIO Insight Magazine* named Red Hat its Most Valued Vendor for the third year in a row. The magazine also recognized the firm as the No. 1 IT vendor doing business with companies in Japan.

Visit Vault at **www.vault.com** for insider company profiles, expert advice, career message boards, expert resume reviews, the Vault Job Board and more.

VAULT CAREER LIBRARY   389

Although Red Hat is a large global enterprise, the firm says it has retained its "small-company spirit." In the United States, full-time employees receive benefits that include medical, dental, vision, disability and basic life insurance. These perks kick in the first day of employment. If Red Hat "associates" (the company doesn't call them "employees") are at least 21 years old, they are eligible to enroll in the company's 401(k) plan on the first of the month following their hire date. Other benefits include paid time off and holidays, as well as an employee referral plan. In addition, the company offers tuition assistance for job-related courses.

How to hang your hat at Red Hat

Departments at Red Hat range from accounting to training. Recently, the firm was hiring for hundreds of positions including a human resources representative in Japan, software engineers in Raleigh, a senior consultant in New York and technical writers in Australia. The company also has an internship program for students, which allows interns to work in a "professional environment learning from Red Hat's experienced management team." The firm says it looks for interns who are motivated, smart and industrious. Interns should be able to work well in high-pressure settings.

Red Hat recruits at college career fairs. In addition, potential applicants can find more information about careers at Red Hat and apply for jobs online at www.redhat.com/about/careers. Applicants can search for openings by location, division (Red Hat or JBoss) or job category. The company keeps resumes in its database for six months. If you have trouble applying for jobs through the web site, you should e-mail the company at careers@redhat.com.

Research in Motion Limited

295 Phillip Street
Waterloo
Ontario N2L 3W8
Canada
Phone: (519) 888-7465
Fax: (519) 888-7349
www.rim.net

LOCATIONS

Waterloo, Ontario (HQ)
Charlotte, NC
Chicago, IL
Dallas, TX
Houston, TX
Irving, CA
Los Angeles, CA
Miami, FL
Redwood City, CA
San Diego, CA
Seattle, WA
Washington, DC

Locations in 16 countries
worldwide, including the UK, Japan,
India, Australia and Canada.

DEPARTMENTS

Corporate
Executive
Hardware
IT
Manufacturing
Marketing
Operations
Sales
Software
Technical Support

THE STATS

Employer Type: Public Company
Stock Symbol: RIMM
Stock Exchange: Nasdaq
Chairman: John Richardson
Co-CEO: Jim Balsillie
President & Co-CEO: Mike Lazaridis
2007 Employees: 6,254
2007 Revenue ($mil.): $3,037

KEY COMPETITORS

Microsoft
Nokia
Palm

EMPLOYMENT CONTACT

www.rim.com/careers

Visit Vault at **www.vault.com** for insider company profiles, expert advice,
career message boards, expert resume reviews, the Vault Job Board and more.

VAULT CAREER LIBRARY   391

THE SCOOP

E-mail in motion

Research in Motion is best known to legions of on-the-go workers as the manufacturer of the BlackBerry, that pocket-sized tether to the office that provides instant access to e-mail, text messages and phone calls. The company also manufactures radio modems for wireless phones and smart card readers to control access to its devices. The company's revenue and profits both increased dramatically from 2006 to 2007—its revenue from $2 billion to $3 billion, and its profits from $374 million to $631 million. The same year, *BusinessWeek* ranked it just outside the top 10 of its InfoTech 100—it came in at No. 11. The company's employees are enjoying the ride as well, as their numbers increased by over 1,000 from the year before.

Some users refer to the device as the CrackBerry, a nod to its enormously addictive potential—which is not limited to instant e-mail and message access, but also includes such games as Brick Breaker, originally included in the devices in 2002 as a way to show the new color screen, but which now constitutes a boardroom phenomenon. The BlackBerry prayer, meanwhile, is the name given to the bowed head and clasped hands of the employee consulting the device during a meeting. There's even a parody advertisement for a helmet, complete with an antenna and little orange safety flag, poking fun at oblivious BlackBerry users who wander about, nudging their devices with their thumbs. The ubiquity of such twiddling has even spawned the neologism "thumbing" to describe walking whilst keying in a message on the device.

Birth of the 'Berry

Research in Motion did not emerge as a result of Japanese efficiency or U.S. innovation, but of Canadian fortitude. Michael Lazaridis (current co-CEO) and Douglas Fregin (VP of operations) founded the company in 1984, on the cusp of the digital age. By all accounts, Lazardis named the company Research in Motion after hearing a football pro's play described as "poetry in motion." Early on, the company contracted in software development, electronic engineering and radio communications. In November 1991, RIM announced its collaboration with Ericsson GE, RAM Mobile Data and Anterior Technology on a wireless e-mail gateway service, which would transform Ericsson's wireless data network Mobitex into a two-way paging and e-mail system.

By 1997, RIM was ready to start selling its products directly to consumers. Its first commercially available product, the quaintly named Inter@ctive pager, featured a contact manager and scheduling program in addition to its e-mail technology. IBM and Panasonic outfitted their sales forces with the device, and its sales took off; RIM floated its stock on the Toronto exchange the same year. In 1998, the BlackBerry hit the market, and RIM's revenue went from $21 million that year to $47.5 million the next, as corporations bought the devices by the truckload to keep in touch with roving employees.

In 2001, the company hammered out a deal with SAP that allowed workers wireless access to SAP's line of business software via their BlackBerries. In 2002, the BlackBerry evolved into the all-purpose device consumers today know and love when RIM added phone features to it. The company took in $300 million in revenue while posting a loss of $150 million that year, due to an economic downturn and problems with the launch of new products. RIM got its balance sheet back on track by the following year, with a modest profit of $51 million. In 2005, the number of BlackBerry users nearly doubled, from just over 1,000 to 2,510 by the end of the year; the company took in just over $2 billion.

Innovation in design

RIM noticed the cell phone industry's trend of offering sleek, slimmed-down devices with lots of cool (BlackBerry-ish) functions, but the company didn't respond until the launch of the BlackBerry Pearl in September 2006. Not only was the Pearl available in colors (red and white), it also contained a camera—a first for an RIM product, though nowadays practically a prerequisite in cell phones—and a trackball navigation system, replacing the scrolling dial on the side of earlier models. These design innovations struck a major chord with consumers—the company called the Pearl's product launch the most successful in its history.

The next version in the standard BlackBerry line, the 8800, was issued in February 2007, first available through AT&T Wireless (formerly Cingular). Thinner than previous models, it plays music and videos, retains the trackball navigation from the Pearl and the Global Positioning System feature first used in the 7520 model in 2005. However, unlike the Pearl, this latest model still did not feature a camera—the reasoning being that a camera can be a liability in sensitive business environments.

Fun and games

In contrast to BlackBerry's somewhat businesslike demeanor and reputation, users (greatly helped by the trackball) still managed to have some adolescent fun with both the Pearl and the 8800. Game developer Hardmark offers 30 titles that can be played on a BlackBerry; French firm Gameloft offers about 250 games for the BlackBerry Pearl, including poker, and a title based on the hit TV show *Lost*. Another software company, Magmic Games, launched an online site called BPlay dedicated to ringtones and games written specifically for the BlackBerry; it sold more than a million titles in 2006. That year, *The Wall Street Journal* reported on business executives who postponed meetings or delayed calls in order to keep playing to beat competitors' high scores posted online in the classic BlackBerry game *Brick Breaker*.

Even RIM can see the possible need for some diversion now and then—the company recently sent out an open invitation to third-party gaming software coders to develop more games for the platform. (More likely, it just wants to keep its customers constantly engaged.) In early 2007, the company released a large number of new Application Programming Interfaces (APIs) for the BlackBerry Java Development Environment (JDE). The release is significant because 125,000 developers who have already downloaded the BlackBerry JDE can use the APIs as a tool to create applications. RIM hopes these third-party developers will develop e-commerce, social networking and multimedia apps for the BlackBerry, solidifying its position as a tool adaptable for business or pleasure.

A juicy balance sheet

In 2006, RIM posted revenue of $2 billion, with a profit of $382 million. That year, the company also rolled out the BlackBerry in Trinidad and Tobago, Greece, United Arab Emirates, Saudi Arabia, Slovakia, Korea and 13 Latin American countries including Panama, Venezuela, Peru and Argentina. RIM also debuted new features courtesy of tech company-of-the-moment Google, including Google Talk and Google Local, enabling BlackBerry users to instant-message each other and to use Google's handy maps.

Busy signals

The biggest challenge on the horizon for RIM may be the Apple iPhone, which was launched in June 2007. Apple's customers are known for their rabid loyalty to the company's product line, and the phone's smooth touch elements, integrated camera and ability to incorporate audio and video from a customer's personal iTunes library could be enough to convert all those iPod listeners into iPhone (rather than BlackBerry) users. RIM is fighting back, however, by allowing some programmers access to its tightly-controlled operating system, so as to develop new programs for the cute little e-mail device.

GETTING HIRED

Add some Motion to your career

RIM's careers site, at www.rim.com/careers, provides information on the company's hiring process, benefits and job openings. Jobs are searchable by region and then by function. Interested parties may fill out a web form to apply.

Students are also welcome to apply to RIM as well. The company hires 300 interns every four months, in the areas of R&D, marketing, tech support, manufacturing, finance, legal, HR, sales and IT. Students pursuing any major are welcome to apply. Jobs are posted in September, January and May. RIM recruits at colleges and universities in the U.S. and Canada, including the University of Seattle and the University of Saskatchewan. In addition to its internship opportunities, the company also offers summer jobs in administration and manufacturing.

RIM usually conducts initial screenings over the phone (your phone need not be a BlackBerry). Once applicants have successfully gotten over that hurdle, they are taken to a RIM office for subsequent interviews; the company will help with any travel arrangements. Interview questions are generally behavioral. RIM's offers of employment are provisional until the candidate passes reference and background checks, and the company does not recommend that job seekers end their previous employment until the checks are complete.

RIM's benefits include salary and incentives, employee assistance plan and travel insurance, summer and holiday parties, and contests and giveaways. The company also subsidizes gym memberships and sports clubs, and provides an on-site clinic with massages, health education, ergonomic assessments and health fairs. And, of course, employees get a free BlackBerry.

Ricoh Company, Ltd.

Ricoh Building
8-13-1 Ginza
Chuo-ku, Tokyo 104-8222
Japan
Phone: 81-3-6278-2111
Fax: 81-3-6278-2997
www.ricoh.com

Ricoh Corporation
5 Dedrick Place
West Caldwell, NJ 07006
Phone: (973) 882-2000
Fax: (973) 673-6934
www.ricoh-usa.com

LOCATIONS

Tokyo, Japan (Corporate HQ)
West Caldwell, NJ (US HQ)
Atlanta, GA • Chicago, IL •
Cupertino, CA • Dallas, TX •
Houston, TX • Menlo Park, CA •
Miami, FL • Montville, NJ • San
Jose, CA • Tustin, CA •
Washington, DC

DEPARTMENTS

Administration • Development •
Engineering • HR •
Marketing/Engineering Support •
Sales • Service & Support

THE STATS

Employer Type: Public Company
Stock Symbol: 7752
Stock Exchange: Tokyo
Chairman & CEO: Masamitsu Sakurai
2007 Employees: 81,939
2007 Revenue ($mil.): $17,547

KEY COMPETITORS

Canon
Hewlett-Packard
Xerox

EMPLOYMENT CONTACT

www.ricoh-usa.com/careers

THE SCOOP

Company of many colors

Though best known as a manufacturer of printers, copiers, and fax machines, the Ricoh Company also has interests in semiconductors, PCs and networking equipment, measuring devices, heat-sensitive paper for fax machines and receipt printers, and optical sensors. Other arms of the business also handle leasing and credit for the customers of its office equipment division. Ricoh's products are sold in 130 countries. In 2006, Ricoh took in $16 billion in revenue.

Office automata

In 1936, Kiyoshi Ichimura founded Ricoh as a maker of photographic paper in Japan. (Ichimura, a self-made entrepreneur who was born into a poor farming family, would helm the company until his death in 1968.) Ricoh's success in photographic paper naturally led to it manufacturing cameras, in 1938; the company floated shares on the Tokyo Stock Exchange in 1949. It introduced a diazo copier, used for reproducing blueprints, in 1955, and became the first Japanese company to mass-produce cameras two years after that.

Ricoh rang in the next decade with the introduction of an offset copier for the office in 1960, while an automated data processing system followed in 1965. In the meantime, the company opened its first U.S. office in 1962 with a handful of employees and $100,000. The company continued producing automated office products, introducing a desk-sized computer in 1971. More products were introduced at a brisk clip: a high-speed fax machine debuted in 1974, a plain-paper copier in 1975 and a word processing device in 1976. In 1979, the company set up an R&D shop in California.

Ricoh-naissance mission

In 1982, Ricoh scored a coup when Gannett Co., the owner of USA Today, installed a Ricoh system, allowing it to send images of the newspaper to its printer via satellite. By 1984, Ricoh was keeping abreast of the competition when it started researching speech and optical character recognition. It soon purchased another company's semiconductor division and set up a semiconductor design facility in California in 1987. Ricoh then introduced its first plain paper color copier three years later.

Visit Vault at **www.vault.com** for insider company profiles, expert advice, career message boards, expert resume reviews, the Vault Job Board and more.

VAULT CAREER LIBRARY   397

However, due to a slumping Japanese economy in 1992, the company posted its first profit loss in over 50 years of doing business. In order to lower manufacturing costs, the company moved production to facilities in Korea and China. That same year, Ricoh came out with CDs; it introduced a CD-RW, which could be erased and rewritten, five years later. In 2001, Ricoh expanded its distribution arm with the acquisition of office products distributor Lanier. It opened a Chinese subsidiary the next year and moved its headquarters to Chuo-ku, Tokyo the year after that.

You can print in any color you want, as long as it's green

Ricoh is committed to making its effects on the planet as benign as possible. In 2006, the Japanese government awarded the company its Ecology Design Prize for a design for packaging toner for copiers and printers—the toner's box was created to pack efficiently into a small space, the toner package is designed to be used multiple times, and it is easy to clean and refill.

Ricoh is also experimenting with manufacturing its printers and copiers out of plastics derived from plants. In 2006, the company was testing copiers made of 50 percent plant material in Japan. If the green plastics can stand up to the rigors of the office copier environment, Ricoh plans to take them global. Ricoh credits its founder, Ichimura, for its environmental awareness; the company web site proclaims that "(Ichimura) developed a sense of corporate social responsibility that is talked about a great deal today, but was almost unheard of 50 years ago," and that "he made an early and genuine commitment to social and environmental sustainability in ever aspect of Ricoh's business activities."

Blue period

In 2006, Ricoh agreed to acquire IBM's interest in a joint venture, InfoPrint Solutions. Big Blue is seeking fatter margins on its products and looking to free up capital to invest in higher-growth areas, following some reductions in its printing division's revenue. Ricoh paid $725 million for a controlling interest in the division, and acquire the outstanding portion by 2009.

Speaking of blue (or is it Blu?), in 2006 Ricoh introduced a device that can read both HD DVDs and Blu-ray discs. These two formats, previously thought to be mutually exclusive (not to mention the new VHS vs. Beta), may yet find common ground in players equipped with the devices.

Golden opportunity

By the end of fiscal 2007, Ricoh had taken in revenue of $17 billion and a profit of $947 million. Earlier that year, the company announced that it had restructured its American operations as Ricoh Americas. Lanier, the office products distributor acquired in 2001, would be folded into an arm of the organization called Ricoh Business Solutions.

GETTING HIRED

Picture yourself at Ricoh

Ricoh's U.S. careers page, at www.ricoh-usa.com/careers, provides information about benefits, recruiting and job openings. Ricoh's benefits include dental, medical, flexible spending accounts, 401(k) with company matching, a wellness program and tuition assistance. Dress at the company is business casual. Open positions are listed at www.ricoh-usa.com/careers/listing.pl?all, and are searchable by location or type. To apply, interested parties may submit their resumes online or e-mail them to a recruiter.

Visit Vault at **www.vault.com** for insider company profiles, expert advice, career message boards, expert resume reviews, the Vault Job Board and more.

VAULT CAREER LIBRARY   399

The Sage Group plc

North Park
Newcastle Upon Tyne
NE13 9AA
United Kingdom
Phone: +44-191-294-3000
Fax: +44-191-294-0002
www.sage.com

LOCATIONS

Newcastle Upon Tyne, UK (HQ)
Austin, TX • Beaverton, OR •
Chicago, IL • Clearwater, FL •
Dallas, TX • Gwinnett, GA •
Herndon, VA • Irvine, CA •
Knoxville, TN • Mayfield Heights,
OH • Nashville, TN • Philadelphia,
PA • Pittsburgh, PA • Rocklin, CA •
Scottsdale, AZ • Williamsburg, VA

Operations in 18 countries
worldwide.

DEPARTMENTS

Accounting
Customer Service
Customer Support
Finance
Human Resources
Information Services
Legal
Marketing/Public Relations
Product Management
Product Marketing
Production
Quality Assurance
Research & Development
Sales
Training & Education

THE STATS

Employer Type: Public Company
Stock Symbol: SGE
Stock Exchange: LSE
Chairman: Anthony Hobson
CEO: Paul Walker
2006 Employees: 13,400
2006 Revenue ($mil.): $1,832

KEY COMPETITORS

Intuit
Microsoft
Oracle
SAP

EMPLOYMENT CONTACT

www.sage.com/careers.php

THE SCOOP

Words for the wise

In 20 years, the U.K.-based Sage Group has evolved from a one-solution business to the region's largest software firm and the FTSE 100's only technology stock. That's all the more remarkable since it markets its business-to-business products (formatted for PC/servers and Internet access) chiefly to small- and medium-sized enterprises (SMEs) of up to 500 employees. Of course, a whole lot of companies fit into that category—which goes a long way toward explaining Sage's elephantine customer base of 5.2 million companies spread over 19 countries.

The Sage Group offers both out-of-the-box software and custom solutions, which meet client needs in accounting/financial, HR and payroll, customer relationship management and payment processing. (Its branded programs include Peachtree in the U.S., Simply Accounting (Canada), Ciel (France), SP PymePlus (Spain), Softline Pastel (South Africa) and Sage 50 in the U.K.) A somewhat newer area for the firm presents industry-specific solutions, which in various geographic regions include health care, food distribution, transportation, manufacturing, real estate, construction, retail and not-for-profit agencies. The company also provides support services to 1.5 million of its clients, and its call-in customer service centers respond to 30,000 calls each day. These days, software only accounts for 35 percent of company revenue—the lion's share is amassed through service contracts.

The gold man of software

The Sage Group was just another start-up in the early 1980s. David Goldman, a printer by trade, thought that the process for estimating job costs might be simplified using the developing tools of computer programming. He recruited local college students to work on the problem, and the resulting program was such a success he sold it to other printing companies.

The subsequent accounting product took off with the help of some venture capital financing and some well-timed good fortune. When the first low-cost personal computer (Amstrad's PCW) came on the market in the U.K. around 1985, Goldman's PC-compatible (and similarly inexpensive) accounting package was ready less than eight weeks later.

To build the company, the entrepreneurial Goldman and his managers took advantage of growing momentum by regularly plowing 50 percent of revenue back into company operations and marketing. And the long-term fate of the company was solidified with one decision: Goldman decided to charge for technical support for his product; though now common practice, it was an innovative strategy at the time. As a result, Sage got literal and figurative buy-in from its customers, and established itself as the No. 1 low-cost accounting software vendor in the market.

Sage strategies

The key to Sage's success so far has been its organizational model. Each of its four regional businesses (North America, U.K. and Ireland, Europe and Asia, and South Africa and Australia) operates with relative autonomy, determining its own local strategies, acquisition possibilities and product line. Not only that, Sage tailors each offices' products—mainly the ones focused on accounting, payroll and human resources—to meet local requirements for tax structures, legal notification and other fiscal necessities. Sage is therefore considerably more flexible and responsive than its competitors, whose main products are mostly standardized and depend on company-specific upgrades.

Sage has also profited through the extensive network of partners and consultants pushing its products. More than 40,000 accountants and financial advisors worldwide recommend Sage to their customers, and approximately 23,000 resellers and business partners with more local and company-focused knowledge also promote the company's products; these resellers also feed information back to the company on potential product or service opportunities. One huge contact, the U.K.-based Barclays Bank, prefers selling Sage business management expertise to its small business clients.

About 50 percent of Sage's business is derived from its accounting and financial products, and human resource and payroll options make up another 11 percent. Industry specific applications such as healthcare, manufacturing and construction make up another 28 percent. However, in the next few years, the company expects to see increased revenue for the higher-margin support segment (which carries an 80 percent customer renewal rate) as companies more frequently request the bundling of custom solutions and support.

Growth spurt

The company spent heavily on acquisitions between 2005 and 2007, expanding both horizontally and vertically. The accounting segment was the beneficiary of most of the takeovers, which added capacity in the U.K., Spain, Switzerland, France and

Malaysia. However, the firm invested the most in its U.S. operations, and in early 2006 it bought Verus, a credit-card processing company, for $330 million.

Near year's end, it purchased Emdeon Practice Services (which, as part of Emdeon, was formerly known as WebMD) for $565 million, the largest deal in company history, which launched Sage into a new industry sector, as Emdeon produces software that manages patient records, prescriptions and billing. The company then attempted an even larger deal for the purchase of Visma, which would have taken it into Scandinavia, but it fell through at the last minute.

Acquisition fever?

All this activity has led some industry watchers to speculate that Sage relies a bit too heavily on growing its business by swallowing other firms. Sage took over 104 companies since its birth—15 of them in the past three years—and Paul Walker, the company's CEO, has admitted that some recent internal issues have slowed Sage's growth in the U.S. and that the division plans to streamline and hone its competitive edge.

Nonetheless, it's hard to argue with a 23 percent increase in revenue, a similar boost in earnings per share during 2006, and organic growth averaging 5 percent to 7 percent across North America, the U.K. and mainland Europe. (In South Africa, Australia and Asia it's a different story—total revenue in those regions was up 17 percent.) Tech insiders on that side of the fence agree that Sage's acquisition process is not a problem, since it permits the company to diversify both its product line and the markets it serves.

Corporately conscientious

The organization's social responsibility policy is quite extensive, with chapters that touch upon employment, the marketplace, the local community and the environment. Besides stressing the importance of diversity in the workplace, Sage looks to develop the entrepreneurial spirit and sense of teamwork among its staff, and provide social and intellectual growth. It calls itself an "integral part of the communities" in which it's based and encourages its employees to volunteer locally and donate to worthy causes. Moreover, Sage has taken a great many steps to reducing the operation's environmental impact by using recycled water for landscaping, minimizing the use of printed materials and packaging wherever possible, and advocating limiting power usage in its offices in the U.K. and Europe. Sage prides itself on encouraging and supporting the development of its employees, with strong internal promotion strategies and reportedly low staff turnover.

GETTING HIRED

Work, wisely

In keeping with its decentralized structure, Sage doesn't keep a listing of all international vacancies (although its web site states that one is in the works). The company's online presence (at www.sage.com) does, however, include links to the country-specific Sage web sites in America and Canada, the U.K., Ireland, France, Germany, Switzerland, Australia, South Africa, Spain, Portugal, Poland and Belgium. (Note that some are not in English.) As might be expected, openings are generally concentrated in the areas of sales, customer support and software design and development.

Sage employs about 1,400 at its Newcastle, U.K., headquarters, and another 3,000 in various U.S. locations. Each office shapes its own culture, with "social events in all regions to promote and embed" corporate principles. The company is said to be strong on training and recognition programs; in addition, some regions provide health awareness programs, and office workers in mainland Europe can take advantage of English lessons.

Samsung Group

Samsung Plaza
263 Seohyeon-dong
Bundang-gu Songnam, 463-721
Phone: 82-2-751-7114
Fax: 82-2-727-7892
www. samsung.com

Samsung America, Inc.
105 Challenger Road
Ridgefield Park, NJ 07660
Phone: (201) 229-5000
Fax: (201) 229-5080
www. samsungamerica.com

LOCATIONS

Ridgefield Park, NJ (US HQ)
Houston, TX
La Mirada, CA
Los Angeles, CA
New York, NY
San Jose, CA
Seattle, WA
Secaucus, NJ

DEPARTMENTS

Accounting & Financial
Administrative
Business Development
Chemical
HR
Legal
Petrochemical
Semiconductor

THE STATS

Employer Type: Public Company
Stock Symbol: SMSD
Chairman & CEO: Lee Kun-Hee
2006 Revenue ($mil): $64,324

KEY COMPETITORS

LG Group
Hynix Semiconductor
SK Group

EMPLOYMENT CONTACT

All branches except CA:
Samsung America Inc.
Human Resources Dept.
105 Challenger Road
Ridgefield Park, NJ 07660
Fax: (201) 229-5083
E-mail: careers@samsungamerica.com

Branches within CA:
Samsung America Inc.
Human Resources Dept.
14251 East Firestone Boulevard
La Mirada, CA 90638
Fax: (562) 921-6384
E-mail: hr@la.sai.samsung.com

Visit Vault at **www.vault.com** for insider company profiles, expert advice, career message boards, expert resume reviews, the Vault Job Board and more.

VAULT CAREER LIBRARY   **405**

THE SCOOP

We're No. 1!

The Samsung Group is far more than flat-screen TVs and cell phones. As South Korea's largest conglomerate—it recently overtook the floundering Hyundai Group—it is best known for its electronics, but produces a number of other services and products as well. This is especially true in South Korea, where on any given day you could attend a Samsung Lions game (the group's Korean professional baseball team), tour the Samsung Museum of Art; ride a rollercoaster at Samsung Everland, the largest amusement park in the country; or go on a shopping spree for men's wear, women's wear, sportswear and accessories made by Samsung's Cheil Industries, Inc.

Wait, there's more. Consumers can grab a room at The Shilla Seoul (owned by Samsung affiliate, The Shilla Hotels & Resorts) and charge it to Korea's largest independent credit card issuer, Samsung Card Co. Ltd. ($3.14 billion sales in 2003). If in need of medical attention, head over to the Samsung Medical Center. You can even take out a life insurance policy at Samsung's Life Insurance Company (2003 sales of $19.3 billion). The diverse company, which posted 2003 sales of $1.6 billion, also sells chemicals and electronic chemical materials.

These are just a few of Samsung's roughly 35 affiliates, although the number is significantly down from the 61 affiliates Samsung boasted before the Korean economic crisis of the late 1990s. The company emerged from near bankruptcy to reconfigure itself into a powerful conglomerate with a total of more than $100 billion in 2003 sales. While the group's largest holding, Samsung Electronics, had a record year in 2004, a host of worrying issues remain ahead of the company, including Korea's shaky economic future. Whether Samsung can weather the storm remains to be seen.

It all began ... with dried fish

Lee Byung-Chul, born into a wealthy landowning family in Korea, founded the company that would become Samsung in 1938. Japan had been occupying the country since Lee's birth in 1910, and it heavily influenced the development of the Korean economy at the time, establishing an infrastructure for the ease of exporting materials and foodstuffs to other Japanese territories. Lee's company began under these conditions as an exporting company, trading in dry fish, but it diversified its offerings soon afterwards.

The Korean War wiped out almost all of Samsung's assets in the 1950s, but the company was soon back on its feet, thanks in part to a favorable government policy that aided family controlled conglomerates like Lee's in order to help rebuild the economy. Indeed, many of these family-operated businesses still dominate the Korean economy and are called "chaebols." Although professional executives conduct the day-to-day business of such firms, members of the founding families often still have a stake in the direction of the company, Samsung included—Lee Kun-Hee still helms Samsung, the son of founder Lee Byung-Chul.

Gimme some sugar ... government

Lee Byung-Chul formed the highly profitable Cheil Sugar Company—then the country's only sugar refinery—in 1953, and textile, banking and insurance ventures followed. In 1961, however, Major General Park Chung-Hee became South Korea's president in a bloodless military coup, on an anti-corruption platform. As Lee was the richest man in the country, he was an easy target for government prosecution, and promptly fled.

He returned to Seoul and struck a deal with Park that typified the government-chaebol relationship in the postwar period (Park remained president until 1979). South Korea was still a predominantly agricultural country at the time, but Park was intent on industrializing the economy (as was his new ally, the United States). One of Park's first moves upon gaining power was to nationalize the country's banks, therefore controlling all of the country's capital. Companies such as Samsung, therefore, had to lobby the government for the money to pursue business opportunities, and this was precisely the nature of Lee's deal with Park. If Samsung would occasionally pursue business in the government's interest, he would have all the necessary startup capital.

Chaebols were therefore often less successful than they appeared, as they only grew relative to their government funding, not the same kind of success a similar-sized firm would enjoy in an American-style open market. Many chaebols consequently suffered from debt once the government withdrew its monetary support. From a foreign perspective, this relationship between chaebols and the government can also resemble collusion, thus numerous scandals have lingered around Samsung and the Lee family ever since. But through perseverance—and the occasional bribe—Samsung continued to grow. In fact, Samsung has been one of Korea's largest chaebols in every decade since the 1950s.

Visit Vault at **www.vault.com** for insider company profiles, expert advice, career message boards, expert resume reviews, the Vault Job Board and more.

VAULT CAREER LIBRARY   **407**

Going electronic in the '70s

Samsung capitalized on low wage rates and a favorable export policy by forming Samsung Electronics in 1969. During the 1970s Samsung became a truly global organization, producing inexpensive TVs, VCRs, and microwave ovens, and then manufacturing them under private labels for corporations like General Electric and Sears.

The government also proved a valuable ally in the company's ascent during this decade, awarding it a number of lucrative grants, an "export prize" in 1976 was worth $300 million and another grant in 1978 bloated the company's coffers by $100 million. By the 1980s Samsung was exporting electronics under its own name and spending heavily on R&D to develop new technology. Samsung Electronic became the world leader in chip production in 1990.

Sputtering engines and shuttering operations

When Lee died in 1987, his son Lee Kun-Hee assumed control of Samsung, and he has increasingly come under fire for making unwise investments on pet projects during his tenure as chairman. A notorious car lover, he announced plans to form Samsung Motors in 1994, and didn't revise his position when the Asian financial markets crashed in 1996—pouring $3 billion more into the venture and rolling out its first cars in 1998. In 2000, Samsung was forced to sell the bulk of the unsuccessful operation to Renault in France.

Whatever the cause of the late 1990s Asian economic crisis, it exposed instability amongst the chaebols—11 of the 30 largest firms collapsed between 1997 and 1999, including Daewoo, which went bankrupt with a staggering $80 billion worth of debt. At this point, Lee demonstrated that he possessed some good judgment, probably saving his family's company from bankruptcy. With the help of Yun Jong-Yong, appointed as co-CEO in 1996, Lee orchestrated a massive restructuring, selling more than 100 nonessential businesses, cutting around 30 percent of the workforce, and slashing inventories by closing factories for weeks at a time. At the same time, he worked on brand building, recognizing that Samsung could produce products technologically comparable to Sony's, but that Samsung's bargain-basement reputation ensured they would collect dust on the back of shelves. He doubled the marketing budget and employed tactics such as product placement, in films like Matrix Reloaded employed.

Ten billion dollar baby

The efforts paid off with Samsung Electronic's entrance into the exclusive (and elusive) "10 billion dollar club" in 2004, when the company achieved eye-popping net earnings of just over $10 billion for the year (roughly double those for 2003). It was just short of repeating the feat in 2005, when it pulled in $9.4 billion.

Cell phones were a large part of the success, keying the transformation of Samsung's reputation from a retailer of dodgy, second-rate VCRs to a peddler of premium-tech, premium-price cell phones. A July 2004 *Forbes* article noted that "The cell phone is to Samsung what the Walkman was to Sony—a slick growth engine and an icon of innovation. Suddenly Samsung is number two worldwide, and it aims to supplant Nokia by 2010." It still has plenty of catching up to do in that department—Samsung has a 13 percent share of the global market to Nokia's 36 percent—but the company is actively mounting an aggressive attack. Lack of brand recognition is still a major issue, especially in the U.S., and Samsung Electronics has allocated $800 million a year for ads, promotion and marketing.

Uncle Samsung wants you!

Samsung's American offshoot (Samsung America Inc.) is split into three divisions: international commodities trading, venture capital and product marketing and distribution. It is still working hard on spreading brand awareness, especially stateside, but its efforts reach beyond what one would might think of as exclusively Samsung. Its engineering marketing offices negotiate with semiconductor companies about equipment to send back to the Korean headquarters while its textile distribution offices actively market the clothes of FUBU, with whom Samsung has had a corporate partnership since 1994.

Tech another look at Samsung

Samsung is also intent on funding new research and focusing on attractive, eye-catching design. The company's most recent mobile phone innovations include one that automatically scans business cards and inserts the details into the user's address book and another that has, believe it or not, a joystick for navigating menu options.

Chip shot

Samsung Electronics is a global leader in memory chips, but it has to watch its margins, as the Korean won recently soared to a seven-year-high against the dollar, reducing the value of overseas earnings. But chairman Lee stresses the importance of "preemptive investment," and the company plans on spending around $24 billion on new chip-making facilities by 2010.

The chip industry realized how important Samsung is to its current overall health, and Samsung realized how badly it needs to modernize its chipmaking, when one of its DRAM chip plants was hit with a surprise power outage in August 2007. The outage shut down chip-making operations for about a day, and the company predicted it had lost about $40 million worth of chips. Samsung has a share of about 40 percent of the market for memory chips, so if its factories are outdated, the company has a lot to lose—losing the single day of operations was enough to bump it from the top spot in the industry. Later that month, it announced in that it would be spending an additional $788 million on upgrading its memory chip production.

Reform school

As South Korea's once and future No. 1. chaebol, Samsung has faced many calls to reform its way of doing business. In December 2006, the South Korean government announced that it was investigating Samsung and LG, another big chaebol, over allegations of price-fixing. The previous year, Samsung had pled guilty to just such price-fixing, paying the U.S. government $300 million in October 2005 for colluding to fix the price of DRAM chips with Hynix and Infineon Technologies (who also paid to settle the lawsuit). In August 2007, though, the Hankyoreah Economic Research Institute analyzed sustainability reports of major chaebols, and Samsung received a better-than-average score in terms of financial transparency and honesty.

Looking ahead

Samsung's plans include possibly seeking a listing for Samsung Electronics on the New York Stock Exchange in the hopes of increasing the company's value (many feel the company may be undervalued because nervous investors shy away from the Korean Exchange, referred to as the "Korean Discount").

Daddy dearest

The powerful Lee family (No. 157 on *Forbes*' 2005 list of the World's Richest people) is also busy battling familiar charges of nepotism: For years, civic groups have accused Lee of questionable attempts to transfer management control to his son, and in January 2005 prosecutors sentenced two Samsung Group executives to jail for illegally selling Lee's son and his sisters convertible bonds well below their price. In

May 2007 an appeal court in Seoul upheld the conviction and increased each executive's fines to $3.2 million.

The court offered no word on whether the elder Lee instigated the entire ordeal. He might not have time to dwell on it, though, as he is busy dealing with another scandal, as a wire-tapped phone conversation from before South Korea's 1997 election went public in late 2005, detailing Lee's illegal campaign funding of a losing presidential candidate. The ties between government and chaebols are not a thing of the past, it seems.

GETTING HIRED

A city unto itself

A truly global empire, Samsung Group has a workforce the size of a small city— roughly 195,000 employees—operating in over 50 countries. Samsung Electronics alone employs 75,000 in 87 facilities in 47 countries. The career section allows interested job seekers to click on openings by global region, and search by each specific Samsung company in the region. Register to submit to positions quickly online and receive e-mail notification of matching positions.

Programs for MBAs

MBAs should check out the site for a description of Samsung Electronic's intensive 10-week MBA internship, offered to first- and second-year students from top MBA programs in the areas of R&D, strategy development, product marketing, sales operations and corporate planning in the digital consumer electronics, semiconductor and telecommunications industries.

Lending a hand

The group also prides itself on its strong philanthropic focus. Samsung spent over $1 billion on community outreach in 2001 alone and encourages employees to become personally involved through the SAMSUNG Community Service Team. According to the web site, nearly 71,000 employees are currently involved in volunteer activities, around 61 percent of employees have participated in these endeavors at some point, and outstanding volunteers are recognized with awards. Check out the Community Activities page for more details on various programs.

OUR SURVEY SAYS

Want it the Samsung Way?

"Samsung is a tough ride," one insider declares. It's a "very demanding place to work," "Samsung can be a very difficult environment," other employees chime in. "I was very surprised that teamwork was not a priority," observes one hire. The company has a "Korean military structure. No questions, no chitchat—just follow your orders and get your work done quickly." Not surprisingly, "Turnover is extremely high. I would guesstimate that 80 percent of non-Korean born employees leave within a year," speculates one source. "Even Koreans who lived and worked abroad find it hard to adjust to strongly hierarchical structure, or complete lack of formal rules and procedures," adds a colleague.

Want to take a break from this office? Not a chance. "The phrase 'work/life balance' evinces a sour expression from all … Samsung lifers and this is a non-expectation for employees," says one insider. "Working hours can be long, and taking holiday is regarded by Koreans (including many bosses) as inappropriate," agrees a co-worker. "If you left one night at say 10 p.m. and so got in the next morning at say, 10 a.m. (rather than the rigidly enforced 9 a.m.) you would be sent a warning letter by the president! No consideration would be given to the fact that you had gone above and beyond the call of duty the previous night," says a third. "The dress code was strongly enforced and it was part of my unenviable job to 'tell off' those who wore jeans on Fridays," recalls another insider.

"Diversity was limited—white and Korean and then very, very few Korean women and even fewer female managers. The hours were long due to the presenteeism culture. The opportunities for advancement were fine as long as you were male," adds one hire. Is it all so bleak? "Benefits were very good depending upon your level: gym subsidy, life insurance, long-term disability cover, 25 days holiday, staff canteen," says one source. His colleague cautions, "heretics, beware."

SanDisk Corporation

601 McCarthy Boulevard
Milpitas, CA 95035
Phone: (408) 801-1000
Fax: (408) 801-8657
www.sandisk.com

LOCATIONS

Milpitas, CA (HQ)
Anyang-City, Korea • Bangalore •
Dublin • Edinburgh • Hsinchu,
Taiwan • Juvisy Sur Org Cedex,
France • Ofuna, Japan • Riemerling,
Germany • Shanghai • Shenzen,
China • Shin Sugita, Japan •
Singapore • Sydney • Taichung,
Taiwan • Taipei • Tefen, Israel • Tel
Aviv • Wanchai, Hong Kong •
Yokkaichi, Japan • Yokohama,
Japan

DEPARTMENTS

Administration • Engineering •
Facilities • Finance • HR • IT • Legal
• Marketing • Operations • Safety •
Sales • Sales Operations • Security

THE STATS

Employer Type: Public Company
Stock Symbol: SNDK
Stock Exchange: Nasdaq
Chairman & CEO: Dr. Eli Harari
2006 Employees: 2,586
2006 Revenue ($mil.): $3,257

KEY COMPETITORS

Micron Technology
Renesas
Samsung Electronics

EMPLOYMENT CONTACT

www.sandisk.com/Corporate/Careers

Visit Vault at **www.vault.com** for insider company profiles, expert advice,
career message boards, expert resume reviews, the Vault Job Board and more.

VAULT CAREER LIBRARY   413

THE SCOOP

No flash in the pan

SanDisk is a top purveyor of flash memory, the solid-state drive on a chip that allows a gigabyte of storage to fit on a fingertip's worth of space. The company's products include USB drives, removable data storage cards for cameras, video recorders, cell phones and PDAs, embedded chips for long-term memory in computers and memory for MP3 players. SanDisk also makes the Sansa, the second-most purchased MP3 player (behind guess what fruit-named firm's device), with a market share of 10 percent.

Remembrances of things past

Flash memory technology was only four years old when Eli Harari founded SanDisk in 1988 (he is still the company's CEO today). Harari is a physicist and expert on solid-state memory, a way of storing data without using moving parts (like the platter in a hard drive). The company was originally known as SunDisk, but changed to SanDisk when it went public in 1995, as customers were confusing it with the other Sun in computing.

SanDisk released its first flash memory chip in 1991, but the company's big breakthrough came in 1996 when it debuted a way to double the capacity of a flash chip. In 2000, the company formed a joint venture with Toshiba, called FlashVision, for the production of advanced flash memory. By 2002, all of the company's flash manufacturing was consolidated into Toshiba's facilities in Yokkaichi, Japan. The two companies teamed up to further manufacturing capacity even further in 2004, with the creation of Flash Partners. The next year, SanDisk demonstrated its commitment to new kinds of chip technology with the $300 million purchase of Matrix Semiconductor.

Mnemonic expansion

In 2006, SanDisk acquired Israel-based M-Systems, another manufacturer of flash products, for $1.5 billion. The new acquisition increased SanDisk's control of the market for flash memory to about a third, and complements its product line quite well, since msystems was in the business of manufacturing flash drives for sale under the names of other companies, while SanDisk primarily sells memory under its own name.

Price elasticity

Despite its previous ability to stay one step ahead of the cyclical drop in prices for memory chips, SanDisk proved susceptible to the recent freefall in memory chip prices that gripped the entire sector. In early 2007, the company announced that it would be cutting 10 percent of its workforce, or about 250 jobs, in the face of a bigger dip than anticipated in the price of its products. Industry analysts suspect that the drop in price is a sign that there is a glut of memory chips in the marketplace. SanDisk's remaining employees will have their wages frozen and executives will take a pay cut of around 15 percent. The company is dropping the prices of its products by as much as 40 percent in order to maintain its market share.

Manufacturers hold out hope, however, that the price of their goods will bounce back on demand for memory in the new crop of iPhones and their inevitable imitators. Although SanDisk pulled in $28.5 million in sales in the second quarter of 2007, down 70 percent year-over-year, it was an improvement after two consecutive quarters of losses. CEO Harari interpreted the results as the start of a turnaround in the market.

SanDisk and the land of the rising sun

Despite its job cuts, SanDisk charged ahead in September 2007 with the opening of a new $170 million microchip plant in Shanghai, China. The plant bodes well for Chinese engineers, as it is located close to both Jiaotong and Huadong Universities and will employ an initial number of 700 employees. China, of course, is the world's largest market for cell phones, which are powered by Sandisk's No. 1 product—flash memory chips. CEO Harari said that the new plant will install the company close to its customer base and hopes that it will serve as SanDisk's strategic base of Asian operations in years to come.

GETTING HIRED

Get a flash-y career

SanDisk's careers page (www.sandisk.com/Corporate/Careers) provides information on job openings at the company, both for graduates and experienced hires, as well as benefits on offer for employees. At press time, most of the openings seemed to be in engineering or finance. To apply, job seekers must first create a profile. SanDisk does not provide much information on what its college graduate career program entails, but

Visit Vault at **www.vault.com** for insider company profiles, expert advice, career message boards, expert resume reviews, the Vault Job Board and more.

V/\ULT CAREER LIBRARY   **415**

recent grads are invited to submit their resumes at www.sandisk.com/Corporate/Careers/CollegeOpportunities.aspx.

Benefits (which may vary by location) include health, dental, vision and employee assistance, health and dependent care, spending accounts, stock options, stock purchase plan and 401(k), discount tickets to amusement parks, free gym membership, events like book fairs, picnics and holiday parties, and discounts on SanDisk merchandise.

Sanmina-SCI Corporation

2700 North First Street
San Jose, CA 95134
Phone: (408) 964-3555
Fax: (408) 964-3636
www.sanmina-sci.com

LOCATIONS

San Jose, CA (HQ)
Allen, TX • Clinton, NC • Costa
Mesa, CA • Fountain, CO •
Fremont, CA • Huntsville, AL •
Kenosha, WI • Livermore, CA •
Manchester, NH • Newark, CA •
Owego, NY • Phoenix, AZ • Plano,
TX • Rancho Santa Margarita, CA •
Rapid City, SD • Research Triangle
Park, NC • Salt Lake City, UT • San
Jose, CA • Turtle Lake, WI •
Woburn, MA

37 locations throughout Europe and
Asia.

DEPARTMENTS

Administrative Support • Defense &
Aerospace • Engineering • Facilities
• Finance/Accounting • HR • IT •
Legal • Logistics/Transportation/
Exporting/Importing •
Manufacturing/Operations •
Marketing • Materials/Planning/
Procurement • Program
Management/Customer Service •
Quality • Sales/Business
Development/Mergers &
Acquisitions • Technicians

THE STATS

Employer Type: Public Company
Stock Symbol: SANM
Stock Exchange: Nasdaq
Chairman & CEO: Jure Sola
2006 Employees: 54,397
2006 Revenue ($mil.): $10,955

KEY COMPETITORS

Flextronics
Hon Hai
Solectron

EMPLOYMENT CONTACT

www.sanmina-sci.com/Info/HR/
career_op.html

Visit Vault at **www.vault.com** for insider company profiles, expert advice,
career message boards, expert resume reviews, the Vault Job Board and more.

VAULT CAREER LIBRARY   417

THE SCOOP

Sanmina is never board

Sanmina-SCI is a leading contract manufacturer of electronics, whose services and goods can carry an electronics manufacturer from the drawing board to finished product. In addition to making circuit boards, cables and all manner of electronic viscera, the company also offers precision machining and system assembly and testing.

Its factories are located essentially wherever electrons can be found, in 20 countries on five continents. The company's customers include such tech industry heavyweights as Lenovo, Alcatel, Cisco Systems, Hewlett-Packard, IBM, Nokia and Ericsson. Sanmina's circuit boards find their way into all manner of electronic devices, from the tiny to the not at all tiny, from cell phones and computers to cars and even airplanes.

Child's play

Two Bosnian immigrants, Jure Sola and Milan Mandaric, started Sanmina (named for one of the founder's children) in 1980 to manufacture circuit boards. (Mr. Sola, who was 25 years old at the company's inception, is still the company's CEO.) However, when margins on their products showed signs of falling, they moved into the more profitable realm of backplane assemblies (that is, constructs which link computer chips together). The company went public in 1993, and used the funds raised by its offering to purchase a number of manufacturing plants from other electronics companies.

In 1998, Sanmina graduated from buying factories to buying competitors. It snapped up rival Altron in 1999, and barely had time to swallow before gobbling up another circuit board manufacturer, Hadco, in 2000. The burst of the tech bubble didn't slow Sanmina's brisk purchasing pace either, as another competitor, the eponymous SCI Systems, became the object of acquisition integration for $4.5 billion in 2001. Thereafter, the newly dubbed Sanmina-SCI rolled in the deals with IBM and boosted its headcount with the purchase of a number of European plants in 2002. Things continued apace through 2004, until the circuit board industry hit a slowdown. In 2005, the tech equipment firm's net sales decreased by nearly 4 percent, while net losses exceeded $1 billion. The company attributed its disappointing results to a drop-off in sales to personal and business computer users, as well as multimedia sectors of the electronics market.

Down Mexico way

Little marketplace hiccups are no match for the likes of Sanmina, however. In 2006, the company opened a 330,000 square foot facility in Guadalajara, Mexico, adding to its other facilities there; attendees at the opening included such luminaries as Mexican President Vicente Fox. The opening expanded the company's enclosure manufacturing capacity to over half a million square feet, making it the biggest exporter in the state of Jalisco and the largest enclosure manufacturing entity in North America. The company has had a manufacturing presence in Guadalajara for more than 18 years, with services including new product introduction (NPI), complete system-level build-to-order (BTO) and configure-to-order manufacturing support, logistics and distribution; and high-volume manufacturing.

Electrifying developments

Inhabitants of dry climates and owners of wool carpets are familiar with the phenomenon—a mundane walk across a carpeted room is suddenly punctuated by a shock. Though sometimes hair-raising, these instances of triboelectric effect—discharges of static electricity—can do a good deal of damage to the dainty circuitry required by today's ever-smaller and ever more feature-packed electronics. A big discharge can destroy delicate junctions on circuit boards, while numerous small electrostatic events can gradually degrade the functionality of circuit boards over time. An industry group, helpfully named the Electrostatic Discharge Association, claims that the electronics industry loses billions every year as a result of electrostatic damage.

A great deal of thought in the circuit board industry has thus gone into preventing the damage caused by electrostatic discharge, or ESD. Sanmina, along with a company called Shocking Technologies, are working on a way to incorporate protections against ESD into the circuit board itself, instead of relying on devices on the board surface or separate grounds, which impose limitations on chip design and performance. ESD protection in the chip substrate allows all the circuits on the board to be protected, yet removes obstacles to chip configuration and performance imposed by traditional methods. Sanmina hopes that this new technology will give it a leg up over its competitors in the increasingly commoditized circuit board industry.

Commodity blues

Sanmina needs to dig itself out from the black hole of a commodity industry in order to survive. In 2006, its sales slid an alarming 6.6 percent, to $11 billion, in the face of decreased demand for circuit boards for computers. The company posted a loss of $141,000 for the year, its fifth consecutive year with negative profits. However, Sanmina is taking steps to diversify its products beyond computer boards—in 2006, two of the company's factories were certified to provide components for aerospace applications.

Going up

The aerospace factories got their first orders in 2007 when Sikorsky, a manufacturer of helicopters for the military, opted to include one of Sanmina's communications systems in a helicopter for the Marines. The communications system will allow soldiers to send voice and high-speed data communications to and from the helicopter.

GETTING HIRED

Sign me up for Sanmina

Sanmina-SCI's careers site (www.sanmina-sci.com/Info/HR/career_op.html) provides information about all the benefits of life at the company, which include the standard medical, vision, dental and drug prescription coverage. Employees also have access to a discount stock purchase plan, 401(k) with company matching, tuition reimbursement, a work/life balance program and a credit union.

There is a database of available jobs, searchable by location and category, which can include administrative, defense and aerospace, engineering, facilities, finance, HR, IT, legal, logistics, manufacturing, marketing, materials, mergers and acquisitions, program management/customer service, quality and sales. To apply, job seekers can fill out an e-mail form online.

SAP Aktiengesellschaft

Dietmar-Hopp-Allee 16
69190 Walldorf
Germany
Phone: (49) 6227-74-7474
Fax: (49) 6227-75-7575
www.sap.com

SAP United States
3999 West Chester Pike
Newtown Square, PA 19073
Phone: (610) 661-1000
Fax: (610) 661-1896

LOCATIONS

Walldorf, Germany (Corporate HQ)
Newtown Square, PA (US HQ)
Atlanta, GA • Austin, TX •
Bellevue, WA • Boston, MA •
Bristol, PA • Brookfield, WI • Bryan,
TX • Burlington, MA • Charlotte,
NC • Chicago, IL • Cincinnati, OH •
Cleveland, OH • Dallas, TX •
Detroit, MI • Durham, NC •
Englewood, CO • Exton, PA •
Foster City, CA • Fremont, CA •
Houston, TX • Irvine, CA •
Kennesaw, GA • Los Angeles, CA •
Miami, FL • Minneapolis, MN •
Morristown, NJ • New York, NY •
Overland Park, KS • Palo Alto, CA •
Parsippany, NJ • Phoenix, AZ •
Pittsburgh, PA • Pleasanton, CA •
Scottsdale, AZ • Seattle, WA • St.
Louis, MO • Washington, DC

Locations in over 50 countries
worldwide.

THE STATS

Employer Type: Public Company
Stock Symbol: SAP
Stock Exchange: NYSE
Chairman: Dr. Hasso Plattner
President & CEO: Henning Kagermann
2006 Employees: 39,355
2006 Revenue ($mil.): $12,413

DEPARTMENTS

Administration
Consulting
Development
Facilities
Finance
HR
IT
Legal
Management
Marketing
Pre-Sales
Project Management
Purchasing
Sales
Support
Training

KEY COMPETITORS

IBM
Microsoft
Oracle

EMPLOYMENT CONTACT

www.sap.com/usa/careers

THE SCOOP

Systems for business

Germany's SAP is a provider of software for businesses. (The firm's name used to be the delightfully agglutinative SAP Aktiengesellschaft Systeme, Anwendungen, Produkte in der Datenverarbeitung, which translates loosely in English to "Corporations Systems, Uses and Products in Data Processing.") SAP's software, scaled for large, midsize and small businesses, has a facet for nearly every line of business under the sun. It manages customer and supplier relations, product lifecycle, human capital, supply chain and manufacturing—it even makes sure that accounting stays on the right side of Basel II and Sarbanes-Oxley.

SAP's software is tailored for nearly 30 industries, including banking, insurance, defense, health care, research, postal services, media, retail, utilities, railways, mining and consumer products. The company rounds out its suite of offerings with training, support and consulting services. As of 2006, SAP is the world's No. 3 independent software vendor and has 1,500 partner companies and more than 38,000 customers in over 120 nations—everyone from the McLaren-Mercedes Formula One team to Microsoft runs SAP.

"You won't get fired for buying SAP"

SAP was founded in 1972 by five employees from another three-letter acronym in computing: IBM. The Big Blue vets set up shop in Mannheim, Germany, and rolled out their first product—an accounting program. Six years later, SAP had racked up 40 businesses as customers, including BMW and Siemens.

In 1979, SAP introduced its next product, R/2, a customizable accounting program that could cope with the demands of companies operating with multiple offices, languages and currencies, and could keep all of a company's data in one place. By the early 1980s, SAP software was being used by such diverse clients as Bayer and BASF, Dow Chemical, General Mills, Shell Oil and Heinz. In 1988, SAP floated shares on the Frankfurt and Stuttgart Stock Exchange and opened its first office in the U.S.

R2-D2, meet your long-lost cousin

In 1992, in response to the move toward more decentralized computer networks, SAP launched R/3, the next iteration of its accounting software. Despite the high costs of customizing the product for a particular business, many believed that "you don't get fired for buying SAP," such was the company's reputation for quality, helpful products. As such, companies in America rapidly cottoned to R/3, and more than half of the company's revenue comes from outside Germany during the year of its release. SAP joined with Microsoft in 1994 to integrate R/3 with Microsoft's database and operating system. Just two years later, R/3 was being run on systems at 9,000 companies. In 1998, SAP issued some of its stock on the New York Stock Exchange.

Do you think I'm some kind of SAP?

As the workstation-and-server architecture of business network design was giving way to a more Internet-like design around 1999, SAP rolled with the punches and came up with MySAP, a way for employees to log into whatever programs they needed for their jobs over the Web. By the year 2000, though, waning sales, stiff competition and troubled U.S. operations had prompted SAP into some major restructuring.

The development division was drastically overhauled, as development was linked to marketing strategies and attacking e-business markets. SAP also broke with its in-house development tradition and its tendency to develop without the aid of alliances and acquisitions, partnering with companies including Sun, Commerce One and Nortel. In an aggressive move, in 2005 SAP bought a start-up called TomorrowNow, which provides cut-rate tech support for the users of Oracle products, in the hopes that it can convince those users to migrate to SAP.

Twosome

In 2006, SAP and its old buddy Microsoft launched the harmoniously named Duet, a program which provides a way for employees to gain access to SAP data via more familiar Microsoft Office programs. Duet can help keep track of an employee's billable hours and vacation time through Microsoft Outlook's calendar, while additional add-ons include programs to manage recruitment, sales, purchasing and business travel. The idea behind Duet is to get employees up and running with the SAP system, all without letting them know they've left the comfortable confines of Microsoft Office.

Going, going, Gurgaon

SAP also didn't waste time thinking about its global identity, and in 2006 opened an office in Gurgaon, near New Delhi, India. The new facility has 250 employees and provides one fifth of the company's R&D output. The company is planning on spending $1 billion in India by 2011. Also that year, it unified its North and South American departments into a single division.

Are the numbers up enough?

SAP posted respectable numbers for fiscal 2006. Revenue from products was up 11 percent over 2005's totals, while software brought home 10 percent more than the year before. Overall, revenue for the whole company increased 10 percent to €9.4 billion, while profits were up 25 percent to €1.8 billion. Wall Street analysts noted that the numbers were less than expected, given the strong rate of growth in the software market and the robust growth experienced by smaller companies in the industry. Indeed, these revenue figures were less than SAP directors themselves were hoping for, as 2006 revenue grew by 11 percent instead of the 15 to 17 percent predicted by the board.

Being big, thinking small

These numbers convey larger forces at work in the industry. SAP, like Oracle and IBM, have an old-school, increasingly outdated business model, wherein a customer (usually a fairly sizeable, established business), buys a customized package from the software vendor for an upfront cost; this package is then added onto the company's in-house servers and computers. Companies like salesforce.com and Google (in the form of Google Apps) are rendering this process obsolete by giving businesses the option of buying software licenses as needed for each user on a per year or per month basis, keeping data off site and allowing employees access to their data and projects from anywhere on the Internet. Whereas companies once had to upgrade software and maintain mainframes and server farms—not to mention keep enough IT guys on staff to keep the machines humming—now all they need are workstations for employees and a fast Internet connection.

To ensure that it adapts to this trend, SAP launched a product along the same pay-to-play lines in 2007. The company also formulated a new strategy in late 2006 to aggressively pursue customers among small and medium-sized businesses. SAP combined these two aims in September 2007, when the company unveiled Business ByDesign, an online subscription service targeting small business customers (at the small business-friendly price point of $149 per month) that gives them access to SAP's software online. The product is available to selected early customers in the United States and Germany and will be available to the company's clients in China,

France and the United Kingdom by 2008. Though favorably reviewed by tech insiders, some industry watchers worry that the lower-priced options may cannibalize some of SAP's business for its more expensive product lines.

Sayonara Shai

SAP suffered a boardroom shakeup in August 2007, as current CEO Kagermann opted not to retire, and his contract was extended until 2009. This led to the departure of his heir apparent, Israeli chief software architect Shai Agassi, whose goal had been to move the company away from its old-school business model. Indeed, Agassi spearheaded SAP's new, web-based strategy, and the company had specifically appointed him to help transition from an essentially German corporate culture to a more global, modern one. Kagermann has since stated in an interview with *The New York Times* that the chances of the next CEO coming from outside of Germany are "pretty high."

Business object ... ions

Following the launch of Business ByDesign, SAP announced its intentions to acquire Paris-based Business Objects, one of the last remaining independent business intelligence software companies, in a deal valued at some $6.6 billion. The announcement was seen as a direct response to Oracle's acquisition of Hyperion, another small competitor in the business intelligence market. In previous years, SAP avoided making big acquisitions, but the temptation of Business Objects (its client base overlaps with SAP's by 40 percent) was too great to pass up. The French firm will remain an independent subsidiary of SAP after the acquisition.

GETTING HIRED

Don't be a sap; join SAP!

SAP's careers site, at www.sap.com/careers, provides hiring information for the 50 or so countries and regions in which the company operates. In 2007, SAP plans to roll out a new recruiting page in the U.S., where wannabe employees can create pages for themselves in the company's talent registry and keep it apprised of any new qualifications, degrees or long-term plans. However, as it stands at press time, SAP's job search page is fairly standard. There are separate search pages for those interested in working at SAP and at its development office based in Silicon Valley, SAP Labs. The HR department can be contacted at careers@sap.com with any questions or comments.

Benefits offered by SAP include medical, dental, vision, tuition reimbursement, flexible spending accounts, a stock purchase plan and 401(k) with company matching. Career development plans in place at the company include succession planning, a career performance measurement tool and company-sponsored training from SAP University.

SAS Institute Inc.

100 SAS Campus Drive
Cary, NC 27513-2414
Phone: (919) 677-8000
Fax: (919) 677-4444
www.sas.com

LOCATIONS

Cary, NC (HQ)
Arlington, VA • Atlanta, GA •
Austin, TX • Beaverton, OR •
Boston, MA • Charlotte, NC •
Chicago, IL • Dallas, TX • Denver,
CO • Detroit, MI • Hartford, CT •
Irvine, CA • Kansas City, KS •
Miami, FL • Middleton, MA •
Minneapolis, MN • Bedminster, NJ •
New York, NY • Phoenix, AZ •
Pittsburgh, PA • Philadelphia, PA •
Rockville, MD • San Diego, CA •
San Francisco, CA • Seattle, WA

More than 400 offices in 52
countries worldwide.

DEPARTMENTS

Administrative • Consulting •
Corporate Services • Design •
Executive • Finance • HR •
Information Systems • Legal &
Contracts • Marketing •
Procurement • Sales • Software
Research & Development •
Technical Documentation •
Technical Support • Training

THE STATS

Employer Type: Private Company
CEO: Jim Goodnight
2006 Employees: 10,101
2006 Revenue ($mil.): $1,900

KEY COMPETITORS

Business Objects
Cognos
Hyperion Solutions

EMPLOYMENT CONTACT

www.sas.com/jobs

Visit Vault at **www.vault.com** for insider company profiles, expert advice,
career message boards, expert resume reviews, the Vault Job Board and more.

V/\ULT CAREER LIBRARY   **427**

THE SCOOP

A very SAS-y company

SAS, the world's largest privately held software company, is the leader in business intelligence and analytical software and services. SAS software is used by 96 of the top 100 companies on the 2007 Fortune Global 500 list, has analyzed data provided by the U.S. Department of Defense and the U.S. Census, has managed customer loyalty programs at Marriott Hotels and Air France, monitors store performance for Subway and Williams-Sonoma, and aids Merck and Pfizer in new drug development. More than 43,000 customer sites in business and academia use SAS software.

SAS' multitudinous offerings include data warehousing and software for data mining, product line management, customer relations management, software that detects money laundering and insurance fraud and programs that ensure companies are keeping on the right side of banking and accounting reform regulations like Basel II and Sarbanes-Oxley. In 2006, the company enjoyed impressive revenue totals of $1.9 billion.

31 years in the data mines

CEO Jim Goodnight founded SAS in 1976, along with John Sall; both were statisticians from North Carolina State University. By 1977, the company had all of seven employees, who enjoyed such idiosyncratic perks as a family Halloween party, fruit on Mondays, breakfast on Fridays and M&Ms on Wednesdays. In 1978, revenue topped $1 million, as SAS products reached 600 customers. In 1980, after just four years in existence, the company opened its first office in Europe, and five years later added offices in Japan and Hong Kong.

In 1986, SAS bought Lattice, another software company, while the closing years of the 1980s saw the company form partnerships with a number of tech heavyweights, including Intel, Sun Microsystems, Microsoft, Apple and Hewlett-Packard. In 1989, SAS opened its consultancy. Five years later, the company's global employee headcount had reached 3,000, and revenue was up to nearly $500 million. In 1995, SAS added data warehousing to its list of services, making its data mining software all the more attractive. Revenue surpassed $1 billion in 1999. In 2000, rumors abounded that the closely held company might go public; however, in 2006, Goodnight reiterated his commitment to keep the company private. The company made another acquisition to boost its position in retail in 2003. Revenue increased by 15 percent in 2004, and the next year, *Fortune* included the company included on its list Best Companies to Work For.

Data, data everywhere

SAS has a bright future, as companies show no signs of reducing the amount of data that needs to be stored and studied—some analysts say that the amount of data companies are faced with doubles each year. Luckily, computers are becoming increasingly able to cope with growing piles of ones and zeroes; according to SAS's CEO Goodnight, computer power is the latest thing on IT managers' minds. He points out that the new generation of 64-bit processors can bite off a much bigger chunk of data than their 32-bit brethren, opening the door for faster crunching of vast quantities of data.

In addition to faster speeds, the increased power of computers has opened up the field to a whole new breed of company—such as Google. The search giant collaborated with SAS on its launch of the Google Search Appliance, a device that allows employees to search for information on company intranets, while protecting confidential data.

Carpe Veridiem

SAS boosted its oomph in marketing metrics with the March 2006 acquisition of Maynard, Massachusetts-based Veridiem, a provider of software that measures the efficacy of ads and marketing campaigns and can even provide models on the effects of different marketing strategies—e-mail spam versus the dead tree variety, for instance.

As companies keep ever greater amounts of data, SAS has more opportunities to serve their information management needs. In 2006, the company took in $1.9 billion in revenue, a 12 percent rise over the previous year. Data warehousing, doubtless boosted by Sarbanes-Oxley and Basel II regulations, was up by 50 percent, while the company's other product lines also put in a respectable showing.

GETTING HIRED

Have an analyzed career

SAS posts its current job openings on its careers site, at www.sas.com/jobs. To apply, interested parties must first create a profile. Jobs are listed by region and then by country. While SAS' emphasis on research and development requires a continuous supply of technically proficient workers, the company also has opportunities in sales, marketing and finance. When applying, be sure to include the job posting number indicated on the web page. SAS also posts a list of recruiting events, which are held at North Carolina state schools and Georgia Tech.

Beneficial treatment

Competition is stiff for career opportunities at SAS, mainly due to the gobs of perks and on-site facilities available to its employees. Keeping its staff happy is one of the company's secrets of success—the pampered employees save the company about $75 million per year in charges associated with finding and training new employees. The employee turnover rate (4 percent) is well below the industry's average (20 percent), and people want to stay for a reason.

At the company's world HQ in Cary, North Carolina, the swimming pool, fitness center, track, child care facilities, medical facilities and various other amenities encourage long-term careers and commitment while maintaining a high-quality workload and work ethic. In 2007, for the 10th consecutive year, SAS was granted a spot on *Fortune*'s Best Companies to Work For list, and its generous health benefits received special mention.

Mr. Holland's HR

Music lovers seeking employment with SAS might be impressed by this extra perk: the company uses daily performances of live music to keep employees at its Cary location relaxed. Daily during lunch, a food-service employee or a member of the SAS choir (another company-sponsored extracurricular) takes a turn as a pianist. SAS also runs a summer camp, Camp Awesome Adventure, at the HQ site for its employees' children.

ScanSource, Inc.

6 Logue Court
Greenville, SC 29615
Phone: (864) 288-2432
Fax: (864) 288-1165
www.scansource.com

LOCATIONS

Greenville, SC (HQ)
Bellingham, WA • Buffalo, NY •
Lenexa, KS • Memphis, TN •
Mendota Heights, MN • Miami, FL •
Norcross, GA • Tempe, AZ

Bad Homburg,Germany • Brussels •
Crawley, UK • Eindhoven,
Netherlands • Hull, UK • Liege,
Belgium • Mexico City • Olivet,
France • Richmond, British Columbia
• Toronto

DEPARTMENTS

Accounting & Finance
Administration
Corporate Operations
Credit
Customer Service
HR
Legal
Mailroom
Marketing
Merchandising
Partner Services
Payroll
Sales
Technical Services

THE STATS

Employer Type: Public Company
Stock Symbol: SCSC
Stock Exchange: Nasdaq
Chairman: James G. Foody
CEO: Michael L. Baur
2006 Employees: 916
2006 Revenue ($mil.): $1,665

KEY COMPETITORS

Ingram Micro
SYNNEX
Tech Data

EMPLOYMENT CONTACT

www.scansource.com/employment.asp

Visit Vault at **www.vault.com** for insider company profiles, expert advice,
career message boards, expert resume reviews, the Vault Job Board and more.

VAULT CAREER LIBRARY   431

THE SCOOP

Your source for all things scan-related

ScanSource is a distributor of electronic equipment, including bar code readers and other automatic identification and data capture (AIDC) devices, which are mostly used to control inventory in warehouses and stores. It also distributes electronic security products, voice, video and data equipment, and videoconferencing products. ScanSource also provides services like training, education, and pre- and post-sale support to its customers. ScanSource ranked No. 956 on the 2007 Fortune 1000 and, as of that same year, offered its clients over 40,000 products.

Some code walks into a bar ...

ScanSource was founded in 1992 to provide bar code readers and other AIDC items to resellers. Despite the company's oddly fatalistic (and sort of non-sequitur) motto, "What's supposed to happen, happens," it has confidently ridden the wave of increasing demand for its products. A year after its founding, ScanSource struck a deal to distribute products from AIDC leader Symbol Technologies (now a part of Motorola). Initially, ScanSource did not sell its products for use in point-of-sale (POS) systems, which are used to scan bar codes at cash registers, but it expanded into that arena in 1993 with its acquisition of Alpha Data Systems.

The following year proved quite eventful—ScanSource went public, in addition to hammering out deals for the distribution of IBM and Zebra products. In 1997, ScanSource moved into Canada and added telecommunications products to its lineup under the Catalyst Telecom division. Catalyst got a boost in 1998 with the acquisition of another telecom products distributor, and ScanSource started to offer logistics services under the ChannelMax name. It opened an office in Mexico in 2001, and got all the waffles it could eat in 2002, when it opened a European HQ in Liege, Belgium. In 2004, keeping up with the pace of technology, the company started carrying RFID products (a fancy kind of neo-bar code, which picks up radio frequency, hence the name: radio frequency identification). It also started offering security equipment.

T for two

In 2006, ScanSource acquired T2, a provider of videoconferencing equipment, for about $50 million. In addition to its videophones and other devices, T2 offers training and tech support. T2 will boost ScanSource and Catalyst's prowess in the increasingly popular realm of videoconferencing.

Listage

In January 2007, ScanSource was named to the 131-spot on *Forbes'* Platinum 400 list of the best big companies for the third straight year, ranking behind five other technology concerns. The companies on the list are evaluated on several criteria, including the amount of debt the company carries and growth in sales and profits. In May *Fortune* named the company No. 956 on the Fortune 1000.

How observant!

The Charlotte Observer ran an article in October 2006, alleging that ScanSource's accounting was, well, a little dodgy. Jeffery Bryson, executive vice president of investor relations, resigned the next month, and the board immediately commissioned an internal review of its stock options practices, receiving its findings in January. The report detailed certain irregularities dating from 1994 to 1996, right after the company had first gone public. The company then provided its findings to the SEC, which had not filed any sort of legal action against the company as of late 2007. The firm expects to take charges of about $5 million for restating past finances.

Movin' on up

In 2007, ScanSource announced that it would be moving its distribution arm to larger digs, conveniently situated close to the airport in Memphis, Mississippi, once the facility was up and running the following year. The new facility will have 600,000 square feet of space for storage and logistics, as well as even more room to expand the facility.

GETTING HIRED

Let ScanSource be your career's security system

ScanSource's careers page, at www.scansource.com/employment.asp, provides everything you'll need to know to become part of the company's workforce. Job openings are listed at www.prohire.com/candidates/default.cfm?szWID=11115&sz

Visit Vault at **www.vault.com** for insider company profiles, expert advice, career message boards, expert resume reviews, the Vault Job Board and more.

V/\ULT CAREER LIBRARY   **433**

CID=50019, and candidates will have to create a profile to apply. Some of the benefits of working at corporate HQ include an on-site café and gym with personal trainer (for working off all of those snacks), health, vision and dental insurance, 401(k) and stock purchase plan, a generous vacation package, tuition assistance, flexible spending accounts for child care and internships for employees' children.

The site also provides information about HQ hometown Greenville, South Carolina, which is apparently located on "a prime spot on the planet." This prime slice of planetary real estate boasts a temperate climate, proximity to the Blue Ridge Mountains, Atlanta and Charlotte, and some fine barbecue.

Science Applications International Corporation

10260 Campus Point Drive
San Diego, CA 92121
Phone: (858) 826-6000
Fax: (858) 826-6800
www.saic.com

LOCATIONS

San Diego, CA (HQ)
Albany, NY • Albuquerque, NM •
Alexandria, VA • Anchorage, AK •
Ann Arbor, MI • Atlanta, GA •
Austin, TX • Baltimore, MD •
Cheyenne, WY • Chicago, IL •
Cleveland, OH • Dallas, TX •
Denver, CO • Destin, FL • Fort
Worth, TX • Honolulu, HI •
Houston, TX • Jacksonville, FL •
Knoxville, TN • Las Vegas, NV •
Little Rock, AR • Los Alamos, NM •
Los Angeles, CA • Miami, FL • New
Orleans, LA • New York, NY •
Oakland, CA • Pittsburgh, PA •
Portland, OR • Salt Lake City, UT •
San Antonio, TX • San Francisco,
CA • San Jose, CA • Savannah, GA
• Tucson, AZ

400 other locations in the U.S.

THE STATS

Employer Type: Public Company
Stock Symbol: SAI
Stock Exchange: NYSE
Chairman & CEO: K.C. Dahlberg
2007 Employees: 44,100
2007 Revenue ($mil.): $8,294

DEPARTMENTS

Biopharm • Manufacturing/
Development • Business
Development/Marketing • Clinical/
Regulatory • Commercial/Industry
Relations • Communications •
Consulting Employee • Contracts/
Pricing • Defense/Intelligence/
Geopolitical • Engineering •
Environmental • Executive
Management • Facilities/Physical
Security • Finance/Accounting/
Business Management • General Office
• HR • Health Services • IT/Telecom
munications • Internal Audit • Legal •
Lending • Logistics • Management •
Manufacturing • Medical Research •
Operations • Planning • Procurement •
Product Engineering • Project/Program
Management • Proposal/Publications •
Research & Development • Skilled
Trades/Technical • Training

KEY COMPETITORS

Booz Allen Hamilton • Computer
Sciences • Lockheed Martin

EMPLOYMENT CONTACT

www.saic.com/career

Visit Vault at **www.vault.com** for insider company profiles, expert advice,
career message boards, expert resume reviews, the Vault Job Board and more.

VAULT CAREER LIBRARY   **435**

THE SCOOP

Science for a safer world

Science Applications International Corporation (known more briefly as SAIC) has built its business on fulfilling various scientific and engineering contracts with the Department of Defense and other government agencies, including the CIA, the National Security Administration and the U.S. Army. It also sometimes works with major U.S. and foreign corporations, but government contracts currently account for 90 percent of its business, more than half of them are directly related to matters of national security.

The firm's areas of expertise range widely—from communications and computer networking to geologic exploration, from logistics to border security systems. Its current services include software and computer system development, plus integration and tech support for both. SAIC has played a role in high-profile projects like the cleanup of Three Mile Island to the construction of the Hubble Space Telescope and America's Cup racing boats. On the 2006 Fortune 500, the company was ranked a respectable No. 285.

Employee-owned and -operated

At the end of the 1960s, Bob Beyster left a research position at General Atomics to start his own company, intending to perform contract research in nuclear physics. With the help of his longtime friend (and entrepreneur in his own right) Myron Eichen, Beyster began interviewing prospective employees in February 1969, using the concept of employee ownership as a focal point. Soon, SAIC had a home base in a small office in La Jolla, California, and by 1970 it had 20 employees, all of whom owned parts of the company. SAIC's employees would own the company (which could be traded on an internal exchange) until the corporation offered some stock to the wider world in 2006.

The times—and SAIC—have been a-changin'

As national security threats have evolved, from the Cold War specter of nuclear attack to that of any number of threats during the "War on Terror," SAIC's areas of interest have varied accordingly. SAIC had gained contracts with Brookhaven National Labs for work on breeder reactors in 1975; began research on geothermal, wind and solar power in 1977; and improved defense systems for NATO in 1978. By

1979, its number of employees had grown to 3,600 and its revenue to over $100 million. In 1981, SAIC participated in the discovery of chemical contamination in Love Canal, New York. Between 1984 and 2003, the firm upgraded to space vehicles and oceanographic equipment and helped design Stars and Stripes, 1987's victorious America's Cup boat.

In 1989, SAIC created devices to prevent drug smuggling and to scan luggage for evidence of explosives. By this point, the company had nearly 10,000 employees and revenue exceeding $800 million. In 1992, SAIC expanded into providing services for the commercial arena; its diversification paid off when nongovernment contract revenue grew by nearly a third in 1997. The following year, the company acquired Telcordia Technologies, a company that specialized in IT consulting and networking, boosting SAIC's abilities in these areas. But SAIC hadn't abandoned working for the feds. In 2001, it became one of the principal contractors on the Yucca Mountain radioactive waste depository, and three years later it signed a deal to provide IT to NASA's Marshall Space Flight Center. It also provided security for nine seaports during the Olympic Games.

Applied strategy

Some of SAIC's success is attributable to its aggressive policy of acquiring other companies, to increase growth and expand capabilities. Indeed, the company acquired four firms in 2006 alone. The first acquisition, in August 2006, was the purchase of bd Systems, a provider of IT for aerospace engineering. A few days later, SAIC acquired Cornerstone Industry, a company that coordinates military activity across the various branches of the armed forces. In September, SAIC then bought Varec, a manufacturer of pumps and measurement equipment for the aviation and oil industries. A few months after that, in December, SAIC snapped up Applied Marine Technology, a provider of a grab bag of military needs, from disaster-response team training to explosives disposal and more. The company has acquired more companies in 2007, most recently the consulting firm Benham Investment Holdings in August.

Going public

Later in 2006 SAIC issued shares to holders outside of the company for the first time in its nearly 40-year history. Until the offering, SAIC shares could be purchased only by employees and had to be sold when employees left the company—a practice instituted by founding CEO Beyster, it was designed to motivate employees and result in better teamwork and overall success. SAIC moved towards public ownership for financial reasons: to acquire other companies in stock transactions and to free up capital for more expansion. A little more than 80 percent of the company was owned by its employees when it issued 75 million shares in October 2006. However, to keep its workers in charge, the company structured its stock offering so that shares held by its employees had 10 times the votes of any share issued outside of it. Smart!

Containing threats

SAIC is currently involved in a number of projects to beef up security, mostly in ports nationwide. In 2007, the company won a contract with the Department of Homeland Security to install its VACIS P7500 container inspection systems at seaports. These systems scan dense cargo with X-rays in order to detect suspicious contents, and it can scan as many as 150 40-foot containers per hour. Later that year, SAIC was contracted to store chemicals for the U.S. Army's Chemical Materials Agency. Scientists at the company are also working on a system which can identify people by the unique properties of their heartbeat in an ECG—scary.

But as high-tech as the company is becoming, it is still susceptible to human error. Its most egregious recent mistake occurred in July 2007, when it admitted a possible data breach. While processing health care claims for roughly 580,000 military personnel and their families, it transmitted their personal data over the Internet in unencrypted form, leaving each person open to potential identity theft. Although the company claimed its internal investigations had revealed no evidence of any such fraud, it could not rule out its occurrence.

A point in space

Thanks to SAIC, you need never lose your place in the world. In early 2007, the U.S. Air Force awarded the company a contract to manage the development of its next generation of global positioning, or GPS, satellites, due to launch sometime after 2010. The contract, valued at around $250 million, will use SAIC's consulting skills to coordinate efforts between the builder of the satellites and the builder of the ground receivers.

GETTING HIRED

Apply yourself to a new career

SAIC's careers page, at www.saic.com/career, has a searchable list of job openings, as well as a wealth of information about the company. To apply for a job, potential employees must first create a profile. SAIC offers many opportunities for training, including discounted admission to colleges near company locations and on-site graduate degrees in business and systems engineering. Benefits at SAIC include health, vision, dental and life insurance, paid vacation and nine holidays, tuition assistance, retirement plans with company matching, and shuttle buses and flexible hours for commuters.

SAIC also provides information for college students who might be interested in internships or co-ops. Majors in engineering, computer science, finance, accounting, political science, security and intelligence studies, and similar areas, with GPAs of 3.2 or higher, are all welcome to apply. Interested students can either contact SAIC's recruiters by phone at (610) 336-4316 for East Coast opportunities and at (858) 826-7624 for West Coast ones; by e-mail form at www.saic.com/career/college.asp or at a career fair. The career fair schedule is posted at www.saic.com/career/schedule.asp and stops include Texas A&M, Louisiana State University, UC San Diego and Brandeis University.

OUR SURVEY SAYS

Have a career in the applied sciences

Insiders report a straightforward interview process at SAIC. "I had one round of interviews with the company that consisted of ... five separate interviews in one day," says one hire. "I had three interviews in total," said one hire. He reports being asked questions like "Why do you want to work? What do you bring to the table?"

"The office I work in has a diverse mix of people, mostly well-educated academic types who get along well," says one insider. "There is no training or support at all. Most new hires leave within two years," says one member of the finance department. "The work and atmosphere for on-site government contractors matters much more on the government client than on SAIC corporate culture or interaction," notes one respondent.

Visit Vault at **www.vault.com** for insider company profiles, expert advice,
career message boards, expert resume reviews, the Vault Job Board and more.
VAULT CAREER LIBRARY   **439**

"The hours are flexible. As long as you get 80 hours in, most managers don't care when you do it … That said, certain positions will require a lot of hours," warns a member of the finance department. "The workplace isn'tt bad. The hours are good, but there is little to no opportunity for advancement at the current location," says one source at a satellite office. SAIC offers its employees "a lot of opportunity, especially for ex-government personnel with security clearances. If you have a top secret clearance, you are golden," according to an insider.

Seagate Technology LLC

920 Disc Drive
Scotts Valley, CA 95066
Phone: (831) 438-6550
Fax: (831) 429-6356
www.seagate.com

LOCATIONS

Scotts Valley, CA (HQ)
Fremont, CA
Longmont, CO
Milpitas, CA
Minneapolis, MN
Oklahoma City, OK
Pittsburgh, PA
Shrewsbury, MA
Sunnyvale, CA

17 locations throughout Europe and Asia.

DEPARTMENTS

Administrative • Customer Service •
Engineering • Engineering Services •
Finance/Accounting • General
Management • HR • IT • Legal •
Manufacturing • Marketing •
Materials • Operations Support •
PR/Communications • Quality •
Sales • Sales Operations • Six
Sigma

THE STATS

Employer Type: Public Company
Stock Symbol: STX
Stock Exchange: NYSE
Chairman: Stephen J. Luczo
CEO: William D. Watkins
2007 Employees: 54,000
2007 Revenue ($mil.): $11,360

KEY COMPETITORS

Fujitsu
Hitachi Global Storage
Western Digital

EMPLOYMENT CONTACT

www.seagate.com/www/en-
us/about/jobs_at_seagate

Visit Vault at **www.vault.com** for insider company profiles, expert advice,
career message boards, expert resume reviews, the Vault Job Board and more.

VAULT CAREER LIBRARY   441

THE SCOOP

Save me, Seagate

Seagate is a major manufacturer of computer hard disk drives, controlling around a third of the market for the devices in 2007. It sells them to OEM manufacturers of PCs and servers such as Dell, Hewlett-Packard and IBM, along with DVR and consumer product manufacturers; through the distribution channel to system builders, resellers and distributors; and at retail, under the Seagate name, to consumers who wish to pump up their drive capacity independently.

Seagate hard disks come in a variety of sizes and capacities, from 1.8-inch drives for use in smaller consumer applications, to 2.5- and 3.5-inch drives for use in computers, on-demand television systems and data storage servers.

Seagate's got your back

As the largest independent manufacturer of disc drives, and the share leader in most market segments, Seagate has exhibited strong financial and operational performance over the years. In 2007, sales were up nearly 25 percent over 2006, to $11.4 billion, and profit topped $900 million.

In December 2006, Seagate acquired eVault, a provider of online data backup (no relation to this company), a complement to its 2005 acquisition of a data recovery company, to create the Seagate Services group. This strategy makes the company a source of all things data-related. Seagate's strategy of parlaying its hard drive expertise into data backup and recovery services is a smart move. These days, hard drives are mostly treated as commodities, and although the landscape of competitors has shrunk dramatically over the past decade, it remains a ruthless and highly competitive environment.

If Seagate succeeds in breaking away from the limits of this marketplace, especially one in which the price of goods is continually reduced in the face of the next larger, faster product, its future is almost assured. People are showing themselves to be excellent accumulators of growing troves of data in the form of pictures, movies, TV shows and music—and as such are proving to be eager consumers of devices with ever-larger storage capacity.

It's a Seator! No, it's a Maxgate!

In December 2005, Seagate announced a deal to purchase its rival, the Maxtor Corporation, for $1.9 billion. The Maxtor acquisition gave the company a second well-known name in rotating storage. Lest consumers be unnecessarily confused by two brands of hard drive, Seagate cleverly solved the problem through advertising in 2007. Seagate-branded drives were hyped with pictures of people moving through their lives, surrounded by a swirling cloud of data saved on their hard drives—pictures, music, tax forms, that unfinished novel, what have you. Seagate has also trotted out a new motto, "Your On." While it makes writers queasy and editors sic, the awkward grammar in the company's slogan is intended to get consumers excited about hard drives. As a perhaps more concrete way to get the public riled up about its product, the company enlisted a design firm to tart up its line of Free Agent external hard drives in cuter covers (which Seagate, in a fit of linguistic originality bordering on the oxymoronic, dubbed "data movers").

The Maxtor line, on the other hand, is being pushed on the data-backup front. Like eating five servings of vegetables per day, the backing up of data is one of those chores that regularly gets put off. Maxtor's campaign endeavors to remind people to protect their data, by showing them what might be lost. A 12-foot pile of CDs and an eight-foot heap of pictures were prominently displayed at the Consumer Electronics Show in Las Vegas, and the company's print campaign showed people whose drives had lost a few bits missing bits of themselves—what better way to drive home the point that we are our data?

Smaller, faster, greener

In early 2007, Seagate announced the Savvio 15K, the world's fastest hard drive (for now, at least). The platter rotates around the spindle at a brisk 15,000 revolutions per minute (rpm), and the average time it takes for the read/write head to find a particular bit of information is a hair under three milliseconds. This drive, in addition to being presented in a spatially efficient 2.5-inch format, is about a third less power-hungry than other 15,000 rpm drives, making it ideal for the eco-friendly data center.

Visit Vault at **www.vault.com** for insider company profiles, expert advice, career message boards, expert resume reviews, the Vault Job Board and more.

V/\ULT CAREER LIBRARY   **443**

GETTING HIRED

Drive your career at Seagate

Seagate's careers site, at www.seagate.com/www/en-us/about/jobs_at_seagate/, provides a searchable list of jobs and information on Seagate's culture, benefits and hiring events. The benefits include several health insurance options, dental insurance, medical and dependent care spending accounts, 401(k) with company matching, profit-sharing, discount stock-purchase plan, tuition reimbursement, adoption assistance and work/life balance assistance. Recruiting events are posted on the site as well (though there were none listed at press time). However, several internship positions were listed in the searchable jobs database. To apply, job seekers must first create a profile on the site.

Silicon Graphics, Inc.

1140 East Arques Avenue
Sunnyvale, CA 94085
Phone: (650) 960-1980
Fax: (650) 933-0316
www.sgi.com

LOCATIONS

Sunnyvale, CA (HQ)
Chippewa Falls, WI
Eagan, MN
Silver Spring, MD

DEPARTMENTS

Administration
Corporate Marketing
Field Operations
Finance
HR
Information Systems
Legal
Manufacturing Operations
Server & Platform
Storage & Software
Technology Solutions

THE STATS

Employer Type: Public Company
Stock Symbol: SGI
Stock Exchange: Nasdaq
Chairman: Kevin Katari
CEO: Robert H. Ewald
2006 Employees: 1,738
2006 Revenue ($mil.): $518

KEY COMPETITORS

Hewlett-Packard
IBM
Sun Microsystems

EMPLOYMENT CONTACT

www.sgi.com/company_info/careers

Visit Vault at **www.vault.com** for insider company profiles, expert advice,
career message boards, expert resume reviews, the Vault Job Board and more.

VAULT CAREER LIBRARY   445

THE SCOOP

See the power

SGI, formerly known as Silicon Graphics, is a manufacturer of supercomputers and servers for scientific, business, engineering and government applications. SGI's computers crunch numbers for Raytheon, model weather patterns in Hungary, power a computational model of the tobacco mosaic virus and enlighten scientists about the origins of our galaxy. SGI's computers helped Speedo design a new generation of swimsuits and let Egyptologists see inside a mummy without having to unwrap it.

Clark & co.

James Clark, a computer science professor at Stanford University, worked with six students in the 1970s and early 1980s to come up with a new way to build an inexpensive 3-D graphics computer. They developed a powerful new semiconductor chip, which separated the chips and circuits for graphics, thereby relieving the burden on the computer's central microprocessor, producing processor-intensive 3-D models more quickly, and enabling small computers to produce graphics simulations (formerly the domain of mainframes). By 1982, Clark left Stanford to form Silicon Graphics, and the company first marketed a 3-D workstation based off of his design for $75,000 in 1984.

Keeping it simple

SGI's future successes followed a similar model: take complicated computer issues involving graphics and make them simple. For its continued record of innovation, many have credited the entrepreneurial spirit of its founder, James Clark. He hit some road bumps along the way, such as when he dropped out of high school in Plainview, Texas, after the school suspended him for igniting a smoke bomb on a school bus; but he became a capable scientist and eventually an academic, teaching at various schools in addition to Stanford. He also knew his strengths—in the early 1980s he hired the former HP exec Edward McCracken to focus on SGI's business end, allowing him to focus on the technology (and serve as company chairman).

Opening an IRIS

SGI introduced a new graphics workstation in 1987, called IRIS 4D/60. The company stayed faithful to Clark's penchant for simplicity, incorporating the simple and innovative RISC (reduced instruction set-chip) system into the workstation's architecture. The next year, IBM purchased the IRIS graphics card for its own graphics system, helping IRIS on the way to becoming an industry standard.

Criticisms soon emerged, however, that the IRIS was too expensive, and SGI duly introduced the Eclipse later in 1988. Although not as technically advanced as the IRIS, it was much, much cheaper—about one-fifth the cost of higher-end machines. It soon received a boost when Chrysler announced it would buy some Eclipses to design new cars.

JEDI of a different sort

In 1987, LucasFilm's famous Industrial Light and Magic movie/special effects division began using SGI's technology, and the partnership proved even more fruitful for both in the 1990s. In April 1993, the two announced a joint venture called the Joint Environment for Digital Imaging (yes, JEDI). It was a pioneer in digital special effects, and its work eventually resulted in the revolutionary special effects in films such as *Star Wars Episode 1: The Phantom Menace* and *The Lost World: Jurassic Park*.

SGI was a player in movies throughout the 1990s; Silicon Graphics computer created effects for *The Hunt for Red October* (1990), *Terminator 2* (1991) and *Jurassic Park* (1993). In the meantime, James Clark had the itch to move on and he left the company; he popped up elsewhere in the tech industry as co-founder of Netscape in 1994 and was thereafter involved in various capacities with the startups Healtheon, WebMD, myCFO and Neoteris. It seems that entrepreneurship will run in his family, too—his daughter Kathy is married to Chad Hurley, co-founder of YouTube.

Missing Mr. Clark

Silicon Graphics soon missed its founder, as it endured its bleakest period ever in the late 1990s, when PCs threatened their market share by growing ever cheaper and more advanced (which must have been painful for SGI, as cheap and advanced had long been its calling card). The company's lowest point arrived in late 1997: after company officials warned that earnings and sales would fall far short of expectations, SGI stock lost a third of its value in one day. The company then laid off nearly 10 percent of its workforce.

In 1999, Silicon Graphics changed its name to SGI, but name changes notwithstanding, things continued to go badly for the company. In 2000, SGI's

Visit Vault at **www.vault.com** for insider company profiles, expert advice, career message boards, expert resume reviews, the Vault Job Board and more.

VAULT CAREER LIBRARY   **447**

annual sales were $2.3 billion, but that number has since declined, dropping to $842 million in 2004. By the time the company announced its 2004 results, the company hadn't had a profitable year since 1997.

A new leader steps in

In 2006, in order to improve its financial position, SGI got a new CEO, Dennis McKenna, who replaced Robert Bishop in the hot seat. McKenna had experience in dragging a tech firm from the red into the black, which he did when he was CEO of SCP Global Technologies. A month after becoming CEO, McKenna made aggressive moves to cut costs, eliminating 12 percent of company headcount at a cost of $20 million. The company's COO and CFO also left the company at that point.

The inevitable

McKenna's cuts could not stave off the inevitable: just a few months later, in May 2006, in the face of nine years of negative profits, SGI declared bankruptcy. It emerged from Chapter 11 in October. The new, improved SGI has fiddled with its balance sheets, trimming $150 million in annual costs and retained some operating capital.

Its business plan calls for growth of 2.5 to 5 percent per year, with profits around 40 percent, with a revenue target of $671 million in 2011. McKenna's ambitious plan calls for shifting the company focus to servers for data mining and data storage—an area where competition has only increased in recent years, and independent data miners live in fear of major takeovers—as well as turning a profit by the close of fiscal 2007, which seems unlikely.

Bo comes home to roost

In 2007, with McKenna having barely warmed the CEO's chair, SGI announced that it would be getting another new head honcho. The new guy, Robert H. "Bo" Ewald, isn't really new at all: he was in management at Cray Research, and was an executive at SGI before leaving to be CEO of Scale Eight, Ceridian and Linux Networx. But Ewald has his work as CEO cut out for him: his job is to put into practice the ambitious business plan detailed above.

GETTING HIRED

Visualize your career at SGI

SGI's careers page, at www.sgi.com/company_info/careers, gives job seekers the lowdown on everything from company culture and benefits to university recruiting. SGI's benefits include a choice of health insurance plans, dental and vision, 401(k) with matching options, sponsored sports activities and a vacation-day donation program. In order to seek your position at SGI, walk your fingers over to the job search page at www.sgi.com/company_info/careers/search.html. To apply for a position, applicants must create a profile. Students are invited to submit resumes, and are generally hired for hardware and software design, manufacturing, tech support and customer service, and in marketing, HR and finance.

Visit Vault at **www.vault.com** for insider company profiles, expert advice, career message boards, expert resume reviews, the Vault Job Board and more.

V/\ULT CAREER LIBRARY   449

STMicroelectronics N.V.

39, Chemin du Champ des Filles
C.P. 21
CH 1228 Plan-Les-Ouates
Geneva
Switzerland
Phone: (41) 22-929-29-29
Fax: (41) 22-929-29-00
www.st.com

LOCATIONS

Geneva (World HQ)
Carrollton, TX (US HQ)
Aliso Viejo, CA • Austin, TX •
Bensalem, PA • Cary, NC • Edina,
MN • Houston, TX • Huntsville, AL
• Indianapolis, IN • Kansas City, MO
• Kokomo, IN • Lexington, MA
La Jolla, CA • Lake Oswego, OR •
Lancaster, PA • Lawrenceville, GA •
Lexington, MA • Livonia, MI •
Longmont, CO • Midvale, UT • New
York, NY • Parsippany, NJ •
Phoenix, AZ • Portland, OR •
Quakertown, PA • Redmond, WA •
San Jose, CA • Schaumburg, IL •
Voorhees, NJ

DEPARTMENTS

Applications • CAD & Library •
Design & Test Development •
Equipment • Finance, Control &
Audit • HR • IT • Logistics •
Marketing • Planning • Process &
Product Engineering • Production &
Maintenance • Product & Test
Engineering • Purchasing • Quality •
R&D Process & Technology
Development • Sales • Site Services

THE STATS

Employer Type: Public Company
Stock Symbol: STM
Stock Exchange: NYSE
Chairman: Gerald Arbola
President & CEO: Carlo Bozotti
2006 Employees: 51,770
2006 Revenue ($mil.): $9,854

KEY COMPETITORS

Infineon Technologies
NXP
Texas Instruments

EMPLOYMENT CONTACT

jobs.st.com

THE SCOOP

It's a small world, after all

STMicroelectronics is one of the top European chip manufacturers, and is one of the firms nudging the world toward smaller and smarter gadgets—as consumers demand more secure credit and ID cards, smaller cell phones, cars that can operate more reliably, give directions, and entertain occupants, and play movies, ST is stepping up to bat. The company makes semiconductors and microchips for the automotive, communications, consumer, industrial and computer industries, in addition to being a leading manufacturer of smart cards (those nifty credit cards with the microchip inside).

Its major customers include Cisco, Delphi, Ericsson, Nokia, Maxtor and Siemens—in other words, all of the big-name consumer and OEM manufacturers. ST manufactures the motion-sensing chips that help enable Nintendo's popular Wii video game system, as well as some of the RFID chips that control Wal-Mart's respected logistics chain and the chips that deliver the video and sound in several recognized cellular handsets. In 2007, ST developed a "lab on a chip" with the ability to test for the presence of avian influenza.

An unlikely merger

ST traces its roots back to 1987, when the French and Italian governments decided to merge their semipublic semiconductor endeavors. One of the merger partners, Italian SGS Microelettronica, was headed by Pasquale Pistorio, who in the early 1980s had turned the company into a lean, mean profit-making machine. Before Pistorio arrived, SGS Microelettronica had been an unprofitable behemoth, churning out commodity chips on outdated equipment and rapidly losing share to Japanese and American competitors; Pistorio transformed it into a firm with efficient facilities putting out specialized microprocessors.

The other merger partner, Thomson-CSF, was also outdated, debt-ridden and unprofitable, not to mention partly owned by the French government. Undaunted, Pistorio worked his management magic on the newly merged entity, now known as SGS Thomson. Needless to say, the merger put the combined companies' balance sheets way out of whack—SGS-Thomson Microelectronics see a profit until 1989, despite generating over $1 billion in revenue in 1988. Revamped manufacturing facilities, innovations in chip design and deals with major semiconductor consumers

Visit Vault at **www.vault.com** for insider company profiles, expert advice, career message boards, expert resume reviews, the Vault Job Board and more.

V/\ULT CAREER LIBRARY   **451**

(like Nokia, in 1990) gave the company's balance sheets their necessary boost, and by 1993 the company was in fighting form once again—as the firm showed, acquiring rival U.S. firm TAG Semiconductors, and having a double-fisted stock offering in both New York and Paris the year following.

... then we take the world

In the late 1990s, the company took its IPO proceeds and invested them heavily in new factories in Europe, China and Singapore. In 1998, it changed the company name to STMicroelectronics and floated an offering on the Italian stock exchange. ST's highly diversified product mix allowed it to weather the tech bubble's burst in 2000 with aplomb, posting profits when other companies in the sector were faring poorly.

In 2002, ST acquired two semiconductor companies, cementing its position high in the global pecking order of chip manufacturers. In 2005, Pistorio, architect of ST's rise to prominence, rode off into the sunset, replaced by longtime ST executive Carlo Bozotti.

Chipper results

ST had a good year in 2006. It posted $9.8 billion in revenue, a hair short of a 4 percent increase in revenue over the previous year's figures, driven primarily by performance in its wireless and industrial segments. Profits also got a sizeable bump, from $266 million to $782 million. The company foresees a slowdown in the chip market in 2007 and is moving to optimize its asset utilization.

Ciao, Flash!

To that end, the company extended its longtime business strategy of teaming up with other industry big boys. In early 2006, ST announced it was seeking a buyer for its flash memory division. Flash is a form of nonvolatile, solid-state memory used in MP3 players, cell phones and digital cameras, and it was a highly volatile and challenging business. ST fully reconfigured the flash division by January 2007, and announced a joint venture with Intel, who also had an unprofitable flash department they were looking to fix, in May. The two firms pooled their flash memory divisions into a new business called Numonyx, which, upon achievement of regulatory approvals, will instantly become the second-largest provider of flash chips in the industry, and will specialize in chips for cell phones and music players.

The Numonyx deal will give ST the flexibility it sought to deal with the slowing microchip market; second quarter sales revealed a 3.1 percent decrease in sales over the year before. (The company fared better than microchip competitor Texas Instruments, though, whose quarterly sales dropped 7 percent in the same quarter.) Soon, ST joined with other major tech firms and announced yet more big deals. In July, it announced a pact with IBM to collaborate on (big surprise) chip manufacturing. The next month, it revealed that Nokia would be transferring its cell phone chip design team to ST (comprising about 200 employees), and that the two firms were now even better buddies, trumpeting "deeper ties" for future exploits. (For more on Nokia's change in cell phone chip design strategy, please see the QUALCOMM profile.)

GETTING HIRED

Put some ST in your career

Interested in joining ST? Walk your fingers over to jobs.st.com, where the company provides information on the organization of the company, diversity, career paths and training through its unique ST University. Benefits vary by location. The company hires about 6,000 people per year in 140 different jobs in such fields as engineering, logistics, marketing, quality and R&D. Internship opportunities are also available for undergrads in France, Tunisia and Morocco. To apply, candidates must first create a profile online.

Sun Microsystems, Inc.

4150 Network Circle
Santa Clara, CA 95054
Phone: (650) 960-1300
Fax: (408) 276-3804
www.sun.com

LOCATIONS

Santa Clara, CA (HQ)
Austin, TX
Beaverton, OR
Broomfield, CO
Burlington, MA
Menlo Park, CA
San Diego, CA

Locations in 44 countries
worldwide.

DEPARTMENTS

Administration
Corporate Information Systems
Customer Service/Field Service
Software Engineering
Hardware Engineering
Manufacturing
Marketing
Sales
Technical Consulting

THE STATS

Employer Type: Public Company
Stock Symbol: JAVA
Stock Exchange: Nasdaq
Chairman: Scott McNealy
President & CEO: Jonathan Schwartz
2007 Employees: 34,219
2007 Revenue ($mil.): $13,873

KEY COMPETITORS

Dell
Hewlett-Packard
IBM
Microsoft

EMPLOYMENT CONTACT

www.sun.com/corp_emp

THE SCOOP

Let the Sun shine in

Need something computer-related? Sun can provide it, as there are few things in the computing universe that are not dreamt of in its philosophy. The Santa Clara, Calif.-based company manufactures all manner of products, including software, servers, data storage centers, workstations, computer chips (Sun's line of chips is known as SPARC) and an operating system (Solaris), as well as networking and security hardware. Sun also offers consulting and support services.

The company is the leading maker of UNIX-based servers and it also developed Java, a popular programming language that can run software on everything from desktop computers to cell phones and smart cards. The company has embraced open source formatting of UNIX, Solaris and Java, letting computer programmers around the world access them for free, with Sun charging for updates and services. Sun can be found in tech hotspots like the Silicon Valley area and up-and-coming, tech-friendly areas like Kazakhstan and Poland.

The dawn of Sun

Sun was founded in 1982 with all of four employees, including former CEO Scott McNealy. International expansion swiftly followed, with a location in Europe in 1983, Canada in 1985 and Asian and Australian outposts the next year. Sun's big break was its introduction of NFS technology in 1984, which became the standard for network file sharing operations. The company expanded its NFS technology to PCs in 1986, the same year it first offered its stock to the public.

Sun partnered with AT&T in 1987 to develop the UNIX operating system, and it had annual revenue of over $1 billion by the next year. It reached a major milestone in 1993, when it was added to the S&P 500. The company introduced Java, the any-platform programming language, in 1995, the same year that the first full-length computer animated movie, *Toy Story*, was partly created on Sun computers. The company agreed to license Java to all hardware and software manufacturers in 1996, and Java soon found its way onto the 1997 Mars mission.

Sunset?

The 2000s were marked by Sun acquiring a number of companies to broaden expertise in networking and data storage. The early part of the decade was marred,

Visit Vault at **www.vault.com** for insider company profiles, expert advice, career message boards, expert resume reviews, the Vault Job Board and more.

V/\ULT CAREER LIBRARY   **455**

however, by a protracted legal battle with Microsoft over the inclusion of Java in Internet Explorer, and by the bursting of the dot-com bubble, which hit Sun especially hard.

Microsoft settled for $1.6 billion in 2004, but Sun had suffered through two straight unprofitable years by then. The company scrambled to cut costs, firing 9 percent of its workers in 2004. At the time, many analysts considered these cuts the absolute minimum of how Sun needed to restructure, and they would be proven right when the company announced further cuts a few years later—see below. During one stretch from 2004 to 2005, Sun posted financial losses in five consecutive quarters.

Get your Java while it's hot!

Sun increasingly views open source formats, that is, free access to its technology, as the key to future success. It released more than 1,500 patents into the public domain in 2005, among them the source code for Solaris, Sun's offering for the UNIX operating system. Sun will still charge for Solaris tech support and software updates, however.

As the next step in its free software strategy, Sun released the Java source code for free in November 2006. Clearly, the company has realized that its best opportunity for future growth lies in offering services to people already using its systems. The hope is that open source formatting attracts more programmers to UNIX and Java, who will then need to purchase services from Sun.

Tap into the Grid

In another 2006 move away from a software business platform, the company announced Sun Grid, a system whereby companies can upload data to be crunched or stored by Sun's computers. Sun pitches the service as a way for IT departments to keep a lid on costs—there are no pricey server farms to maintain when demand for storage space or processor power is low. Thus far, Sun's rates are $1 to store a gigabyte for a month, and $1 per CPU hour of computing time. The company is confident that Sun Grid is the leading edge of computer technology, and that the idea will soon catch on.

1. Free software 2. ??? 3. Profit!

Sun has little choice but to depend on the adoption of Sun Grid, since it has the profitability that one might expect from a company that gives things away for free.

In 2006, the company posted its fourth consecutive year of losses since 2002—and those losses were over 6 percent higher than the previous year. In fiscal 2007 (the year ending June 30, 2007), the company returned to profitability. Sun reported revenue for the full fiscal year of $13.873 billion, an increase of 6.2 percent over fiscal year 2006. Net income was $473 million as compared with a net loss of $864 million for fiscal 2006.

New things under the sun

In April 2006, Sun founder and CEO Scott McNealy bowed down, replaced by former COO Jonathan Schwartz. McNealy isn't jumping off a sinking ship, though; he will remain as chairman. Since Schwartz's appointment, Sun has enjoyed a financial bounce, reporting three consecutive profitable quarters to close the 2006 fiscal year. Figures for 2007, recently released at the end of June, suggest that Sun could be turning its fortunes around as revenue increased even further, to $13.8 billion and net income of $473 million, or $0.13 per share.

As analysts expected, Sun announced more job cuts later in 2006, eliminating 13 percent of its headcount (around 5,000 jobs), and in another move to cut costs, Sun sold its Burlington, Mass. outpost. The 100-plus acre site was bought by the Nordblom Co. for $212 million; Sun in turn leased back the space and still utilizes the space. Most importantly, shedding this asset prevented further layoffs.

In April 2006, Sun founder and CEO Scott McNealy bowed out, naming former COO (and right-hand man) Jonathan Schwartz as his replacement. McNealy isn't jumping off a sinking ship, though; he will remain as chairman. Since Schwartz's appointment, Sun has enjoyed a nice little bounce, reporting three consecutive profitable quarters to close the 2006 fiscal year. Figures for 2007 suggest that Sun could be turning its fortunes around as revenue that year came in at $13.8 billion.

GETTING HIRED

Your time to shine

Sun's careers site, at www.sun.com/corp_emp, tells you everything you'll need to know in order to find your own place in Sun. Sunny hopefuls can check out job opportunities hither and yon, and get the latest on benefits, culture and student recruiting. Benefits include a choice of health plans, dental, life and disability

Visit Vault at **www.vault.com** for insider company profiles, expert advice, career message boards, expert resume reviews, the Vault Job Board and more.

VAULT CAREER LIBRARY   **457**

insurance, 401(k), fitness centers, adoption assistance, emergency child care and discounts on everything from company stock to recreational outings to Sun products—you can start your own server farm!

With a program unique to Sun called Open Work, employees have the ability to work anytime, anywhere, using any device. Over 55 percent of the workforce currently participates in the program. Sun promotes a healthy work/life balance by allowing its employees the work flexibility they need in order to maintain a family, hobbies and other work-unrelated responsibilities. As long as high-quality work is being produced, it doesn't matter when or where Sun employees do it.

Company culture, as befits a California company, is laid-back. Sun is willing to be flexible about hours and telecommuting, and the dress code is casual. The company's employees also have a tradition of pulling pranks on April Fool's Day. One year, the VP of research found his Ferrari floating in a pond, while the next year, several executives' offices were joined to accommodate a par four golf course. Other notable pranks have involved peacocks, sharks and a 60-foot wooden arrow. Since the company's early days, on Friday afternoons employees are invited to have pretzels and beer with the executives. Sun embodies a culture of transparency, encouraging its employees to voice their opinions, air complaints and make suggestions for a better work environment. Much like a professor hosting office hours, Sun's top execs have adopted an "open door policy," where all employees are invited to interface with Sun's "top dogs." Have a new idea? Share it with the CEO!

If all this sounds a great deal like college to you, then you can apply for as many years of postgraduate hijinks as you'd like at www.sun.com/corp_emp/zone/index.html. Sun is looking for students with degrees in business, marketing, CS or computer engineering and allied disciplines. Sun recruits at a large number of schools, including Brigham Young, Boston University, Syracuse, Carnegie Mellon, MIT and Yale. It also gladly accepts resumes from students attending schools outside its recruiting circuit. Engineering students are encouraged to acquaint themselves with the SEED program, Sun Engineering Enrichment & Development, for young engineers. The SEED program is one of Sun's largest mentoring programs and pairs promising college recruits (as well as established employees) with distinguished engineers, Sun fellows, VPs and directors who have volunteered to be mentors. Sun also offers internship opportunities for college and MBA undergraduates pursuing certain areas of study. Interns are paid (they even get overtime!) and have access to the fitness facilities where they work as well as the opportunity to attend networking events with executives.

Sybase Inc.

One Sybase Drive
Dublin, CA 94568
Phone: (925) 236-5000
Fax: (925) 236-4321
www.sybase.com

LOCATIONS

Dublin, CA (HQ)
Albany, NY • Alpharetta, GA •
American Fork, UT • Bethesda, MD
• Boise, ID • Boulder, CO •
Chantilly, VA • Charlotte, NC •
Chicago, IL • Clearwater, FL •
Concord, MA • Concord, NH •
Corvallis, OR • Dallas, TX • Detroit,
MI • Englewood, CO • Houston, TX
• Irvine, CA • New York, NY •
Orem, UT • Parsippany, NJ •
Philadelphia, PA • Phoenix, AZ •
San Diego, CA • Seattle, WA •
Mississauga, Canada • Vancouver •
Waterloo, Canada

40 locations in 25 countries
worldwide.

DEPARTMENTS

Business Operations • Consulting &
Professional Services • Corporate
Development • Customer Service &
Support • Engineering • Finance •
Human Resources • iAnywhere •
Information Technology • Legal •
Manufacturing & Logistics •
Marketing • Purchasing • Real
Estate & Facilities • Sales • Senior
Management • Sybase 365

THE STATS

Employer Type: Public Company
Stock Symbol: SY
Stock Exchange: NYSE
Chairman & CEO: John S. Chen
2006 Employees: 4,067
2006 Revenue ($mil.): $876

KEY COMPETITORS

IBM
Oracle
Microsoft

EMPLOYMENT CONTACT

www.sybase.com/about_sybase/
careers

Visit Vault at **www.vault.com** for insider company profiles, expert advice,
career message boards, expert resume reviews, the Vault Job Board and more.

VAULT CAREER LIBRARY   459

THE SCOOP

Make information work for you (wherever you are)

Sybase, once a leader in software for database applications, is reinventing itself as a provider of software for mobile communications. Currently, the company is somewhere in the midst of these two areas. Its database unit specializes in servers and software for handling and analyzing large amounts of data in a short amount of time, principally for companies in the financial, government and health care sectors. Its clientele includes 80 companies in the Fortune 100. However, while its database software division still brings in the lion's share of revenue, 70 percent in 2006, Sybase trails the other big players in the DB game, IBM, Oracle and Microsoft. So, like its competitors in the relatively stagnant database market, it has latched onto what it hopes is the next big thing.

In addition to Sybase's traditional information management business, the company has two major mobility-focused subsidiaries, Sybase iAnywhere and Sybase 365. The iAnywhere business unit develops mobile middleware software that allows wireless devices to communicate seamlessly with each other, for instance, allowing laptops to talk to BlackBerry devices and vice versa. Other iAnywhere tasks including managing mobile and embedded databases on devices and combining tasks like e-mail, IM, collaboration and security on mobile devices for enterprises. Sybase is expected to roll out more and more gadgets that chat with other gadgets in the coming year and it is also working on a number of RFID initiatives.

Sybase 365 is one of Sybase's newer subsidiaries, focused on mobile services, messaging technology and applications (SMS and MMS) and m-commerce (that's mobile commerce—or buying things via cell phone. The subsidiary processes more than six billion messages a month for more than 2.2 billion mobile device users worldwide—about 77 percent of the world's subscribers. With some mobile marketing campaigns and m-content deals already under its belt with financial institutions and retail organizations, it is expected that Sybase 365 will continue to expand its portfolio of services.

Database darling

In 1984, tech whizzes Mark Hoffman and Robert Epstein came together with the goal of marketing a relational database management system (RDBMS), calling their new venture Sybase (short for "System Database"). Relational databases, the most common form of databases, allow data (for instance, a customer's name) to be hooked to, or related to, other relevant pieces of information (like his address, which widgets he has ordered and when, or which stocks he owns, and so forth). When the product was released in 1987, its use was limited to IBM and a handful of other hardware platforms.

In 1988 Sybase formed an alliance with Microsoft, which had previously licensed Sybase technologies, to co-develop and sell versions of Sybase's RDBMS with its operating systems. In 1990 the company began pursuing an aggressive growth strategy, purchasing SQL Solutions, a database consulting company, then going public in 1991, and buying multimedia application tool designer Gain Technology in 1993. By emphasizing service, pushing international growth, providing quick delivery and specializing in databases for UNIX systems, Sybase had become the second-largest supplier of software for relational databases by that time, trailing only Oracle.

Not a perfect 10

Then, the company introduced the ill-fated System 10 in 1993, which simply proved too slow. The flop provided rivals Oracle and Informix with a golden opportunity to eat into the company's market share. Sybase quickly hammered out another deal with Microsoft the following year, hoping to staunch the loss of business. The deal, wherein Microsoft agreed to co-develop and sell the company's tools, provided a boost, but Sybase still lost money in 1995.

Co-founder Hoffman attempted to right the ship in 1996, removing David Peterschmidt as CFO and eliminating the COO position, but the Sybase board ousted him later in the year. (Hoffman would go on to found Commerce One.) Later that year, Sybase bought another company and got a new CEO in one fell swoop, when it purchased Powersoft for $950 million and installed its co-founder, Mitchell Kertzmann, as Hoffman's replacement. Kertzmann quickly began to streamline operations by laying off staff and selling non-core businesses, but left in the middle of the process to head an Oracle spin-off in 1997. John Chen immediately took the vacant position, and has served as chairman, CEO, and president ever since.

Changing with Chen

Chen had his work cut out for him. At a time when industry database sales instantly slowed, Chen continued restructuring, laying off workers and reorganizing operations. The company finally returned to annual profitability in 1999, after experiencing four consecutive years of losses.

In 2000 Sybase again looked toward expansion, buying Home Financial Network (Internet financial services) for $130 million. The next year, it bought software integration specialist New Era of Networks in a deal valued around $373 million. Sybase then poured some money into advertisements to spread some brand recognition and catch up with Oracle, Microsoft and IBM. It was to no avail, as Sybase's database software sales stagnated to 3 percent growth by 2006, and the company found itself entrenched in fourth place in the market. But Chen's reorganizing would soon start to bear fruit, steering Sybase in a new direction.

"Unwired" is the new black

Ladies and gentlemen, start your thumbs: Sybase expects you'll be using them a lot in the near future. Companies are starting to think about the possibility that having the workforce in contact via IM, text messaging, and e-mail will facilitate more efficient collaboration. And Chen believes that's the wave of the future.

Back in 2003, Chen selected the mobility sector as Sybase's next great area of growth, and began looking for acquisition possibilities. He selected iAnywhere Solutions as Sybase's flagship line for mobile products, and soon started beefing up the mobility initiative with acquisitions. The company bought AvantGo in 2003, one of the largest providers of personalized content to PDAs, and Dejima in 2004 a provider of mobile access solutions using natural language interface technology. In 2006, it acquired Solonde, a German provider of software that can access and display data from multiple sources at once. It also acquired assets from iFoundry, a Singapore-based provider of wireless technology. Then, Sybase made a successful $425 million bid for Mobile 365, now Sybase 365, the company's flagship messaging and mobile services unit

The iAnywhere division generated growth of 11 percent in 2006, and Sybase hopes that the growth rate for the mobile tech industry will increase nearly 20 percent year-over-year until 2010. If so, the market will transform from a $600 million market into a $1.5 billion one. The strategy is working so far: Revenue and profits have increased each of the last three years, most recently clocking in at $876 million revenue and $95 million in profits for 2006.

GETTING HIRED

Relate yourself to Sybase

Job seekers can check out the Employment section of Sybase's web site, at www.sybase.com/about_sybase/careers, to either upload a resume or search current job openings by category, location or keyword. Sybase offers a competitive benefits plan featuring medical, dental, vision and life insurance, employee stock purchase plan and stock awards, 401(k) plan, education assistance program, adoption assistance program and even onsite child care and fitness centers at some locations. The company prides itself on its lack of gratuitous layers of management and flexible career paths; there are ample opportunities for career development through Sybase University.

College students can join the Sybase family, at least until their classes start. The company offers internships and co-ops at various locations. Interns get paid holidays, access to fitness facilities (if available) and training on any software they may utilize at work.

Visit Vault at **www.vault.com** for insider company profiles, expert advice, career message boards, expert resume reviews, the Vault Job Board and more.

VAULT CAREER LIBRARY   **463**

Symantec Corporation

20330 Stevens Creek Boulevard
Cupertino, CA 95014-2132
Phone: (408) 517-8000
Fax: (408) 517-8186
www.symantec.com

LOCATIONS

Cupertino, CA (HQ)
Austin, TX • Chicago • El Segundo,
CA • Englewood, CO • Heathrow,
FL • Houston, TX • Irvine, CA •
Mountain View, CA • Newton, MA
• Oak Brook, IL • Redmond, WA •
San Francisco • Santa Monica, CA •
Springfield, OR • Sunnyvale, CA •
Woodbridge, NJ

International offices in Singapore,
Dublin and Tokyo.

DEPARTMENTS

Administration • Brand Management
& Communication • Business
Development • Consulting •
Contingent Worker • Corporate
Security • Development • Executive
Office • Finance & Accounting •
Fixed Term Employee • Global
Services & Support • HR •
Information Systems • IT • Legal •
Manufacturing • Marketing •
Operations • Purchasing • Sales •
Services • Technology • Training •
Web

THE STATS

Employer Type: Public Company
Stock Symbol: SYMC
Stock Exchange: Nasdaq
Chairman & CEO: John W. Thompson
2007 Employees: 17,100
2007 Revenue ($mil.): $5,199

KEY COMPETITORS

CA
McAfee
Microsoft

EMPLOYMENT CONTACT

www.symantec.com/about/careers

THE SCOOP

Safety for data

If the Web is the Wild West, Symantec is the sheriff. In addition to its Norton Antivirus line of PC protection, the Cupertino, Calif.-based software company has made its name developing utility programs, which manage and protect files from computer crashes and viruses. It also produces fax software including the pcANYWHERE, WinFax and ACT! lines.

Symantec produces Internet development tools in the Cafe product lines. It also releases the ominously named Internet Security Threat Report, a biannual analysis that summarizes the changing nature of malicious activity on the Internet, as well as the Internet Threat Meter, which uses the same color-coded advisory system as the Department of Homeland Security, only for the Internet.

Hendrix rocks Stanford! The tech one, that is

Unlike most other Silicon Valley startups, Symantec didn't get its financing by maxing out its founders' credit cards. The company started out in 1982 with brainpower from Stanford and a grant from the National Science Foundation. The fledgling company's first product didn't ship until 1985.

The company built on founder Gary Hendrix's expertise in natural language processing, released as its first product a simple database program that could manipulate data using commands in the form of English sentences instead of formulas. This functionality earned the program the name Q&A. But despite a nifty user interface, it never really caught on to the extent that similar programs did, like Lotus 1-2-3.

Symantec buys up the competition

Luckily, the company had another business model up its sleeve. In 1987, the company embarked on a series of acquisitions that powered it along for over a decade. It quickly snapped up two software companies, a manufacturer of project-management software and a creator of presentation software. Symantec went public in 1989 to further finance its acquisition strategy and snapped up Peter Norton's eponymous company hot on the heels of its IPO. At the time, Norton was known for its data-protection and crash-recovery software, and its name lives on in Symantec's modern versions of the same products.

Symantec's purchasing spree slowed somewhat after the company incurred a $40 million loss as a result of its 1995 acquisition of Delrina. In the early 2000s, however, business rebounded, largely on the strength of the company's Norton utility and antivirus programs—especially as viruses matured into ever more annoying forms, such as the notorious Melissa virus. Symantec beefed up its backup in 2005 with its $10 billion-plus purchase of Veritas, a provider of software for backing up data.

Acquisition position

The acquisitions continued fast and furious through early 2007, with the company snapping up eight more companies after the Veritas purchase. These acquisitions included outfits specializing in security for instant messaging, antifile-piracy programs, networks and server monitoring. Symantec even branched out, boosting its prowess in the IT consulting arena with the acquisition of Company-i in November 2006. A few months later, Symantec was back to its usual ways with the early 2007 purchase of Altiris, a provider of network and device security, for $830 million.

The cost of spending

Despite the brisk pace of acquisitions or perhaps because buying so many companies had been expensive, Symantec scaled back its earnings forecast by about 25 percent in early 2007. The company sought to minimize any losses, though; it announced an attempt to cut $200 million by curtailing spending on consultants and air travel, the cutting of 800 jobs and a hiring freeze.

Trends in security

Symantec's security software is well-placed for future profits as computer viruses grow even more serious. The malefactors of the Internet have progressed from the relative pranks of viruses like Mydoom and Slammer into serious identity theft propositions, which target banks and consumers for data such as credit card numbers and bank account numbers.

From its most recent Internet Security Threat Report, Symantec predicted that nearly 90 percent of all online malfeasance, such as viruses, phishing scams, Trojan programs and similar ventures, are aimed at the home computer, with 15 percent aimed at banks and similar institutions.

GETTING HIRED

Find a secure career

Symantec's careers site, at www.symantec.com/about/careers, provides information on the company's benefits and culture. The culture is marked by the company's four principal values—trust, innovation, value and customer-driven. Benefits include health, dental and vision, as well as 401(k), stock purchase, paid pregnancy and disability leave, generous vacation allowance, and kits for parents of children of all ages. The company also offers employee- and management-nominated bonus programs.

Internships are available for college students in the divisions of finance, information technology, product management, product marketing and software architecture. College recruits are also put to work in those areas. In order to qualify for these positions, students must major in a relevant field. For graduate students, Symantec funds a graduate research fellowship.

To apply, point your browser to www.symantec.com/about/careers/search.jsp. Hiring freeze notwithstanding, there were still a handful of jobs posted on the site. In order to apply, candidates must create a profile.

SYNNEX Corporation

44201 Nobel Drive
Fremont, CA 94538
Phone: (510) 656-3333
Fax: (510) 668-3777
www.synnex.com

LOCATIONS

Fremont, CA (HQ)
Atlanta, GA
Chantilly, VA
Dallas, TX
Glendale Heights, IL
Greenville, SC
Keasbey, NJ
Las Vegas, NV
Los Angeles, CA
Memphis, TN
Miami, FL
Portland, OR
Los Angeles, CA

DEPARTMENTS

Accounting/Finance
Corporate
Credit & Customer Service
Finance
Legal
HR
IT
Logistics & Integration Services
Loss Prevention
Marketing
Purchasing
Sales/Technical Services

THE STATS

Employer Type: Public Company
Stock Symbol: SNX
Stock Exchange: NYSE
Chairman: Matthew Miau
President & CEO: Robert Huang
2006 Employees: 2,647
2006 Revenue ($mil.): $6,343

KEY COMPETITORS

Bell Microproducts
Ingram Micro
Tech Data

EMPLOYMENT CONTACT

www.synnex.com/careers

THE SCOOP

They're IT

Synnex is one of the largest distributors of IT supplies from 100 world-leading IT manufacturers and software publishers ranging from Acer to Xerox. Synnex offers distribution, assembly and marketing services to more than 15,000 retailers and resellers of computer equipment throughout the United States, Canada and Mexico. The company is ranked No. 360 on the Fortune 500.

MiTAC steps in

Founded in 1980 by current President and CEO Robert Huang, the company was known as COMPAC Microelectronics until 1994, when it changed its name to SYNNEX Information Technologies, Inc. Motivated by the desire to simplify the distribution of its products, Taiwanese circuit board manufacturer MiTAC bought a majority of the company in 1992. Six years later, Synnex won contracts with HP and IBM to assemble computers and servers.

The company expanded into Mexico and China by 2002, and started selling its stock on NYSE for the first time in 2003. MiTAC maintains a majority stake in the company, and Matthew Miau currently serves as chairman for the boards of both MiTAC and Synnex.

The NEXCE great thing

Synnex branched out beyond its sales of computer parts with the September 2006 when it created a new consumer electronics division, NEXCE. It hopes to provide consumer electronics retailers with all the latest gadgets, from flat-screen TVs to GPS devices to digital cameras. Synnex will be teaming up with consumer electronics giant JVC on the deal.

Getting and spending

Synnex acquired five companies in 2006 and 2007 to expand its offerings. In April 2006 it purchased Telpar, a distributor of point-of-sale equipment and equipment for auto-ID technology like bar code scanners and RFID, for $5 million. It bought a Canadian supplier of ink for printers and copiers for $17 million the following month. In September, Synnex acquired Concentrix, a marketing services firm.

Visit Vault at **www.vault.com** for insider company profiles, expert advice, career message boards, expert resume reviews, the Vault Job Board and more.

VAULT CAREER LIBRARY   469

In 2007, the company acquired the Redmond Group of Companies and the computer wholesale division of Insight for more consumer electronics offerings. Synnex also obtained the tech support firm, Link2Support, and a controlling interest in HiChina Inc., a well-respected Chinese e-commerce company. The company's 2006 revenue increased from the previous year by 13 percent to $6.3 billion. So, if Synnex intends to expand its wide array of products in the future through new acquisitions, it should have ample cash to do it.

GETTING HIRED

Assemble your career at Synnex

Synnex's careers page, at www.synnex.com/careers/careers.html, provides information about job openings, benefits, training and how to apply at Synnex. Jobs are only searchable by location; to apply, resumes and cover letters may be submitted via e-mail to the specified address, or by fax or snail mail, using the information on the How to Apply page (at www.synnex.com/careers/how_to_apply.html).

The company's benefits include medical, dental, vision, 401(k), discount stock purchase, tuition reimbursement for approved courses, a generous employee referral program—and discounts on merchandise and gym memberships for employees and their families. New sales hires are offered three weeks of training. There's also the Synnex Knowledge Transfer Team for employees' knowledge transfer needs. A college recruiting calendar is posted, but contained no events at press time. The company also has information about lovely downtown Fremont, Calif., where the company HQ is located.

Teradyne, Inc.

700 Riverpark Drive
North Reading, MA 01864
Phone: (978) 370-2700
Fax: (978) 370-2910
www.teradyne.com

LOCATIONS

North Reading, MA (HQ)
Agoura Hills, CA • Allentown, PA •
Austin, TX • Boise, ID • Chandler,
AZ • Deerfield, IL • Detroit, MI •
Essex Junction, VT • Fridley, MN •
Irvine, CA • Plano, TX • Poway, CA
• Richardson, TX • San Jose, CA •
Winston-Salem, NC • Tualatin, OR

Locations in 20 countries
throughout Europe, Asia and South
America.

DEPARTMENTS

Business operations
Financial Analyst • Product Manager
• Sales Engineer
Engineering
ASIC • Hardware • Software • Field
Applications • Test Development •
Applications • Mechanical •
Manufacturing

THE STATS

Employer Type: Public Company
Stock Symbol: TER
Stock Exchange: NYSE
Chairman: Patricia S. Wolpert
President & CEO: Michael A. Bradley
2006 Employees: 3,800
2006 Revenue ($mil.): $1,376

KEY COMPETITORS

Advantest Corporation
Credence Systems Corporation
LTX Corporation

EMPLOYMENT CONTACT

www.teradyne.com/hr

Visit Vault at **www.vault.com** for insider company profiles, expert advice,
career message boards, expert resume reviews, the Vault Job Board and more.

VAULT CAREER LIBRARY   **471**

THE SCOOP

Testing, testing

Teradyne is the world's top manufacturer of automated test equipment (ATE), used to analyze semiconductor chips and circuit boards. These bits of electronics, once vetted by Teradyne, wind up in everything from aerospace equipment, automobiles, computers and defense electronics to telephone and data systems. Notable consumers include Boeing and the U.S. government. The insatiable consumer appetite for electronics and telecommunications in recent years has Teradyne's sales climbing, but Teradyne's core business is dependent on semiconductor production, which is a cyclical business indeed.

Hungry? I just ATE

Teradyne was founded in 1960 by MIT graduates Nicholas DeWolf and Alexander d'Arbeloff, who headquartered the company in a downtown Boston loft, conveniently located over a hot dog stand. DeWolf and d'Arbeloff's idea was to take high-tech test equipment out of the lab and put it on the production line, increasing the speed and accuracy of the tests performed on newly-manufactured electronics. The company initially produced testers for integrated circuits, resistors, transistors and diodes. But when Teradyne began using computers to speed up the testing process in the late 1960s, it created the ATE industry it leads today. The company started Teradyne Components (later Teradyne Connection Systems) in 1968 to produce electronics connection assemblies, and the firm went public in 1970. In 1972, Teradyne began developing a telephone system testing device called 4Tel.

The company formed its computer-aided engineering (CAE) group in the 1980s by purchasing and combining Aida Corporation and Case Technologies. Decreased military spending in the early 1990s hurt sales, but a $63 million contract from the German national telephone system put Teradyne back on top in 1993. The company started its software and systems test unit the next year. As demand for PCs grew in 1995, so did sales of the company's semiconductor testing equipment, lifting Teradyne to the $1 billion sales mark for the first time. In late 1999, the company announced that it would spin off this software division. The following year, the company acquired two competitors, and opened its first location in China in 2003.

Vocation relocation

In 2006, Teradyne announced it would be moving its headquarters from Boston to North Reading, a town in Massachusetts about a half-hour north of the city. The company's spacious, five-building, 45-acre campus accommodates 1,800 workers in newly renovated offices. In order to keep its employees from fleeing due to the move, the new buildings offer a number of cruise-ship perks, like room to play horseshoes, a shuffleboard court, a hair salon and a gym. Employees will also be offered mortgages at attractive rates should they wish to relocate in order to be nearer the new HQ, and there will be shuttle busses for those who do not. Teradyne's former offices in downtown Boston were sold, and will be converted into office space.

Bright ideas

In addition to dragging its employees hither and yon across the Massachusetts countryside, the company posted 2006 revenue of $1.38 billion (a slight dip from a 2004 peak of $1.4 billion), with a profit of $198 million. Shortly thereafter, in 2007, the company acquired ATE assets for $17 million from Mosaid, an Ottowa-based semiconductor company, which will bolster Teradyne's R&D for memory testing.

GETTING HIRED

Test your limits

Teradyne's careers pages provide information on benefits and job openings, as well as diversity at the company and its involvement in the community. Benefits include profit sharing and 401(k), health insurance for you, your spouse of whatever gender, and your pet, as well as tuition reimbursement and company-sponsored classes. All that on top of shuffleboard! To apply for a position, job seekers must first create a login.

If you're of the student persuasion, Teradyne offers everything from interview tips to internships. Student opportunities are available through the career center on campus. Interviews are initially given on campus; subsequent rounds happen at Teradyne locations. Internship opportunities are also distributed through college career centers.

GO FOR THE GOLD!

GET VAULT GOLD MEMBERSHIP AND GET ACCESS TO ALL OF VAULT'S AWARD-WINNING LAW CAREER INFORMATION

◆ Access to **500+ extended insider law firm profiles**

◆ Access to **regional snapshots** for major non-HQ offices of major firms

◆ Complete access to **Vault's exclusive law firm rankings**, including quality of life rankings, diversity rankings, practice area rankings and rankings by law firm partners

◆ Access to **Vault's Law Salary Central**, with salary information for 100s of top firms

◆ Receive **Vault's Law Job Alerts** of top law jobs posted on the Vault Law Job Board

◆ Access to complete **Vault message board archives**

◆ **15% off** all Vault Guide and Vault Career Services purchases

For more information go to
www.vault.com/law

VAULT
> the most trusted name in career info

Texas Instruments Incorporated

12500 TI Boulevard
Dallas, TX 75243-4136
Phone: (972) 995-2011
Fax: (972) 927-6377
www.ti.com

LOCATIONS

Dallas, TX (HQ)
Alpharetta, GA • Attleboro, MA •
Austin, TX • Baltimore, MD •
Boston • Burlington, MA • Cary, NC
• Chicago • Detroit • Duluth, GA •
Fishkill, NY • Ft. Lauderdale, FL •
Germantown, MD • Houston, TX •
Huntsville, AL • Indianapolis, IN •
Irvine, CA • Jackson, MS •
Kokomo, IN • Laurence Harbor, NJ
• Longmont, CO • Manchester, NH
• Mansfield, MA • Milwaukee, WI •
Minneapolis, MN • Palo Alto, CA •
Philadelphia, PA • Pittsburgh, PA •
Plano, TX • Portland, OR • Raleigh,
NC • Rochester, MN • Rochester,
NY • San Diego, CA • San Jose,
CA • Santa Barbara, CA • Santa
Rosa, CA • Seattle • Sherman, TX •
St. Louis, MI • St. Petersburg, FL •
Sunnyvale, CA • Tempe, AZ •
Tucson, AZ • Waltham, MA •
Warren, NJ • Warrenville, IL •
Warwick, RI • Woodland Hills, CA

More than 40 locations throughout
Europe, Asia and Canada.

THE STATS

Employer Type: Public Company
Stock Symbol: TXN
Stock Exchange: NYSE
Chairman: Thomas J. Engibous
President & CEO: Richard K.
  Templeton
2006 Employees: 30,986
2006 Revenue ($mil.): $14,255

DEPARTMENTS

Application Specific Integrated
Circuit (ASIC) • Broadband • Digital
Consumer Applications • Digital
Control Applications • Digital Signal
Processing (DSP) • DLP® Technology
• Educational and Productivity
Solutions (E&PS) • Engineering •
Facility Services • High-Performance
Analog • HR • Logic • Management •
Marketing • Quality • Radio
Frequency Identification (TI-RFid™) •
Technology and Manufacturing
Group • Wired Digital Media •
Wireless • Wireless Infrastructure •
Wireless Networking

KEY COMPETITORS

Analog Devices, Inc.
National Semiconductor Corporation
STMicroelectronics NV

EMPLOYMENT CONTACT

focus.ti.com/careers

Visit Vault at **www.vault.com** for insider company profiles, expert advice,
career message boards, expert resume reviews, the Vault Job Board and more.

VAULT CAREER LIBRARY   475

THE SCOOP

Smaller, faster, better

The company that brought you the handheld calculator (in 1967) is now helping to make computers and cell phones even smaller and faster. 87 percent of Texas Instruments' sales come from computer chips, making it the third-largest manufacturer of chips in the world.

Among Texas Instruments' developments is a 0.18-micron transistor, which is so small that 125 million of them can fit onto a thumbnail-sized chip. It has also produced a digital-signal processor (DSP) that processes information 10 times faster than its predecessors—it's a key component of a number of high-tech audio and video processing technologies. Perhaps more importantly, DSPs now account for two-fifths of TI's semiconductor business.

In the beginning

Texas Instruments was initially founded as an oil prospecting venture in 1930, called Geophysical Service, Inc. (GSI). It separated itself from many of its Texan peers through its high-tech method of reflection seismography, which involved exploding small amounts of dynamite aboveground and analyzing the resulting shockwaves to predict underground oil and gas deposits. The technique would soon benefit the U.S. government, who contracted GSI during World War II and thereafter to track the underground activity of enemy forces (it located submarines during the 1940s, and it later pinpointed Russian nuclear testing facilities during the Cold War).

In the meantime, the firm divested its oil prospecting division, selling it in 1941 to the oil company that would become Amoco. The technological branch of GSI changed its name to Texas Instruments a decade later to reflect its focus on electronics. In 1952, the company bought a license which allowed them to manufacture germanium transistors and began its now historical work in the semiconductor industry.

Let there be silicon!

The company created the first practicable silicon transistors in 1954, and a pocket transistor radio later that year. In 1958, engineer Jack Kilby created the first microchip on a strip of germanium in a Texas Instruments laboratory (he was awarded a Nobel Prize for his efforts in 2000). Some of the first uses to which microchips were put included the computer in the Minuteman II missile in 1964 and a handheld calculator in 1967.

The company developed the single-chip microprocessor and the single-chip microcomputer in the 1970s. More recently, Texas Instruments' success in developing new and better chip technology has led to a number of joint ventures with electronics manufacturing giants such as Hewlett-Packard and Hitachi.

Chips dip

Of course, a company that grows fat on profits from silicon goes hungry when silicon prices fall, which occurs often in the highly cyclical chip industry. As newer, smaller, faster chips are developed, expensive factories for their manufacture are built. The field fills with competitors, and prices drop off sharply until the next greater, tinier chip comes along. Rinse, repeat. In the mid-1990s, chip prices fell, and Texas Instruments refocused itself in the face of industrywide instability by completely divesting GSI in 1991 (it had sold 60 percent of its stock to Halliburton, another Texan defense contractor, in 1988). It also sold off its notebook computer operations by the end of the decade.

TI then concentrated on what it does best—its calculators, which had begun to set new sales records, and its microchips for all kinds of electrical devices, notably cell phones. Nevertheless, a collapse in memory chip prices hurt the company in 1998; its memory chip division lost almost half a billion dollars that year. The company sold its memory business to Micron Technology in September and focused more intently on producing DSPs.

Picking up revenue and other companies

By the end of the 1990s, the company seemed to have weathered the industry-wide drop in sales quite well. Revenue had increased, due in part to a rise in demand for cell phones. TI continued to shed non-core businesses and made a number of strategic acquisitions, including Unitrode, a manufacturer of analog chips (critical for turning voice input into digital signals, and vice versa), and Toccata Technology, which advanced its digital-speaker technology. The company also teamed with SCS Corporation to develop its UHF RFID technology, which is a small, identifying circuit that can be tagged onto an object and recognized by radiowaves, often used by businesses for tracking and managing inventory.

As a step into the burgeoning VoIP (voice-over-Internet protocol) market, which allows phones to make calls over an Internet connection instead of a phone line, TI formed an agreement with Clarent Corporation in 2000 to develop new VoIP solutions. The biggest acquisition came in August of the same year, when TI purchased analog chip manufacturer Burr-Brown for more than $7.5 billion. The market didn't look up for long, though, as a slumping chip market forced the company to lay off some 2,500 employees in 2001. TI eventually turned itself around, with help from a reviving market for semiconductors, in 2002 and 2003.

TI phone home

In January 2005, the company announced it had integrated most of the computing functions that power wireless phones onto a single chip. These chips, which initially became available in mid-2006, can incorporate features like an MP3 player and ringtones, an FM radio receiver, color display, and are even more efficient with power than phones built with multiple chips. They are so small that they can become a part of a cell phone the size of a matchbox. But what's smaller than their size is their price—TI has cleverly dubbed a line of them LoCosto—a phone made with them can be sold for around $20. TI is already selling these phones in developing countries, where the adoption of cellular technology is booming.

The numbers back up TI's strategy: cell phone use in China is expanding at a cracking 15 to 20 percent per year, and India is posting similar growth numbers. Market researchers forecast that by 2010, more than 50 percent of the world's population will own cell phones—and TI wants to be at the heart of them. The company already boasts an impressive 58 percent share of the phone chip market. A new generation of the chip, slated to go into production in 2008, adds even more features, like 30 frame-per-second video and 3-D video game graphics.

Many analysts predict that, in the near future, cell phones will be the main gateway to the Internet for developing economies. TI agrees: it is heralding the beginning of the "cell phone revolution," and sees many possibilities for Internet communications moving from laptops and desktops to pocket-sized devices that can handle music, video, e-mail, web browsing and video games. Wherever cell phone technology goes, TI and its chips plan to be along for the ride.

Cell well

TI's strategy seems to be paying off—its business was buoyed by its involvement with cell phone chips in 2006. Revenue was $14.25 billion, up 16 percent over 2005.

The analog chip division turned in particularly strong numbers, with sales increasing by over 30 percent year-over-year, while the DSP division also acquitted itself admirably. Company profits overall clocked in at a robust $4.3 billion.

Also in 2006, the company sold off its sensors division (not including RFID) for $3 billion. TI then bolstered its RFID division with the acquisition of Norway-based Chipcon, a company that makes low-powered radio frequency chips. The first quarter of 2007 bore good tidings for TI as well. Although profits decreased by 4 percent, they still exceeded Wall Street forecasts, and the company expects growth to continue in the second quarter.

GETTING HIRED

Instrumental careers

TI's informative careers site (at www.ti.com/recruit) has information for job seekers of every age and experience level. The site has nifty facts about the company, as well as lists of benefits and job opportunities the world over. To apply, applicants must create a profile. Employees must also pass a drug test.

High achievers wanted

TI offers co-op and paid internship programs that require a commitment of two academic terms. Setting the bar high, TI requires that students have a 3.0 GPA and preferably be pursuing a degree in engineering, science or finance. The company offers paid relocation, relocation assistance and employee benefits. MBA internships are available as well.

Recent graduates can find out about job opportunities on the site, too. TI posts a college recruiting schedule (forthcoming at press time). Positions are available for graduates with engineering and finance degrees, and for those in doubt, there's a Fit Check tool that can tell if the company is looking for particular skills.

Once you've jumped through all the hoops in the TI hiring process, the company amply rewards employees with generous benefits, like stock purchase plans, 401(k)s, bonus programs, and on-site cafeterias and fitness facilities. In 2007, TI's efforts in this area were recognized for the seventh time in *Fortune*'s list of the Top 100 Best Places to Work. The company was received accolades for its diversity program, as well as its employee retention.

OUR SURVEY SAYS

Add a little Texas to your resume

Insiders report that "TI is surprisingly conservative for a high-tech company." Its conservatism is surprising, considering that it is "a youth-oriented company, not the best place for people over 40," according to one hire. "Work hours are long with 50-60 per week being the norm. Frequent weekend [and] holiday work," notes an insider. "The company talks about work/life balance, but that's all talk. You are expected to work very long hours; something a lot of the company's working moms struggle with," agrees a colleague. "Pay and opportunities for advancement are open to all and based primarily on showing good results," adds a source.

Total System Services, Inc.

1600 1st Avenue
Columbus, GA 31901
Phone: (706) 649-5220
Fax: (706) 644-8065
www.tsys.com

LOCATIONS

Columbus, GA (HQ)
Alpharetta, GA
Atlanta, GA
Boise, ID
Golden, CO
Tempe, AZ

DEPARTMENTS

Assessment & Consulting •
Business Continuity Management •
Card Services-Personalization •
Client Acceptance • Client
Relations-PCS • Conversion Project
Management • DBS Support • Dialer
Support • DTS Quality Assurance •
DTS Technology • FCS
Implementation Services • Finance •
Hosting • Implementation Team •
Information Security • Internal Audit
• IT Card Applications • IT
Statements/TS2 Correspondence •
IT-Project Office • Marketing •
Platform Development • Press Room
• Product Support • QA • Risk
Management • SS—Statement
Production • Strategic Planning •
System Development • Systems
Management • TLP Group • TSYS
Analytics • TSYS Business Support
Services • TSYS Prepaid Services

THE STATS

Employer Type: Public Company
Stock Symbol: TSS
Stock Exchange: NYSE
Chairman & CEO: Philip W. Tomlinson
2006 Employees: 6,644
2006 Revenue ($mil.): $1,787.17

KEY COMPETITORS

BA Merchant Services
First Data
NOVA

THE SCOOP

Pay up

Total System Services, Inc., better known to its closer friends as TSYS, is one of the world's largest credit card payment processors, serving banks and private-label card issuers the world over. The company offers its clients credit authorization, payment processing, merchant acquiring, account management, e-commerce services, card issuance, call-center operations, customer loyalty programs, prepaid programs and fraud monitoring.

Charging ahead

The credit card system was in its infancy in 1959 when The Columbus Bank and Trust Company (now part of Synovus Financial Corp.) became the second bank in Georgia to issue cards to its customers. Within five years, the bank's card holders increased by more than 10 times, and CB&T installed a computer to process transactions in 1966. With time, the bank became increasingly interested in creating a computer system to further automate the work. By 1969, it had written credit-processing software, which would become known as the Total System in 1973.

In 1974, CB&T began offering to process transactions for other banks, and business took off when credit cards became more popular in the 1970s. Among its achievements, the Total System became the first of its kind to process cards from both Visa and MasterCard (then known as BankAmericard and MasterCharge, respectively). The division's revenue was $6 million in 1980. Two years later, when CB&T decided to spin it off into a subsidiary known as Total System Services (for obvious reasons), revenue stood at $10 million. TSYS, as it would soon be familiarly known, was then processing over one million accounts for 60 banks in 28 states.

Onwards and upwards

The next year, 20 percent of TSYS was offered as stock; CB&T (known as Synovus after 1989) has maintained its ownership of more than 80 percent of the company's stock until the current day. (In July 2007, Synovus began discussions about divesting its stock in TSYS to its shareholders.) In 1992, TSYS added mass mailing capabilities to its transaction processing services, for billing and marketing purposes. Three years later, the company inked a deal with the Mexican bank Controladora Prosa, expanding abroad under the name TSYS de Mexico.

Plans for world domination

Geographic expansion continued in 2006 for TSYS. In July, the company acquired Card Tech, a U.K.-based company that provides card services and payment processing in 70 countries. The deal will strengthen TSYS' position in Asia, Africa and the Middle East. TSYS made further inroads into Asia with a deal to handle transactions for the credit arm of Toyota in September.

Shortly thereafter, in November, the company announced a joint venture with Merchants, a South African subsidiary of the IT firm Dimension Data, which would incorporate call centers and payment processing for card issuers in Europe and the Middle East. Growth overseas is proving to be lucrative for the company, as revenue from international transactions increased by nearly 25 percent between 2005 and 2006.

A healthy line of business

TSYS isn't just after consumer spending in other countries. The advent of American consumer-driven health care (not to mention meteorically increasing costs for health care in general, which consumer-driven plans should do little to mitigate) has banks and credit issuers jockeying to get into the game. UnitedHealth, one of the largest American insurance companies, has hatched a plan to give consumers a card linked to their health savings accounts, complete with a built-in line of credit. TSYS won the right to process these payments for United in April 2006. The card can accept payments from insurance companies, so that reimbursement for covered procedures is automatically added to the account; this reduces administrative costs for the insurer.

Hey, big spender

In October 2006, Bank of America, which accounted for nearly a quarter of TSYS' total revenue that year, opted to move its credit card processing over to MBNA's system, following the two companies' merger. TSYS charged BoA nearly $70 million for ending the contract early, but the loss of business put a kink in the bottom line for 2007. 2006 was still a good year for TSYS, though; revenue was $1.7 billion, and profits were nearly $250 million—up 30 percent over the previous year. TSYS doesn't predict that losing business with Bank of America will cost a 25 percent decline in revenue for 2007, though; the company forecasted only a decrease of between 3 and 5 percent for the fiscal year.

GETTING HIRED

Charge your career at TSYS

TSYS provides a wealth of career-related information on its web site, located at www.tsys.com/careers. There is a searchable list of job openings (as usual, interested candidates must create a profile in order to apply). The company accepts unsolicited résumés. There's also information on the wealth of benefits offered by the company, which include career development opportunities, including tuition reimbursement, a cafeteria-style benefits plan with health, dental and vision options, 401(k), discount stock purchase and profit sharing plans, and work/life balance programs, including time off to volunteer at a child's school.

Chattahoochee choo-choo

TSYS helps employees balance work and life in other ways, as well. The company's campus, conveniently located on the beautiful Chattahoochee River, offers such sumptuous amenities as a café for lunch or takeaway dinner, 15-mile park for running or biking, an on-site gym and a cozy library and child care facilities. There's also a full-service bank, mini mall with a dry cleaner's, florist and beauty salon. And don't forget the wonderful attractions of Columbus, Georgia, just 100 miles south of Atlanta.

Trimble Navigation Limited

935 Stewart Drive
Sunnyvale, CA 94085
Phone: (408) 481-8000
Fax: (408) 481-7781
www.trimble.com

LOCATIONS

Sunnyvale, CA (HQ)
Austin, TX • Cambridge, MA •
Chantilly, VA • Chicago, IL •
Corvallis, OR • Dallas, TX • Dayton,
OH • Deerfield Beach, FL • Folsom,
CA • Fremont, CA • Long Beach,
CA • Plano, TX • Salem, MA •
Santa Clara, CA • Tempe, AZ •
Westminster, CO

23 international locations.

DEPARTMENTS

Accounting & Finance
Administration
Business Development
Engineering
HR
IT
Legal
Manufacturing & Production
Operations
Sales & Marketing
Service, Support & Training

THE STATS

Employer Type: Public Company
Stock Symbol: TRMB
Stock Exchange: Nasdaq
Chairman: Ulf J. Johansson
President & CEO: Steven W. Berglund
2006 Employees: 2,842
2006 Revenue ($mil.): $940

KEY COMPETITORS

Leica Geosystems
Motorola
QUALCOMM
Topcon

EMPLOYMENT CONTACT

www.trimble.com/employment.shtml

Visit Vault at **www.vault.com** for insider company profiles, expert advice,
career message boards, expert resume reviews, the Vault Job Board and more.

VAULT CAREER LIBRARY   485

THE SCOOP

Wherever you go, there you are

Trimble is in the business of marrying GPS, laser and wireless technology so that its customers need never feel out of place. The company's products are used in agriculture to level fields and guide tractors; in construction and mining to guide heavy machinery for earthmoving operations, keep track of trucks and aid in surveying; and in automobile navigation systems. The company also provides products for precise timing in cellular networks, wireless ticketing for law enforcement agencies and the management of mobile workers.

Location, location, location

Trimble has been the guiding light to the lost—or rather, the guiding electromagnetic radiation—since 1978, when Charlie Trimble and some of his associates from Hewlett-Packard set up shop in California. Trimble began by working with navigation devices that used the LORAN (Long Range Navigation) system, powered by terrestrial radio stations. But when GPS hit the scene in 1982, Trimble made good use of the new technology in products for surveyors, mapmakers and geologists. The company expanded into marine navigation in 1984, and purchased a business to begin developing differential GPS (DGPS) technology to provide increased accuracy for the fleet management market. In 1990, the company was publicly listed on the Nasdaq.

During the 1990s, Trimble worked to combine communications and GPS technology into the same device, and in 1995 developed a GPS device that could be plugged into laptops and PDAs. In the early 2000s, the company continued to expand its product offerings through research and development and via acquisitions—Trimble formed a partnership in 2002 with construction and mining vehicle manufacturer Caterpillar, to install positioning technologies in its wares.

Acquisition position

Trimble seems to have a simple company-building strategy: buy and hold. Since the start of 2006, the firm has purchased nine companies involved in GPS, fleet management and related industries. A highlights reel of its acquisitions that year would include a shot of Trimble cementing a deal to acquire the intellectual property of XYZs of GPS, a company that specializes in devices for GPS measurements of volcanoes and oilfields; Trimble purchasing Quantm, a company which works to identify the most cost-effective and environmentally friendly routes for roads, pipelines and canals; as well as the purchase of retail goods tracking firm Eleven Technology. Trimble kept the acquisitions ball rolling with its purchase of BitWyse, a developer of software for engineers, in May of that year, while the final months of 2006 saw the company purchase concerns in the law enforcement, construction project workflow software and mobile mapping fields.

Come together

But that's not all! On Valentine's Day 2007, the company celebrated its union with INPHO GmbH, a German firm specializing in 3-D imaging and mapping. Two days later, the company acquired the uniquely named fleet management company @Road.

Despite its spate of purchases, the company is well positioned to take advantage of the growing market for knowing where stuff is, whether it be deposits of oil or gold or a garbage truck. In 2006, Trimble's revenue increased more than 20 percent over the year previous to $940 million, while profits were up 8 percent over 2005, to $135 million.

Death in the family

Trimble's forward progress halted for a moment in July 2007, when the company's chairman since 1989, Dr. Robert Cooper, passed away. Ulf J. Johansson, a former chairman at Swedish tech giant Ericsson, filled the post immediately; he had served on Trimble's board since 1999. Seven days after this announcement, Trimble posted its quarterly results: revenue had grown by 34 percent and profits by 23 percent from the year before.

GETTING HIRED

Find your place at Trimble

Trimble's careers site (at www.trimble.com/employment.shtml) provides up to date and comprehensive job information worldwide for experienced hires, recent graduates and college students seeking internships. The site also allows job seekers to create profiles enabling them to be automatically alerted to future new jobs in their area of interest. Trimble is particularly interested in graduates (of any degree level)

Visit Vault at **www.vault.com** for insider company profiles, expert advice, career message boards, expert resume reviews, the Vault Job Board and more.

VAULT CAREER LIBRARY   **487**

who specialized in accounting, computer science, IT, marketing, and agricultural, civil, electrical and mechanical engineering, among other majors. Internships are available for qualified students. To apply, send your resume as a text document (no fancy formatting, please) to college_recruiting@trimble.com. To submit your resume for consideration for a specific position, you must first create a profile on Trimble's site.

Unisys Corporation

Unisys Way
Blue Bell, PA 19424
Phone: (215) 986-4011
Fax: (215) 986-2312
www.unisys.com

LOCATIONS

Blue Bell, PA (HQ)
Primary locations in California,
Georgia, Michigan, Minnesota, New
Jersey, Pennsylvania, Utah and
Virginia.

77 locations in 35 US states.

DEPARTMENTS

Accounting/Finance •
Administrative/Secretarial •
Communications • Distribution/
Logistics/Procurement •
Engineering—Hardware •
Engineering—Software • Facilities •
General Management • HR •
Information Systems Consulting •
Legal • Manufacturing • Marketing •
Medical • Network Management/
Architecture • Programming/
Development • Project Management
• Quality Assurance/Testing • Sales •
Supply Chain • Systems Architecture
• Technical Support/Help Desk
Services • Technical
Writing/Documentation •
Telecommunications/Telephony •
Training • Web
Design/Graphics/Content • Web
Development

THE STATS

Employer Type: Public Company
Stock Symbol: UIS
Stock Exchange: NYSE
Chairman: Henry C. Duques
President & CEO: Joseph W. McGrath
2006 Employees: 31,500
2006 Revenue ($mil.): $5,757

KEY COMPETITORS

EDS
Hewlett-Packard
IBM

EMPLOYMENT CONTACT

www.unisys.com/about_unisys/
careers

Visit Vault at **www.vault.com** for insider company profiles, expert advice,
career message boards, expert resume reviews, the Vault Job Board and more.

VAULT CAREER LIBRARY   489

THE SCOOP

The Unisys universe

Unisys is a provider of technology services—that is, everything from servers to outsourcing to consultants—intended to transform an average company into a lean, mean, profit-generating machine. The company offers a full menu of services, including software development, support and administrative functions, IT infrastructure, business security consulting and the company's flagship product, the 3D visible enterprise, which allows management to see the consequences of a business decision before it's put into motion—kind of like a business "flight simulator" for CEOs.

Unisys works with clients in the financial services, communications and transportation sectors, among other industries. It counts among its clients the U.S. Army, the government of Malaysia and the *Irish Times*, for which it set up an RFID cargo-tracking system, created a national ID card system, and helped bring into the digital age, respectively. Unisys systems are used at 22 of the top 25 banks worldwide, and half of all checks written each day are cleared by the company. Unisys' security consultants help in keeping passengers and cargo secure at 600 of the world's airports by tracking luggage and cargo, and the firm processes nearly 200 million health insurance claims each year for Medicaid.

From typewriters to high tech

Unisys, like the computer, sprang from the marriage of an adding machine and a typewriter. The company was founded officially from the 1986 merger of Sperry (which had absorbed the Remington typewriter company in 1955) and Burroughs, a manufacturer of adding machines. (It must be noted, though, that both companies' computer divisions were well established at the time of the merger.)

Unisys can trace its antecedent companies back to the 1870s, when Remington manufactured one of the first mass-produced typewriters. The company introduced the first commercial business computer, UNIVAC 1, in 1949, and went through a merger in 1955 to become the Sperry Rand Corp. Sperry was acquired by adding machine and mainframe computer giant Burroughs in 1986, forming the Unisys we know and love today.

Making papa proud

The firm carried on its parents' legacy of groundbreaking computer inventions. In 1988, Unisys acquired Convergent Technologies and its proprietary Convergent Technologies Operating System (CTOS). In 1989, Unisys developed Micro-A, the first single-chip desktop mainframe, and in 1993 the 2200/500 mainframe, the first mainframe based on CMOS (complementary metal oxide semiconductor) technology. From there, it began to specialize in services and "solutions," developing complex technologies like ClearPath Heterogeneous Multi-Processing (allowing applications and databases to run with UnixWare and Windows NT applications and databases on a single platform) and Cellular Multi-Processing, which can handle up to 32 processors at once. To date, the firm owns more than 2,600 technology patents.

Integrating pains

From a financial perspective, however, the formation of Unisys didn't exactly go smoothly. In fact, many observers labeled the merger disastrous, prompting customers to flee—mainly to IBM. Larry Weinbach, formerly of Andersen Consulting, was brought on board to calm the waters in 1997. He initiated a restructuring plan that directed the company's focus towards what he knew best: consulting, which represented about 80 percent of the company's resources by 2004. Profits, however, fell in the short term. Unisys got another new CEO in January 2005, Joseph W. McGrath, who started cutting jobs in the face of losses. McGrath also adjusted the company's business focus onto its top customers.

The computer industry's new frontier

Unisys has been an exemplar of outsourcing since 2004, when it invested $180 million in a facility in Bangalore. The following year, the company planned to expand its ranks of employees in India by 2,000 people by December, and to 4,000 by 2006, to keep up with demand for its outsourcing services. In 2006, Unisys opened outsourcing locations in China and Hungary, and announced two more facilities near Bangalore and Hyderabad, which will bring its headcount of Indian employees to 6,000 by 2008.

Labor relations

In 2005, following a year of losses, Unisys announced that it would be cutting more jobs stateside. In April, the company said it would lay off 3,600 people, some 10 percent of its workforce, with plans to turn a profit beginning in 2008. Further losses that year led inevitably to the trimming of an additional 1,000 jobs in the Americas and Europe, coupled with a strong push to hire people in countries with lower labor costs. By the end of 2006, the company had completed the bulk of its terminations.

The company is moving to save money in labor costs in other ways as well. In March 2006, Unisys announced it would freeze its company-run pension plan as of December. Retirees will continue to receive benefits, but employees will no longer accrue them. Instead, Unisys is giving employees' 401(k)s a healthy dose of company stock. It seems they'll be getting it at discount rates—Unisys' revenue barely budged over 2005, but losses of $1.7 billion in 2005 fell to the significantly less epic negative profit of $278 million for 2006. The company did incur some charges associated with its job cuts and readjusting its pension scheme, which pushed down revenue somewhat. However, it seems to be realizing the gains from moving its workforce to India and other less expensive countries.

Bringing in the deals

It wasn't just profit losses and job cuts for the company in 2006, however. The company also scored some contracts that should help lift it back to profitability. In 2006, Unisys announced a contract potentially worth just under $50 million to develop flight simulation software for NASA and extended its contract with the Brussels airport for the maintenance of its ticket- and baggage-handling systems.

Unisys cemented other contracts with Boeing, for its work on the Department of Homeland Security's Secure Border Initiative, and Starbucks. Finally, the company also teamed up with the Department of Justice in 2007 to maintain the system the department uses to keep track of loot seized in criminal proceedings in a deal valued at over $100 million.

Opening up for the future

Unisys has recently taken a page from other struggling computer firms and turned to open source formatting (or, the use of software for which manufacturers charge no licensing fee) as a way to offer its IT services cheaply and efficiently to its clients. In August 2007, the company demonstrated its utilization of disparate open source technologies at the LinuxWorld Conference & Expo. The Open Solutions Alliance, a tech firms' organization devoted to open source formatting, sponsored the event. Unisys is a founding member of the OSA and was the first IT company of its type to join the organization, so it is deeply invested in open sources as a way forward.

Media res

In March 2007, Unisys sold its media holdings to Atex, a U.K.-based software company, for an undisclosed sum. Unisys' media division provided its customers with editorial, advertising and multimedia content for the internet. A few months later, in June, Unisys inked a deal to provide the Department of Defense with $28 million in parts and support for its RFID supply chain.

GETTING HIRED

Unite yourself with Unisys

Unisys' jobs site, at www.unisys.com/about_unisys/careers, gives information on the company's benefits and job openings. Benefits include such options as health, 401(k), dental and flexible spending accounts. The company also has the Unisys University, to foster everyone's budding managerial talent, at five global campuses. Nice perks include a generous slate of work/life programs, including compressed workweeks, job sharing, telecommuting and part-time employment. At some facilities, there are nurses' stations, gym facilities and aerobics classes, to keep employees in tiptop shape, as well as nifty services like photo developing, dry cleaning and car washes, plus employee discounts on things like flowers, museum admission, restaurants and cars. Job openings are divided by region. In order to apply, candidates must first create a profile.

Unisys has plenty of internship and co-op opportunities for people who haven't officially entered the working world. The company partners with INROADS to provide internships and co-op opportunities to minority students, and offers internships to foreign exchange students. Internship positions are listed under the "Job Openings" link.

OUR SURVEY SAYS

They should call it Multisys

One source praises the company's "hardworking and skilled professionals," while a co-worker notes that "many basic internal business processes simply did not work," and that he had "occasional contentious battles with the internal bureaucracy." The management ranks are "made up of professionals from all over the map—each bringing their own methodology," observes one long-time employee. An engineer agrees. "At least one department used matrix management."

Sources report "flexible hours (come and leave any time you want)" but that they "can be very long—most of the higher level executives work 70-80 hours a week." There is a "flexible dress code (mainly jeans and business casual although on a rare occasion I think I saw someone wearing shorts)," according to a member of the engineering department, while in the upper ranks, "Dress was always suit and tie."

Varian Semiconductor Equipment Associates, Inc.

35 Dory Road
Gloucester, MA 01930
Phone: (978) 282-2000
Fax: (978) 283-5391
www.vsea.com

LOCATIONS

Gloucester, MA (HQ)
Beaverton, OR
Mesa, AZ
Newark, CA
Newburyport, MA
Plano, TX
Portland, OR

29 offices in France, Germany, the Netherlands, Japan, Korea, Taiwan, China, Singapore and Malaysia.

DEPARTMENTS

Administrative • Buying • Customer Service • Electrical Engineering • Finance & Accounting • HR • IT • Legal • Manufacturing • Mechanical Engineering • Quality • Systems Engineering

THE STATS

Employer Type: Public Company
Stock Symbol: VSEA
Stock Exchange: Nasdaq
Chairman: Richard E. Aurelio
CEO: Gary E. Dickerson
2006 Employees: 1,588
2006 Revenue ($mil.): $730

KEY COMPETITORS

Axcelis Technologies
Applied Materials
Nissin Electric

EMPLOYMENT CONTACT

www.vsea.com/hr.nsf/docs/work-vsea

THE SCOOP

Isn't it ionic?

Varian is one of the leading manufacturers of semiconductor production equipment. The company's machines implant semiconductor substrates, like silicon or germanium, with ions (or charged particles) in order to produce transistors with the specific properties required by the circuit in which it will be used. These ion implantation devices work by accelerating ions to sufficient energies such that they lodge in the crystal structure of the silicon. (The process is also known as doping.) Ions of arsenic, phosphorous and boron are most commonly implanted in semiconductors manufactured in large quantities. Each wafer of semiconductive material is individually implanted with ions, so it is critical that the machines be consistent.

Varian's equipment can fit 50 billion identical transistors on a 300mm (about a foot long) piece of silicon. The pieces of machinery that Varian makes aren't like ray guns out of Star Wars, though—they're about the size of a garage and can weigh 50,000 pounds. Varian's equipment falls into the categories of high current (for injecting ions deep into silicon pieces), medium current and high energy. Its position as a maker of equipment for semiconductor manufacture means that it is at the mercy of the volatile chip industry. To stay ahead of the competition, the company spent over $90 million on R&D in 2006.

Doping through the ages

The seed of Varian Semiconductor was planted in 1978 when Varian Associates, one of the firms that made Silicon Valley into a tech mecca, bought Extrion, a Gloucester-based manufacturer of ion implantation equipment. Varian Associates had been a large maker and developer of equipment for the military and scientific markets; the company specialized in particle accelerators, computers and vacuum pumps. In 1999, with an eye to improving profits, Varian Associates split into three independent companies. (The other Varians are Varian Incorporated, which makes vacuum pumps and instruments for biological sciences, and Varian Medical Systems, which concerns itself with radiological oncology.)

Varian Semiconductor took over the semiconductor equipment arm functions of the original Varian Associates. Three years after Varian Semiconductor Equipment Associates was spun off, the market for semiconductors (and all other things tech-

Visit Vault at **www.vault.com** for insider company profiles, expert advice, career message boards, expert resume reviews, the Vault Job Board and more.

V∧ULT CAREER LIBRARY   497

related) abruptly dried up. But the company weathered the adverse market conditions to come back swinging—in 2005, it was ranked No. 2 on the Boston Globe 100. Two years later, *BusinessWeek* ranked it No. 60 on its Infotech 100.

Coming up roses

In fiscal 2006, Varian pulled down a revenue of $730 million, its highest to date, with profits of $94 million, up 30 percent over the previous year. The good news continued into 2007 when Varian's market share, which is calculated by some analysts to be nearly 40 percent, got a bump when rival Applied Materials announced that it would be closing one of its plants in England. Analysts expect Varian to pick up some of the slack when the factory closes in late 2007. Already, third quarter results for the year have yielded record revenue and profits.

Major windfall

In 2007, Varian was awarded a $30 million contract to provide several of its ion implanters to a semiconductor manufacturer. The deals also include service for the devices. The contract represents the largest deal in Varian history.

GETTING HIRED

Cozy up to a semiconductor

Varian's careers page (at www.vsea.com/hr.nsf/docs/work-vsea) provides information on benefits and job openings at the company. Candidates may have to undergo a background check before they can be hired. Benefits of working for the company include tuition reimbursement, stock purchase plan, 401(k), health and dental insurance, and, for those who work at HQ, life in the quaint seaside town of Gloucester, Massachusetts. The careers site has a list of open positions for which job seekers can apply online and a college recruiting schedule (forthcoming at press time). It also accepts resumes for general inquiry.

VeriSign, Inc.

487 East Middlefield Road
Mountain View, CA 94043
Phone: (650) 961-7500
Fax: (650) 961-7300
www.verisign.com

LOCATIONS

Mountain View, CA (HQ)
Ashburn, VA • Atlanta, GA •
Boston, MA • Broomfield, CO •
Chicago, IL • Dallas, TX • Dulles,
VA • New York, NY • Norcross, GA
• Olympia, WA • Overland Park, KS
• Providence, RI • Reston, VA • San
Francisco, CA • Santa Cruz, CA •
Savannah, GA • Seattle, WA •
Tampa, FL • Washington, DC

Locations in 15 countries
worldwide.

DEPARTMENTS

Administration
Facilities
Finance
HR
IT
Legal
Marketing
Public Relations
Purchasing
Quality
Research & Development
Sales

THE STATS

Employer Type: Public Company
Stock Symbol: VRSN
Stock Exchange: Nasdaq
Chairman: Jim Bidzos
President & CEO: William A. Roper Jr.
2006 Employees: 5,331
2006 Revenue ($mil.): $1,575

KEY COMPETITORS

Entrust Technologies
Identrus
RSA Security

EMPLOYMENT CONTACT

www.verisign.com/verisign-inc/
careers

THE SCOOP

For your security

VeriSign is striving to become the de facto standard for Internet security. The company's main product is the SSL certificate, which provides security for transactions that happen over the Internet. The technology is used for e-mail, access control for corporate Intranets and virtual private networks (VPNs), and online retail. The company's digital identification products and verification services are recognized by national authorities all over the world—even the FBI uses them.

VeriSign is involved in the cutting-edge areas of VoIP and RFID supply-chain management, and offers its customers Internet consulting and help in meshing different network standards. In addition, VeriSign is the go-to authority for registering all .com, .net and .tv domain names, which it wholesales to retailers, and it manages part of the system that allows computers to look up domain names on the Internet. The company's products secure 700,000 servers and assist in finding 21 billion web sites and e-mails to keep the Internet running smoothly.

The company's success depends on the wide acceptance of its brand, since its technology doesn't work unless users on both ends of a transaction support it. One of the biggest challenges VeriSign has faced is convincing consumers that SSL certificates are a necessity. The company has worked to achieve that goal by getting online vendors to require these digital IDs for online transactions. This level of security is becoming increasingly important to the average consumer, who has no desire to see his or her credit card number fall into the wrong hands. The IDs are recognized worldwide and deemed as legally binding signatures in a number of countries. To ensure its place in the market, the company has partnered with high tech-heavies like Microsoft, Apple and IBM.

Defender of the bytes

In 1995, the business security firm RSA (now owned by EMC Corp.) spun off VeriSign as a venture in technology for the newly emerging sector of digital security. Right then, computers were just beginning to become commonplace in people's homes—and stores were beginning to realize the retail potential of the Internet. Customers, however, were loath to hand over their bank account, credit card or Social Security numbers without assurance that an unauthorized someone wasn't privy to the transaction. That's where VeriSign stepped in, making the Internet safe for all sorts of secure online transactions—payments, bank transfers, securities trades and the like. By 1996, the company had developed the VeriSign seal—the Good Housekeeping Seal of Approval for Internet Security, if you will—and had put it on 12,000 sites.

VeriSign on the dotted line ...

Then, VeriSign embarked on a series of acquisitions to bolster its position in the Internet industry. In 2000 it acquired Network Solutions, the leading Internet domain name registration company, in a blockbuster $21 billion deal. This enabled VeriSign to offer its customers complete startup services, meaning that VeriSign can register the domain name of a new company and then build its security backbone to facilitate e-commerce transactions.

In 2003, the company acquired Guardent, a service products company, for $140 million in cash and stock, and acquired the assets of Unimobile, a company that sells mobile messaging products, the following year. Also that year, VeriSign secured an agreement with EuroTrust A/S to buy Secure Socket Layer (SSL) certificate assets for $8.5 million, with an eye to expanding its market to Europe.

The other kind of security

While VeriSign has proven itself quite adept at handling online security, it hasn't done as well with handling of its own securities. At the beginning of summer 2006, the company's stock dating practices fell under scrutiny from the U.S. Justice Department. A few months thereafter, the company announced it would be restating its earnings from 2001 to 2005 due to creative options dating, which ultimately resulted in a $160 million charge, as well as the resignation of CFO Dana Evan, in July 2007. As a result of the restatements, VeriSign posted a $4.7 million loss in the second quarter of fiscal 2007. In August the company terminated the contract of its general counsel, James Ulam, citing no reason for his departure; *American Lawyer's Corporate Counsel* linked him to the stock uncertainty in its article, "A Very Quiet Departure."

Progressive technology, conservative business

In early 2007, VeriSign agreed to a joint venture with Rupert Murdoch's News Corporation, the parent company of Fox Mobile Entertainment, to combine Fox Mobile with VeriSign's Jamba. The partnership is slated to bring all sorts of family-friendly Fox entertainment, like *Trading Spouses* and *24*, to cell phones everywhere.

Visit Vault at **www.vault.com** for insider company profiles, expert advice, career message boards, expert resume reviews, the Vault Job Board and more.

VAULT CAREER LIBRARY

501

The deal followed shortly on the heels of an agreement with Axiom TV, a site that provides legally downloadable, family-friendly movies, including such favorites as *First Spaceship on Venus* and *Santa Claus Conquers the Martians*. How much luck mothers have persuading their children to see either of those titles when they're having a tantrum about not being admitted to the latest shoot-'em-up horror thriller remains to be seen, as Axiom has not yet opened downloading to the general public. The deal will take advantage of VeriSign's unique peer-to-peer content delivery system, which reduces bandwidth costs.

Also in 2006 VeriSign acquired GeoTrust and inCode Wireless. GeoTrust was a rival supplier of SSL encryption, while inCode is a consultancy that specializes in wireless and Internet technologies. VeriSign wrapped up 2006 with revenue of $1.58 billion, down somewhat from the previous year, due to hiccups in the mobile content arena.

Building a better Internet

In February 2007, VeriSign embarked on a $100 million project to improve the infrastructure of the Internet by 2010, including a tenfold increase in the bandwidth for domain name lookups, which would enable the company to handle four trillion domain name lookups per day. VeriSign is making the changes to cope with increased demand for the Web from the growing number of portable devices that transfer data over the Internet, in addition to the immense projected number of Internet users by 2010.

The company has given the undertaking the appropriately epic name of Project Titan. The project will also naturally provide more robust Internet security against denial of service attacks and other forms of online hijinks. This is how VeriSign is preparing itself for the "Any Era," the phase of Internet development where people demand information everywhere and at all times.

Option to run

In May 2007, the company's CEO Stratton Sclavos resigned. Sources at the company denied that his departure was related to the aforementioned options investigation. William A. Roper Jr. was taken on as Sclavos' replacement, and Sclavos departed with a $25 million severance package.

GETTING HIRED

You have the key

VeriSign has a career site at www.verisign.com/verisign-inc/careers, with information on college recruiting, benefits and (of course) open positions. The college visit bandwagon makes stops at Carnegie Mellon, Georgetown and the University of Rhode Island. The company offers such nice benefits as health club reimbursement, stock options and a full suite of retirement and health plans. Jobs are searchable by location and keyword. To apply, job seekers must fill out an online form.

Visit Vault at **www.vault.com** for insider company profiles, expert advice,
career message boards, expert resume reviews, the Vault Job Board and more.

VAULT CAREER LIBRARY   503

Vishay Intertechnology, Inc.

63 Lancaster Avenue
Malvern, PA 19355-2143
Phone: (610) 644-1300
Fax: (610) 889-9349
www.vishay.com

LOCATIONS

Bennington, VT • Chicago • City of
Industry, CA • Columbus, NE •
Detroit, MI • Monroe, CT • Niagara
Falls, NY • Sanford, ME • Santa
Clara, CA • Shelton, CT •
Westbury, NY

DEPARTMENTS

Administration
Audit
Engineering
Facilities
Finance
HR
IT
Legal
Logistics
Manufacturing
Marketing
Outsource Management
Project Management
Public Affairs & Communications
Purchasing
Quality
Research & Development
Sales

THE STATS

Employer Type: Public Company
Stock Symbol: VSH
Stock Exchange: NYSE
Chairman: Dr. Felix Zandman
President & CEO: Dr. Gerald Paul
2006 Employees: 27,000
2006 Revenue ($mil.): $2,581

KEY COMPETITORS

Philips
Texas Instruments
Toshiba

EMPLOYMENT CONTACT

hr.vishay.com

THE SCOOP

Behind the electronic scene

Vishay makes the bits that handle the bytes—all those little electrical parts like diodes, transducers, transistors and rectifiers that are integral parts of the computers, cell phones, cars, satellites and other gadgetry that makes the modern world go 'round. The company makes its components in places such as the Czech Republic, China and the Philippines.

Dr. Zandman, bring us a dream ...

Dr. Felix Zandman, holding degrees in math, physics and engineering, came to Philadelphia in 1956 by way of France and Poland. He came armed with a nifty new invention called "Photostress," which visually assessed the stresses on metal objects, such as a bridge or an airplane wing, and eventually revolutionized the way engineers would test for adequate construction on such projects. He was employed by the Budd Company in Philadelphia and also came up with a new kind of resistor, which could conduct and regulate electrical currents without falling susceptible to temperature.

The Budd Co. was not interested in either of Zandman's inventions, and he went into business himself in 1962 with his cousin, Alfred P. Slaner, founding Vishay Technology. They name of the company derives from Zandman and Slaner's ancestral village in Lithuania, to honor those who perished in the Holocaust. The company started manufacturing resistors, primarily for the military and aerospace markets, and had its first stock offering in 1972.

Enter Zandman

In the 1980s and 1990s, Zandman recognized that the company needed to diversify its offerings to continue its pace of rapid growth. Thenceforth, Vishay acquired a number of other companies, thereby expanding its product line beyond resistors to capacitors and semiconductors. The exponential growth of the market for electronic goods during the 1980s and 1990s powered the company along, proving Zandman's strategy correct—and invited a number of competitors to the party. Hiccups in the market sparked spates of restructuring for the company in the mid-1990 and again in 2003.

Happy days are here again

It seems, though, that Vishay's days of restructuring are behind it, as are its days under the direct supervision under Dr. Zandman. Effective January 2005, Zandman handed over the CEO position to then COO Dr. Gerald Paul, who has been with the company since 1978. Zandman is not exactly drifting into retirement, though—he will remain chairman of the company's board and has ascended to two newly created positions—chief technology officer and chief business development officer.

Already, Paul has brought the company to new heights: 2006 revenue was record-setting, at $2.58 billion, having jumped 12 percent over the previous year's sales. Also, profits doubled over 2005's take, to $139.7 million. The further strengthened its position in the semiconductor market that year, acquiring the power control systems division of International Rectifier for $290 million.

Weighty matters

In April 2007, Vishay acquired PM Group, a U.K.-based electronics manufacturer whose products are based on technology created by Vishay. The object of Vishay's purchase is PM's onboard vehicle weighing division, which allows truckers to weigh the contents of dumpsters and dump trucks without having to weigh the entire truck. The deal will provide the company with a way to expand in Europe, and went down for $27 million.

GETTING HIRED

Work in the "ohm" office

Vishay lists open positions at hr.vishay.com. There are opportunities for experienced hires as well as interns, at locations both in North America and the rest of the world. To apply for a job, applicants must first create a profile. Other than that, not much information is extant on the site.

VMware, Inc.

3401 Hillview Avenue
Palo Alto, CA 94304
Phone: (650) 475-5000
Fax: (650) 427-5001
www.vmware.com

LOCATIONS

Palo Alto, CA (HQ)
Austin, TX • Basking Ridge, NJ •
Bellevue, WA • Cambridge, MA •
Dallas, TX • Duluth, GA • Houston,
TX • McLean, VA • San Francisco,
CA

International operations in Canada,
Brazil, Mexico, Belgium, Denmark,
Finland, France, Germany, Ireland,
Israel, Italy, the Netherlands,
Norway, Spain, Sweden,
Switzerland, the UK, Australia,
China, Hong Kong, New Zealand,
India, Japan, Korea, Singapore and
Taiwan.

DEPARTMENTS

Facilities • Finance/Accounting •
Human Resources • HW Alliances •
ISV Alliances • IT Development • IT
Ops • Legal • Marketing •
Operations • Professional Services •
Quality Assurance • R&D • Sales
Support • SI Alliance • Technical
Support

THE STATS

Employer Type: Public Company
Stock Symbol: VMW
Stock Exchange: NYSE
Chairman: Joseph M. Tucci
President & CEO: Diane B. Greene
2006 Employees: 3,000
2006 Revenue ($mil.): $703.9

KEY COMPETITORS

Microsoft
SWsoft
Symantec

EMPLOYMENT CONTACT

www.vmware.com/company/jobs

Visit Vault at **www.vault.com** for insider company profiles, expert advice,
career message boards, expert resume reviews, the Vault Job Board and more.

VAULT CAREER LIBRARY   507

THE SCOOP

Virtual reality

Vmware is a major player in the market for virtualization software, which runs underneath a computer's operating system, controlling the distribution of resources like processor cycles or memory space. Why is that useful? It allows for the creation of multiple virtual 'computers' out of a single machine—or a single 'virtual' machine out of several boxes.

Sort of a server-within-a-server, this software enables a single computer to run different operating systems at the same time, servers to create real-time backups without interrupting users' work, and spare processor cycles and memory space are sent to where they're needed, rather than being tied to an unused machine. It all adds up to a more efficient use of computing power, which, in turn, cuts down on hardware costs and server rooms, as firms with Vmware need fewer computers and space to do the same amount of work. Vmware's customers include 100 of the largest companies in the Fortune 500.

Academia pays off

A Silicon Valley entrepreneur and computer scientist named Diane Green, her husband, Stanford professor Mendel Rosenblum and three other friends founded Vmware in 1998 to create and sell virtualization software. Countless all-nighters, along with $3 million in venture capital, got the first version of Vmware's software ready to ship within a year. Though the company didn't show a profit in 2000, it had already taken on 80 employees. Lack of profits wasn't for lack of market share, however—by 2001, Vmware's software was running on 5,000 computers in 100 countries and the computing community was fast embracing virtualization software.

The growth of the company brought it to the attention of bigger players in the industry, and in 2003 the storage giant EMC made an offer the founders couldn't refuse—EMC and Vmware closed on a $650 million deal the following year. Despite its acquisition, EMC didn't totally assimilate Vmware into its operations; co-founder Diane Green is still the company's president and CEO.

In 2005, Vmware made moves to boost its market share when it released vital source codes to numerous Vmware subscribers, including major tech players Cisco Systems, Hewlett-Packard, IBM, Red Hat and Intel. By releasing its source code, Vmware was allowing these firms to tweak its software to work better on their systems, hopefully bringing them into the fold as Vmware clients. Vmware also developed a version of its software for the Mac OS, which launched the following year; it also distributed a free version of its software for home users to try. The strategy seemed to work well—by 2006, the company had a majority of the market for virtualization software.

Share and share alike

The added market share saw considerable gains when EMC opted to spin off 10 percent of the company in 2007. The offering was a success—Vmware was valued at over $18 billion, and most of its shares sold for nearly double the asking price, making the company's IPO the biggest tech offering since Google's, in 2004. EMC hinted that the stock would be used to lure top programming talent with options, though EMC plans to keep its stake in the business.

Vmware has some stiff competition in its search for talent. Silicon Valley insiders say that the company is aiming to vacuum up the best programmers before a competitor can and is making aggressive moves in the salary arena, plying the Valley's best programmers with hefty salaries and options. But this strategy means that it's up against Google, a perennial member of Fortune's Best Places to Work list.

Room to grow

Why all the investment in a company that produces an obscure bit of code that mostly runs in server farms? The potential market is huge, and Vmware is the current market leader. To minimize the disruption from crashes, viruses and all the other woes that servers are heir to, most companies have a machine devoted to running a single program. While this reduces the risk that two programs running simultaneously might cause the server to crash, it also means that the server isn't used to its fullest potential 80 or 90 percent of the time.

This is where virtualization software steps in—it can allow two machines to work as a single unit, which means that if one crashes, the other acts as an automatic, real-time backup of the other system, reducing the fallout from the crash. There are over 24 million servers that can run Vmware software in 2007—and the number is expected to grow 40 percent by the end of the decade. Not surprisingly, the results from Vmware certainly show a company riding a boom—2006 revenue was double that of the year before.

Keeping the data on the DL

Even the NSA is getting in on the game. In 2007 the agency bought Vmware's software to create multiple virtual machines out of one computer. Some of these virtual computers would have access to top-secret information, while other users on the machine would not be able to see it, despite being able to use the computer's memory and processor. Until Vmware virtualized its computers, the NSA stored secret information on dedicated computers.

GETTING HIRED

Have a real career in virtual software

Vmware's careers site, at www.vmware.com/company/jobs, provides jobseekers with all the details on landing a job at Vmware. College students are invited to send resumes to the campus relations department, or chat with a recruiter on a campus visit—the company makes stops at Columbia, Boston University, MIT, Stanford and the University of Wisconsin, among other places. Experienced hires-to-be can search job openings or submit a resume.

Benefits at the company include annual performance bonuses, plenty of time off, health insurance, life insurance, commuter benefits and 401(k), for which employees are eligible from their date of hire. Hopeful hires might also want to check out what Vmware executives have to say on their blog, located at blogs.vmware.com/console.

Western Digital Corporation

20511 Lake Forest Drive
Lake Forest, CA 92630-7741
Phone: (949) 672-7000
Fax: (949) 672-5408
www.westerndigital.com

LOCATIONS

Lake Forest, CA (HQ)
Fremont, CA
San Jose, CA

DEPARTMENTS

Accounting, Auditing & Finance
Customer Care & Quality Assurance
Engineering
Facilities Maintenance, Installation &
  Repair
HR & Staffing
IT & Support Services
Manufacturing, Operations &
  Assembly
Marketing, Advertising & Public
  Relations
Project/Program Management
Purchasing
Sales & Sales Operations
Supply Chain/Logistics

THE STATS

Employer Type: Public Company
Stock Symbol: WDC
Stock Exchange: NYSE
Chairman: Thomas E. Pardun
President & CEO: John F. Coyne
2006 Employees: 23,161
2007 Revenue ($mil.): $5,468

KEY COMPETITORS

Fujitsu
Hitachi Global Storage Technologies
Seagate Technology

EMPLOYMENT CONTACT

www.westerndigital.com/en/company/
employment

Visit Vault at **www.vault.com** for insider company profiles, expert advice,
career message boards, expert resume reviews, the Vault Job Board and more.

VAULT CAREER LIBRARY   511

THE SCOOP

Got data?

Western Digital has made a name for itself the hard way—by making hard drives. WD's drives are found in everything for the modern consumer at work and play: from TiVo to the laptop, desktop, servers and those items' external storage. Of the company's sales, 54 percent come from PC builders like Dell, but the firm also sells drives to consumers who need a little data storage and backup.

A drive to succeed

Alvin Phillips, a veteran in the semiconductor field with Motorola, founded Western Digital in 1970 as a manufacturer of semiconductors and, as the market for PCs grew, so did the company. In 1983, the company targeted the PC sector as its main area of growth, and began to diversify its product line with data storage devices, and hitching its fortunes to the PC market proved wise. By 1988, the company began manufacturing hard drives and its sales had increased to the point where the company joined the ranks of the Fortune 500. The company hit a ceiling during the next decade, though, as it weathered the overproduction of its signature hard drives, as well as the consequent rapid drop in prices.

WD survived because of its quality—its drives were found in many different computers, and the International Standards Organization rewarded the company in 1994 by affording it ISO 9001 status, the first multinational company headquartered in the U.S. to receive the distinction. Revenue continued to grow steadily through the late 1990s and hit a high of $4.1 billion in 1997, but the company took a loss each of the next three years, due to a flagging tech market. As the market for its products gained strength in the following years, revenue began to grow again, reaching $3.6 billion in 2005.

Raptor rapture

In the summer of 2006, WD got kudos from *PC World* magazine for its WD Raptor X 150 GB hard drive. The drive has a clear top, through which users can watch as it writes and retrieves data. To keep fun drives like the WD Raptor coming, the company then purchased a factory in California and Thailand that had been manufacturing semiconductors for its drives for three years. The factory in California employs around 1,000 people.

Coyne of the realm

In November 2006, Western Digital announced a new CEO, named John Coyne. Coyne had been made president and COO only a few months previously. Arif Shakeel, the former CEO, had only held the position for a year; he remained as an advisor to Coyne until June 2007. Coyne, who was born in Ireland, first joined Western Digital in 1983. He initially worked in the manufacturing division, but expanded his expertise as he advanced through the company ranks. Mr. Shakeel himself had been with the company since 1985.

Minding the data

For 2007, the company posted revenue of $5.5 billion and profits of $585 million. The market for digital storage has been growing at a ripping rate due to customers' propensity to attract data—music files, digital pictures, movies, TV shows—and because these customers are increasingly demanding for this data to be accessible at all times; by some estimate, the market is growing by 14 percent.

In 2001, hard drives outsold PCs by a margin of 50 percent, and the margin increased to 74 percent by 2005. WD is hopping on the bandwagon, with hard drives for applications other than desktop computers making up 29 percent of its revenue in 2006, up eight points from the year before. In June 2007, WD added to its technology offerings by acquiring Komag, a manufacturer of thin-film media for storage in hard drives, in a deal valued at $1 billon.

Now hear this

No, really, your TiVo didn't die, the ending of *The Sopranos* was supposed to be like that. In order to prevent the media-consuming public from being exposed to abrupt endings not originating in the mind of quixotic TV writers, in 2007 WD introduced a new line of hard drives designed to hold audio and video files. They range in size from 80 to 500 GB, are quieter and consume less power than WD's other hard drives, and are intended for use in surveillance systems and set-top video recorders. For people who cannot conceive of life without a soundtrack, the drives are designed to be durable enough to be used all the time.

Visit Vault at **www.vault.com** for insider company profiles, expert advice, career message boards, expert resume reviews, the Vault Job Board and more.

VAULT CAREER LIBRARY　513

GETTING HIRED

Go Western, young man (or woman)

Western Digital's careers site, at www.wdc.com/en/company/employment, has information for both college graduates and more seasoned hires. WD seeks college grads with degrees in mechanical or electrical engineering; students with computer science degrees and programming experience in C and C++ are also encouraged to apply. Interested candidates may apply online at www.westerndigital.com/careers. Western Digital also accepts a small number of interns seeking applicable degrees in computer science, electrical and mechanical engineering, physics and tribology. The company offers its workers a suite of benefits including health, vision and life insurance and a retirement plan with company matching. Western Digital maintains a searchable list of jobs at www.westerndigital.com/careers. To apply to a specific position, hopeful job seekers must first create a profile.

Xerox Corporation

45 Glover Avenue
Norwalk, CT 06856-4505
Phone: (203) 968-3000
Fax: (203) 968-3218
www.xerox.com

LOCATION
Stamford, CT (HQ)

DEPARTMENTS
Customer Service & Support
Document Services
Finance
General Management
Hardware/Software/Other
Engineering
IT
Legal
Manufacturing
Research
Sales & Marketing

THE STATS
Employer Type: Public Company
Stock Symbol: XRX
Stock Exchange: NYSE
Chairman & CEO: Anne M. Mulcahy
2006 Employees: 53,700
2006 Revenue ($mil.): $15,895

KEY COMPETITORS
Canon
Hewlett-Packard
Ricoh

EMPLOYMENT CONTACT
www.xerox.com/careers

THE SCOOP

The original copy cats

Any discussion of Xerox must, of course, begin with photocopiers—and that's worth saying again; the name is practically synonymous with the process of making duplicates. In 2006 alone, the company's multifunction machines (print, copy, fax) churned out an astounding 300 billion photocopies. Xerox isn't just copies, though—the company offers a number of other graphics-related products, like printers for everything from an letter size black-and-white resume, to five-foot-tall full-color posters and digital printing presses for on-demand printing of everything from marketing materials to textbooks (let Gutenberg get a look at that!). Xerox also offers business services like task management software, IT consulting and outsourcing. In 2006, the company drew in revenue of nearly $15.9 billion and was ranked No. 142 on the Fortune 500.

Dry writing

Xerox got its start in 1906 in Rochester, New York, which to this day is the epicenter of the U.S. photographic industry (other visual greats in town include Eastman Kodak and Bausch & Lomb). Initially known as the Haloid Company, the firm manufactured photo paper until 1947, when it licensed a patent for xerographic imaging from Chester Carlson, who developed the technology in Astoria, Queens in 1938.

Xerography—after the Greek words for "dry" and "writing"—produced a duplicate image without any mucking around with icky chemicals; Haloid trademarked the truncated word "Xerox" to describe the process in 1948.

The company wasted no time in bringing this new technology to market, and by 1949 had produced its first copier, the Model A. This was followed, six years later, by the Copyflo. In 1959, Xerox kept the copies coming with the 914, an automated office copier that spit out duplicates at a rate of seven per minute.

After changing its name in 1958 to Haloid Xerox Corporation (to honor its most popular product), the company subsequently decided to drop "Haloid" from its name in 1961.

Xerox's dominance in the photocopying market bankrolled an ambitious program of expansion and acquisition in the 1960s and 1970s. During this period, the company acquired a computer company, three publishing companies and a disk-drive business. It was even, briefly, the owner of Western Union.

But in 1974, federal regulators ruled that Xerox had become too powerful and forced the company to license its xerographic technology to competitors. Undaunted, Xerox continued to innovate, unveiling the first laser printer in 1977 and keeping the acquisitions coming in the 1980s, diversifying into desktop publishing, optical character recognition, scanning, faxing and even banking and insurance.

The copy company—for better and for worse

The 1990s were an up-and-down decade for Xerox because, although business was good in its core copying division, the company had missed some major chances to diversify elsewhere in the tech industry. The company chose to ignore its breakthroughs with the graphical user interface (GUI), which Xerox developed at PARC in the 1970s but Apples' Macintosh first brought to market in the next decade. Microsoft's Windows subsequently made a fortune with it. Back in 1979, Xerox entered into an agreement with Apple Computer which allowed an Apple team to tour PARC in return for some Apple stock. Apple CEO Steve Jobs liked what he saw, subsequently hired half of PARC's staff for his own company and basically created the modern personal computer—with a great boost from Xerox's technology.

Perhaps sensing it had lost its opportunity to branch out into other areas of the tech industry, Xerox focused on its core document business in the 1990s, and appeared to settle into its role as a copying and printing company. It concentrated on upgrading its technology from analog to digital and entered into alliances with computer manufacturers Apple and Compaq to provide them with printing components (in 1992 and 1993, respectively). The company also recognized the growing importance of networking in the business environment and began selling networked color laser printers and software in 1995, to facilitate printing documents from the Web.

In 1997, Xerox bought out the Rank Organization's minority interest in Rank Xerox, its European joint marketing venture, and later renamed it Xerox Limited. Around the same time, the company sold off its various insurance interests. Xerox was once again an active acquirer of other companies, but its purchase choices tended to be more closely related to its core (i.e., copying) businesses, a reflection of the company's intention to consolidate, rather than expand its services. It acquired the technology consulting firm XLConnect, now known as Xerox Connect, in 1998. The following year, it purchased the French company SET Electronique, a maker of high-speed digital printing equipment. In 2000, Xerox bought the color printing division of Tektronix. Also that year, Xerox spun off its digital rights management unit as a separate company called ContentGuard.

From Xerox to zero

But however attractive these gains looked in the 1990s, they were a distant memory by 2001. The company posted losses of $198 million in the fourth quarter of 2000, a 10-year low, and its debts were $2.6 billion at the start of the next year (down from a surplus of $38 billion), forcing the company to take out a $7 billion line of credit. Between 1999 and May 2001, the company's stock price fell from $64 to $7, another decade-long lowlight. In March 2000, the board sacked its CEO, G. Richard Thoman; he had been with the company since leaving IBM in 1997. In its March 2001 article "Xerox: the downfall," *BusinessWeek* called the company "a corporate catastrophe of the first order." In fact, the company's financial status was so dire that it even considered

the sale of its stake in PARC, which was unprofitable, but reconsidered that plan, noting that research centers are not intended to be profit centers. Instead, Xerox spun it off as a separate company in 2001, creating a wholly owned subsidiary that allows Xerox to benefit directly from PARC research, while PARC also performs contract research for other companies. (The company's focus on research and success in developing breakthrough technology was recognized in July 2007, when President George W. Bush presented Xerox the National Medal of Technology.)

More trouble arrived soon for Xerox: by 2001, at least 11 different shareholder groups had filed lawsuits against the company, angry about the company wasting its $38 billion surplus. More ominously, the company fired 13 executives from its Mexican division the same year, citing accounting irregularities and unpaid debt. Xerox stressed that this problem was limited to its Mexican branch only, but the Securities and Exchange Commission (SEC) looked into the matter and concluded otherwise—they claimed that the company had overstated its revenue for the last three years, inflating sales figures by up to $3 billion. Xerox and KPMG settled the matter in 2002 and 2005, respectively (Xerox paid the SEC $10 million and KPMG paid $22.5 million), and Xerox restated its earnings from 1997 through 2001— erasing nearly $2 billion in profit from its books. In October 2001 Xerox fired KPMG, who had handled its auditing for three decades and hired PriceWaterHouseCoopers for its future accounting needs.

The Mulcahy miracle

Anne M. Mulcahy, a former HR executive and 24-year veteran of Xerox, was appointed as Xerox's new CEO amidst the turmoil in 2000 and immediately set goals for regaining profitability—selling $2.2 billion of the company's assets and cutting costs by $1 billion. Mulcahy was promoted to chairwoman of Xerox's board in 2001; it was a bid to regain investor confidence, as the previous CEO Thoman's reign was marred by a power struggle between himself and the chairman Paul Allaire, himself also a former CEO at Xerox. Investors started coming back to the company in the summer of 2003, and its stock price continued to improve through 2004.

Mulcahy restored the company's credit and led it to $91 million profits in 2002, and numbers have been climbing slowly but steadily since then. In 2006, revenue was up 1 percent to $15.9 billion, driven by strong performance in the color and business services segments. But the company also cut nearly 300 jobs that year, when it closed a branch of its capital services department, which handled billing for customers' purchasing and leasing activities.

Easy as XMPie

In addition to getting its financial ship in order, Xerox has continued to acquire companies in order to expand its products and service offerings. In July 2006, it purchased Amici LLC for $154 million. Amici, a provider of electronic discovery services to law firms and lawyers, has been folded into Xerox's litigation services division, which helps companies and lawyers go through the reams (not to mention gigabytes) of data in the event of litigation or audit. In November 2006, Xerox purchased multimedia communications company XMPie, in order to further boost its graphics and custom media offerings.

In May 2007, Xerox acquired Global Imaging Systems, pairing Xerox's rapidly expanding portfolio of document products and services with Global's expertise in the small- and midsized business market and nationwide distribution network. With Global now part of Xerox, the company increased its distribution capacity by 50 percent in the SMB market where the total opportunity for document-related offerings is estimated at $16 billion in the United States.

In September 2007, Xerox announced plans to buy Advectis, provider of one of the mortgage industry's most widely used solutions for electronic document collaboration. The pending acquisition is part of Xerox's acquisition strategy to identify successful companies whose offerings align with Xerox's commitment to innovation and reducing the complexity of document management.

Visit Vault at **www.vault.com** for insider company profiles, expert advice, career message boards, expert resume reviews, the Vault Job Board and more.

VAULT CAREER LIBRARY   519

Appropriate documentation

Xerox is not just "the document company"; it's also become a pretty nifty place to work. CEO Mulcahy, under whose tenure stock prices have recovered much of their former health, emphasizes communication and finding the best person for the job. Ms. Mulcahy is serious about communications—even officers with a "C" in their titles spend time dealing with customers.

One year, in order to cheer everyone up, she gave every employee the day off on their birthday. Such employee-appreciation efforts haven't gone unnoticed. The company is consistently recognized for its commitment to workplace equality and opportunity, and, in 2007, alone has been ranked as among the best companies by *Black Enterprise*, *Latina Style*, *Hispanic Business*, and *DiversityInc* magazines, among others. The National Association for Female Executives lauded Xerox for being a great place for women to work, with a spot on the Top 30 Companies for Female Executives. The Families and Work Institute also recognized the firm with its Work Life Legacy Award in 2006.

Fun with advertising

Xerox rolled out two new ad campaigns in spring 2007 to bolster the fastest growing segments of its business: consulting services and color. Aimed at C-level decision makers and IT executives, the ads were placed in such magazines as *Harvard Business Review* and *The Economist*. One ad spotlighted a team of Xerox consultants under the banner, "We Find Millions." It focused on document assessment services that result in millions of dollars in cost savings for companies. Another ad, titled "We Retrieve," shows a golden retriever with a duck decoy in its mouth, an analogy for how Xerox helps search, retrieve and manage everything from accounting and HR documents to legal contracts. A third ad, headlined, "We Go One to One," emphasized personalized, targeted customer communications that boost response rates.

GETTING HIRED

More than making copies

Xerox employs roughly 53,700 people, with 28,400 based in the U.S. While the company is headquartered in Stamford, Connecticut, it retains a major presence in its original hometown of Rochester. Positions are available in one of 12 departments:

customer service and support, hardware engineering, software engineering, other engineering, document services, IT, finance, legal, general management, sales and marketing, research and manufacturing. To apply for a job, navigate to www.xerox.com/careers and click on "apply to jobs." Job seekers can submit a resume or search jobs by various criteria; in order to apply, they must first create a profile.

Campus crusaders

Xerox recruits enthusiastically from the ranks of college students. For students who believe that the company is perfect for them, Xerox conducts on-campus recruiting visits at schools including Carnegie Mellon, Cornell, MIT, Clarkson University and the University of Michigan. At these events, students have the opportunity to meet and interview with company officials for internships, co-op positions and postgraduation employment. The Xerox careers web site outlines the various internship programs and posts a schedule of upcoming recruiting events.

Giving 'em the benefits

Xerox's standard benefits package includes a full slate of company-paid insurance. Employees are given a choice of medical plans as well as dental, vision, life, accident and disability coverage. For retirement benefits, Xerox provides matching funds for a 401(k) account. Other perks include "lifecycle assistance" funds—for child care expenses or mortgage assistance—an adoption reimbursement program, tuition assistance and access to the company's federal credit union. The company also has 4,000 online courses for employees to spruce up their skills or gain new ones.

Letting 'em give back

Xerox encourages its employees to give back to the community through a program that encourages them to take paid leaves of absence of six to 12 months to help out a charity. Projects undertaken by Xerox employees in 2007 included starting an electronic appointment and records management system at a community health clinic, raising funds for a tutoring center for underprivileged children and creating a database for a charity that aids the mentally ill.

In addition, Xerox employees pitch in by getting involved in robotics competitions at local schools, as well as teaching science classes. Because Xerox believes that a successful corporation must be an active participant in society, the company has created a variety of opportunities to give back to the communities from which it draws its employees.

The Xerox Foundation supports employees' philanthropic interests and endeavors through four primary programs. The Xerox Social Service Leave Program provides employees with fully paid corporate sabbaticals (with full benefits, too) to invest their time and talent full time to support the work of a nonprofit agency in the community. Other employees choose to invest their time outside of work through the Xerox Community Involvement Program. XCIP provides employees with seed money and management-supported opportunities to volunteer their time and energy to better the community.

For those employees who would prefer to share their treasure, Xerox's employee matching gifts program and the Xerox Employee United Way Program provide such opportunities. The matching gift program matches employees' contributions to institutions of higher education up to a maximum of $1,000. Or, employees can make a contribution to benefit the United Way programs and services.

Xerox also operates two special educational programs. Xerox researchers and scientists receive paid time to participate in the Xerox Science Consultant Program (in its 40th year), which provides hands-on science instruction by Xerox employees for students in grades three to six. And, Xerox partners with high schools within its communities on U.S. FIRST, an international robotics competition where teams of students build robots with the assistance of their technical sponsors.

The Xerox Foundation also offers financial grants in such areas as education/workforce preparedness, science/technology, employee/community affairs, cultural affairs and national affairs.

About the Editor

Michaela R. Drapes graduated from the University of Texas at Austin and has degrees in radio/TV/film and english. Before joining Vault, she was an editor at award-winning business publisher Hoover's Inc. and covered an array of industry sectors, including pharmaceuticals, amusement parks, real estate, and international banking and finance. Michaela is one of the founders of fashion startup Kindling & Tinder and is an irreverent music critic; she also occasionally DJ's at independent rock shows around New York City.

Nicholas R. Lichtenberg holds degrees from the University of Syracuse in drama and history. Before working at Vault, he covered and canvassed trial court bureaucracies in select states as an assignment editor at the legal publishing company ALM Media Inc. He lives in New York.